An Introduction to the Mechanical Properties of Solid Polymers

An Introduction to the Mechanical Properties of Solid Polymers

I. M. Ward
IRC in Polymer Science and Technology
University of Leeds, UK

and

D. W. Hadley
J. J. Thomson Physical Laboratory
University of Reading, UK

JOHN WILEY & SONS
Chichester · New York · Brisbane · Toronto · Singapore

Baffins Lane, Chichester,
West Sussex PO19 1UD, England

Reprinted May 1997, November 1998

Other Wiley Editorial Offices

John Wiley & Sons, Inc., 605 Third Avenue,
New York, NY 10158-0012, USA

Jacaranda Wiley Ltd, G.P.O. Box 859, Brisbane,
Queensland 4001, Australia

John Wiley & Sons (Canada) Ltd, 22 Worcester Road,
Rexdale, Ontario M9W 1L1, Canada

John Wiley & Sons (SEA) Pte Ltd, 37 Jalan Pemimpin #05-04,
Block B, Union Industrial Building, Singapore 2057

Library of Congress Cataloging-in-Publication Data

Ward, I. M. (Ian Macmillan), 1928–
An introduction to the mechanical properties of solid polymers / I. M. Ward and D. W. Hadley.
p. cm.
Includes bibliographical references and index.
ISBN 0 471 93874 2 (cloth): ISBN 0 471 93887 4 (paper)
1. Polymers – Mechanical properties. I. Hadley, D. W. II. Title.
TA455.P58W36 1993
620.1'9204292 – dc20 93-3431
CIP

British Library Cataloguing in Publication Data

A catalogue record for this book is available from the British Library

ISBN 0 471 93874 2 (cloth)
ISBN 0 471 93887 4 (paper)

Typeset by Keytec Typesetting Ltd, Bridport, Dorset
Printed and bound in Great Britain by
Biddles Ltd, Guildford and King's Lynn

Contents

Preface

The first and second editions of this textbook were particularly directed at research workers in polymer science and postgraduate students with first degrees in physics, chemistry, engineering or materials science. The mechanics of the behaviour of polymers was developed at an elementary level but with no especial concessions to brevity. It has therefore been considered that there would be merit in writing a textbook which was somewhat less extensive and with less detailed discussion of the more advanced topics such as finite elasticity and non-linear viscoelasticity. In the event, we found that we did not wish to abbreviate any of the material to the extent of losing rigour unnecessarily, and that certain new topics such as polymer composites ought to be included. For this reason the present textbook is almost comparable in length to the first and second editions, but it is hoped that it is a somewhat easier read.

We have provided a more extensive index, and some elementary problems, together with their solutions, in the hope that these will be of assistance to the reader.

We wish to thank Dr D. I. Bower and Dr R. A. Duckett for their comments on the draft manuscript, and Miss W. Watson, Mr C. C. Morath and Mrs C. Laverty for their assistance in the preparation of the final manuscript.

I. M. Ward
D. W. Hadley

Chapter 1

Structure of Polymers

The mechanical properties which form the subject of this book are a consequence of the chemical composition of the polymer and also of its structure at the molecular and supermolecular levels. We shall therefore introduce a few elementary ideas concerning these aspects.

1.1 CHEMICAL COMPOSITION

1.1.1 Polymerization

Linear polymers consist of long molecular chains of covalently bonded atoms, each chain being a repetition of much smaller chemical units. One of the simplest polymers is polyethylene, which is an *addition* polymer made by polymerizing the monomer ethylene, $CH_2{=}CH_2$, to form the polymer

$$\left[-CH_2-CH_2- \right]_n$$

Note that the double bond is removed during the polymerization (Figure 1.1). The well-known *vinyl* polymers are made by polymerizing compounds of the form

$$\begin{array}{c} \qquad X \\ \qquad | \\ CH_2{=}CH \end{array}$$

where X represents a chemical group; examples are as follows:

$$\text{polypropylene} \quad \left[-CH_2-\overset{\displaystyle CH_3}{\overset{|}{C}H}- \right]_n$$

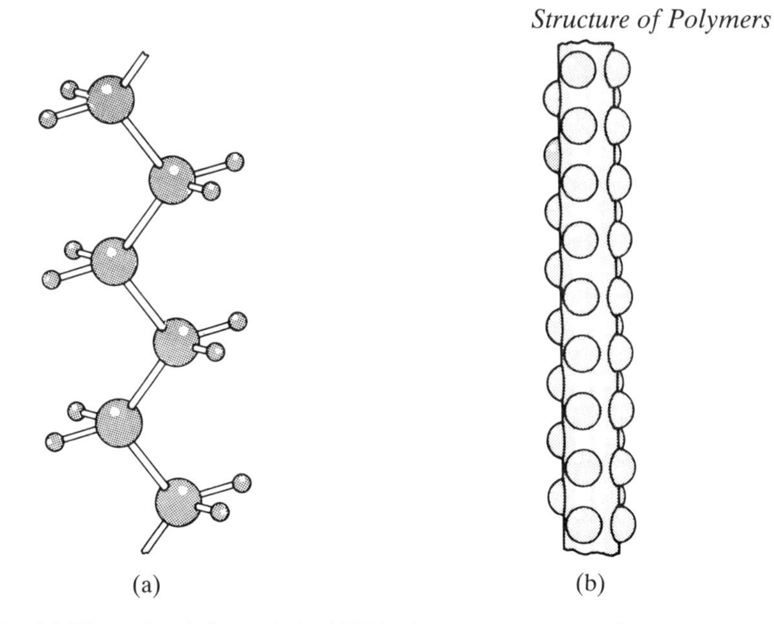

Figure 1.1 (a) The polyethylene chain $(CH_2)_n$ in schematic form (larger spheres, carbon; smaller spheres, hydrogen) and (b) sketch of a molecular model of a polyethylene chain

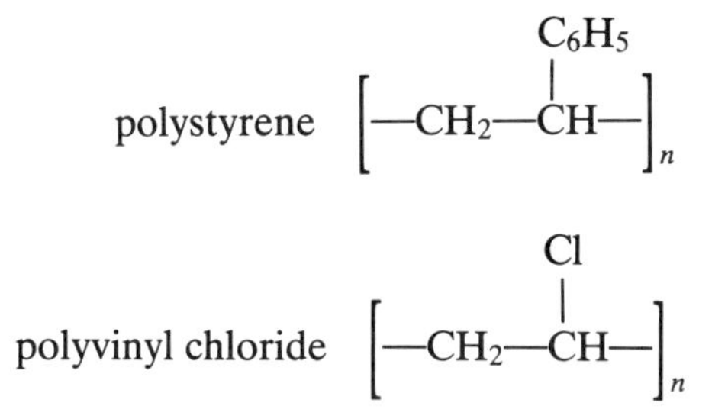

Natural rubber, polyisoprene, is a diene, and its repeat unit,

$$\left[-CH_2-CH{=}\overset{\displaystyle CH_3}{\overset{|}{C}}-CH_2- \right]_n$$

contains a double bond which gives it added flexibility.

Condensation (or step-growth) polymers are formed by reacting difunctional molecules, usually with the elimination of water. One example is the formation of polyethylene terephthalate (the polyester used for Terylene and Dacron fibres and transparent films and bottles) from ethylene glycol and terephthalic acid:

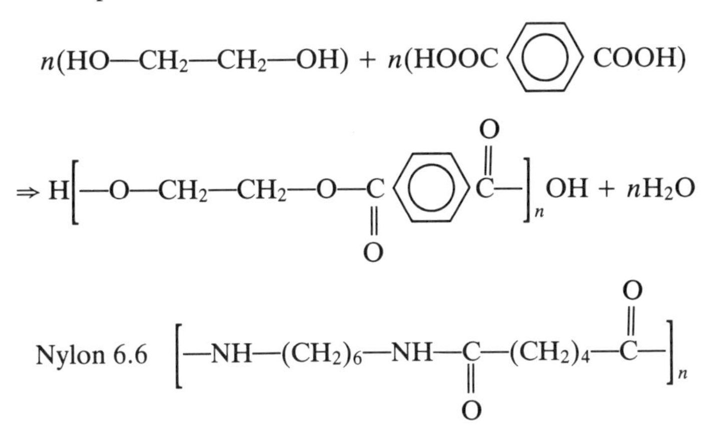

is another common condensation polymer.

1.1.2 Cross-linking and Chain-branching

Linear polymers can be joined by other chains at points along their length to make a cross-linked structure (Figure 1.2). Chemical cross-linking produces a thermosetting polymer, so called because the cross-linking agent is normally activated by heating, after which the material does not soften and melt when heated further; examples are bakelite and epoxy resins. A small amount of cross-linking through sulphur bonds is needed to give natural rubber its characteristic feature of rapid recovery from a large extension.

Very long molecules in linear polymers can entangle to form temporary physical cross-links, and we shall show later that a number of the

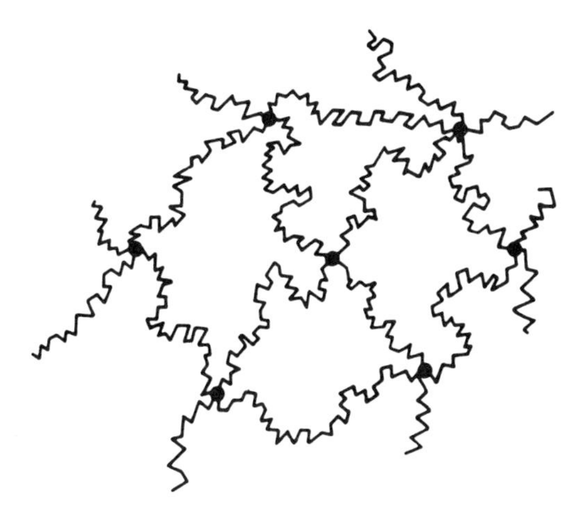

Figure 1.2 Schematic diagram of a cross-linked polymer

characteristic properties of solid polymers are explicable in terms of the behaviour of a deformed network.

A less extreme complication is chain branching, where a secondary chain initiates from a point on the main chain, as is illustrated for polyethylene in Figure 1.3. Low-density polyethylene, as distinct from the high-density linear polyethylene shown in Figure 1.1, possesses on average one long branch per molecule and a larger number of small branches, mainly ethyl [$—CH_2—CH_3$] or butyl [$—(CH_2)_3—CH_3$] side groups. The presence of these branch points leads to considerable differences in mechanical behaviour compared with linear polyethylene.

1.1.3 Average Molecular Mass and Molecular Mass Distribution

Each sample of a polymer contains molecular chains of varying lengths, i.e. of varying molecular mass (Figure 1.4). The mass (length) distribution is of importance in determining the properties of the polymer, but until the advent of gel permeation chromatography [1, 2] it could be determined only by tedious fractionation procedures. Most investigations therefore quoted different types of average molecular mass, the commonest being the number average $\bar{M}_n$ and the weight average $\bar{M}_w$, defined as

$$\bar{M}_n = \frac{\sum N_i M_i}{\sum N_i} \qquad \bar{M}_w = \frac{\sum (N_i M_i) M_i}{\sum N_i M_i}$$

where N_i is the number of molecules of molecular mass M_i, and $\sum$ denotes summation over all i molecular masses.

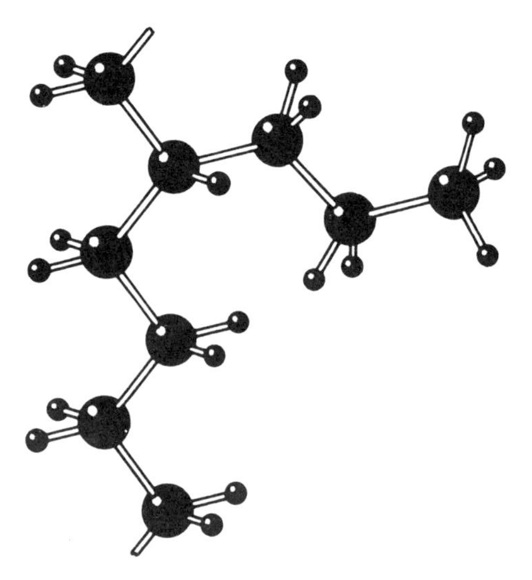

Figure 1.3 A chain branch in polyethylene

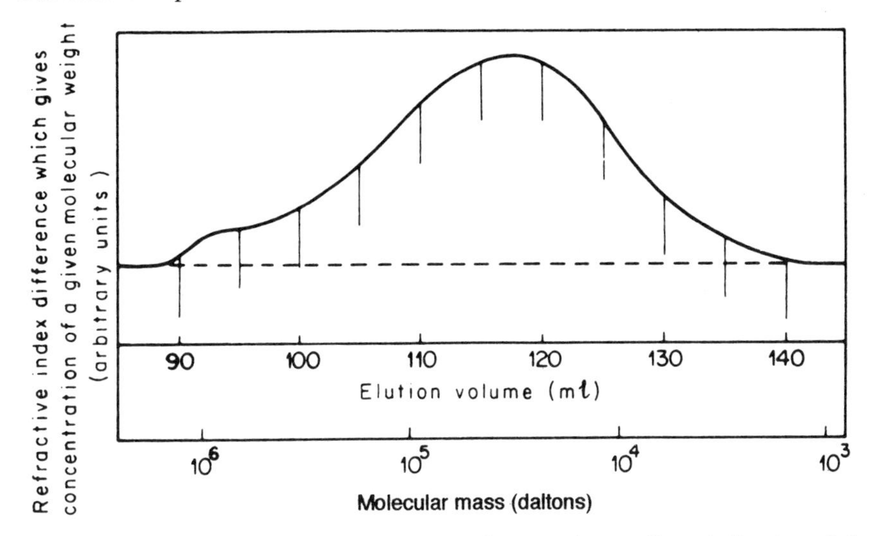

Figure 1.4 The gel permeation chromatograph trace gives a direct indication of the molecular distribution. (Results obtained in Marlex 6009 by Dr T. Williams)

The weight average molecular mass is always higher than the number average, as the former is strongly influenced by the relatively small number of very long (massive) molecules. The ratio of the two averages gives a general idea of the width of the molecular mass distribution.

Fundamental measurements of average molecular mass must be performed on solutions so dilute that intermolecular interactions can be ignored or compensated for. The commonest techniques are osmotic pressure for the number average and light scattering for the weight average. Both methods are rather lengthy, so in practice an average molecular mass was often deduced from viscosity measurements, of either a dilute solution of the polymer (which relates to $\bar{M}_n$) or a polymer melt (which relates to $\bar{M}_w$). Each method yielded a different average value, which made it difficult to correlate specimens characterized by different groups of workers.

The molecular mass distribution is important in determining flow properties, and so may affect the mechanical properties of a solid polymer indirectly by influencing the final physical state. Direct correlations of molecular mass to viscoelastic behaviour and brittle strength have also been obtained.

1.1.4 Chemical and Steric Isomerism and Stereoregularity

A further complication of the chemical structure of polymers lies in the possibility of different chemical isomeric forms within a repeat unit or between a series of repeat units. Natural rubber and gutta percha are chemically both polyisoprene, but the former is the *cis* form and the latter the *trans* (see Figure 1.5). The characteristic properties of rubber are a consequence of the loose packing of molecules (i.e. large *free volume*) that arises from its structure.

Vinyl monomer units

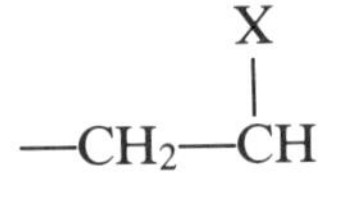

can be added to a growing chain

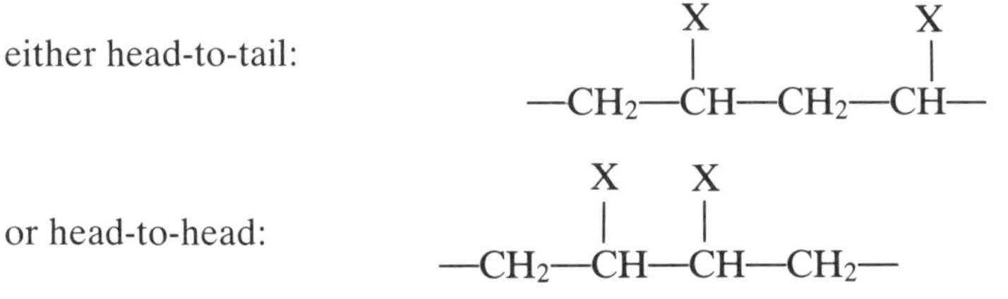

Head-to-tail substitution is usual, and only a small proportion of head-to-head linkages can produce a reduction in the tensile strength because of the loss of regularity.

Stereoregularity provides a more complex situation, which we will examine in terms of the simplest type of vinyl polymer (Figure 1.6), and for which we shall suppose that the polymer chain is a planar zigzag. Two very simple regular polymers can be constructed. In the first (Figure 1.6(a)) the substituent groups are all added in an identical manner to give an *isotactic* polymer. In the second regular polymer (Figure 1.6(b)) there is an inversion

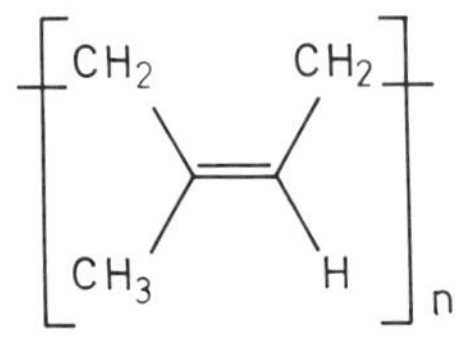

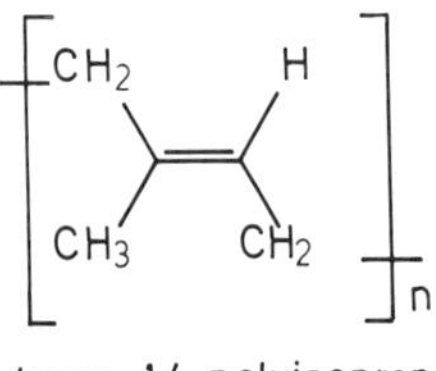

Figure 1.5 (*cis*) 1,4-polyisoprene, (*trans*) 1,4-polyisoprene

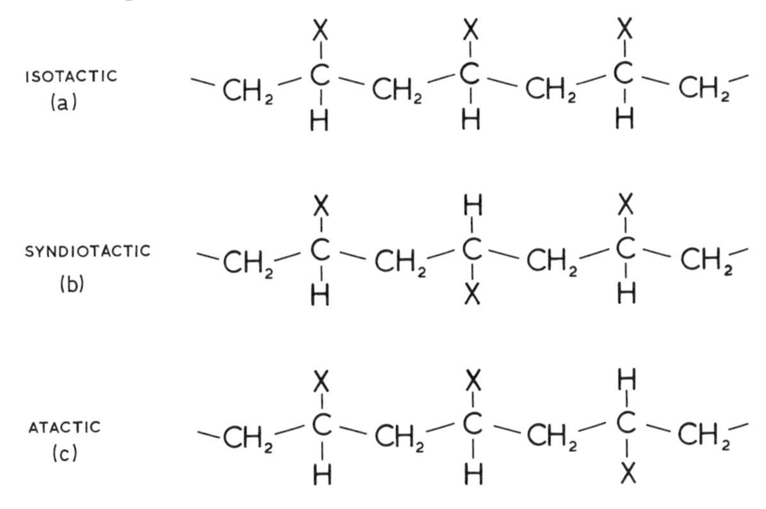

Figure 1.6 A substituted α-olefin can take three stereosubstituted forms

of the manner of substitution between consecutive units, giving a *syndiotactic* polymer for which the substituent groups alternate regularly on opposite sides of the chain. The regular sequence of units is called *stereoregularity*, and stereoregular polymers are crystalline and can possess high melting-points. The working range of a polymer is thereby extended compared with the amorphous form, whose range is limited by the lower softening point. The final alternative structure is formed when the orientation of successive substituents takes place randomly (Figure 1.5(c)) to give an irregular *atactic* polymer which is incapable of crystallizing. Polypropylene $(—CH_2CHCH_3—)_n$ was for many years obtainable only as an atactic polymer, and its widespread use began only when stereospecific catalysts were developed to produce the isotactic form. Even so some faulty substitution occurs, and atactic chains can be separated from the rest of the polymer by solvent extraction.

1.1.5 Liquid Crystalline Polymers

Liquid crystals (or plastic crystals as they are sometimes called) are low molecular mass materials which show molecular alignment in one direction, but not three-dimensional crystalline order. During the last twenty years, liquid crystalline polymers have been developed where the polymer chains are so straight and rigid that small regions of almost uniform orientation (domains) separated by distinct boundaries are produced. In the case where these domains occur in solution, polymers are termed *lyotropic*. Where the domains occur in the melt, the polymers are termed *thermotropic*.

An important class of lyotropic liquid crystal polymers are the aramid polymers such as polyparabenzamide

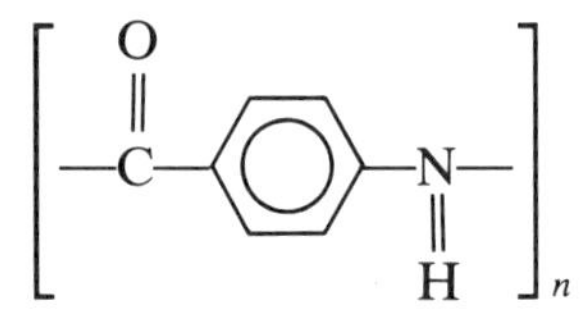

and polyparaphenylene terephthalamide

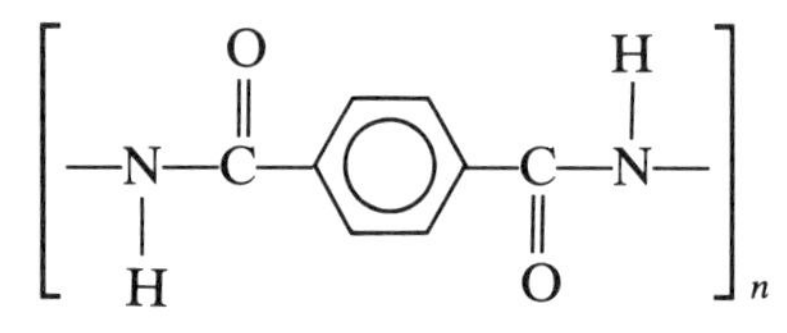

better known as Kevlar, which is a commercially produced high stiffness and high strength fibre. It is important to emphasize that although Kevlar fibres are prepared by spinning a lyotropic liquid crystalline phase, the final fibre shows clear evidence of three-dimensional order.

Important examples of thermotropic liquid crystalline polymers are copolyesters produced by condensation of hydroxybenzoic acid (HBA)

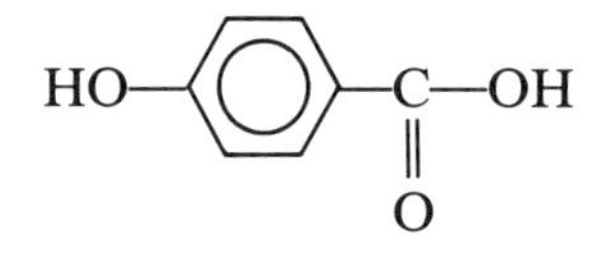

and 2-6 hydroxynaphthoic acid (HNA)

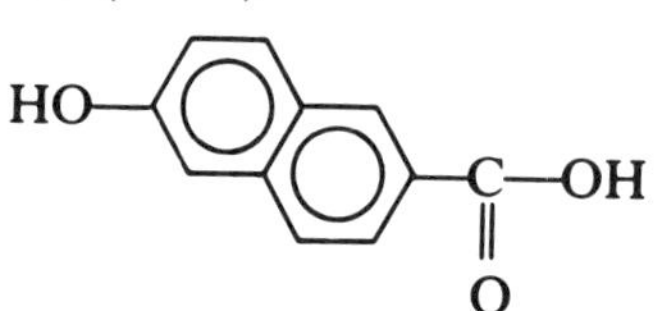

most usually in the proportions HBA:HNA = 73:27.

In addition to these main-chain liquid crystalline polymers, there are also side-chain liquid crystalline polymers, where the liquid crystalline nature arises from the presence of rigid straight side-chain units (called the mesogens) chemically linked to an existing polymer backbone either directly or via flexible spacer units.

The review by Noël and Navard [3] gives further information on liquid crystalline polymers, including methods of preparation.

1.1.6 Blends, Grafts and Copolymers

A *blend* is a physical mixture of two or more polymers. A *graft* is formed when long side chains of a second polymer are chemically attached to the base polymer. A *copolymer* is formed when chemical combination exists in the main chain between two or more polymers, $[A]_n$, $[B]_n$, etc. The two principal forms are block copolymers [AAAA . . .] [BBB . . .] and *random* copolymers, the latter having no long sequences of A or B units

All these processes are commonly used to enhance the ductility and toughness of brittle homopolymers or increase the stiffness of rubbery polymers. An example of a blend is acrylonitrile–butadiene–styrene copolymer (ABS), where the separate rubber phase gives much improved impact resistance.

The basic properties of polymers may be enhanced by physical as well as chemical means. An important example is the use of finely divided carbon black as a filler in rubber compounds. Polymers may be combined with stiffer filaments, such as glass or carbon fibres, to form a composite. We shall show later that some semicrystalline polymers may be treated as composites at a molecular level.

It must not be forgotten that all useful polymers contain small quantities of additives to aid processing and increase the resistance to degradation. The physical properties of the base polymer may be modified by the presence of such additives.

1.2 PHYSICAL STRUCTURE

The physical properties of a polymer of a given chemical composition are dependent on two distinct aspects of the arrangement of the molecular chains in space.

1. The arrangement of a single chain without regard to its neighbour: rotational isomerism.
2. The arrangment of chains with respect to each other: orientation and crystallinity.

1.2.1 Rotational Isomerism

Rotational isomerism arises because of the alternative conformations of a molecule that can result from the possibility of hindered rotation about the many single bonds in the structure. Spectroscopic techniques [4] developed in small molecules have been extended to polymers, and as an example we illustrate (Figure 1.7) the alternative *trans* and *gauche* conformations in the glycol residue of polyethylene terephthalate [5]: the former is a crystalline conformation, but the latter is present in amorphous regions.

To pass from one rotational isomeric form to another requires that an energy barrier be surmounted (Figure 1.8), so that the possibility of the

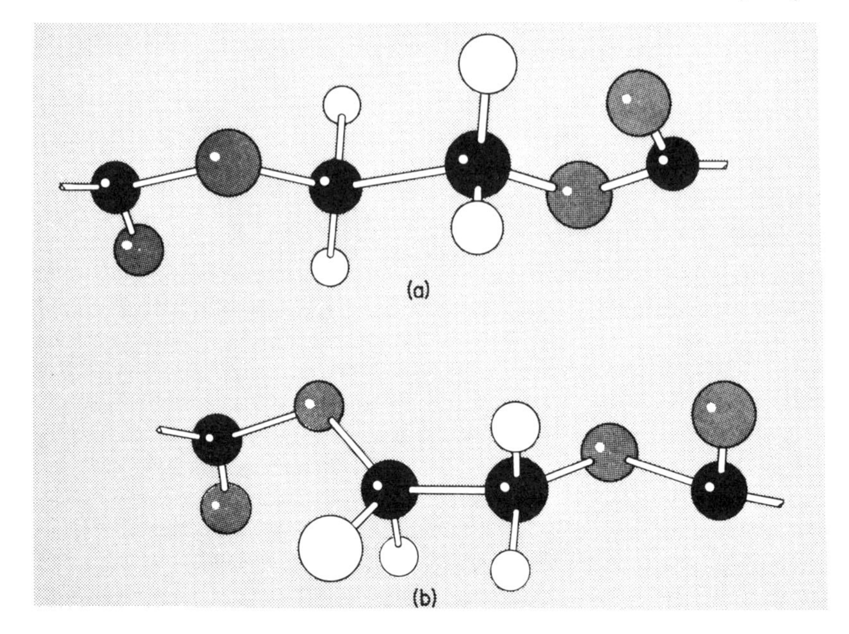

Figure 1.7 Polyethylene terephthalate in the crystalline *trans* conformation (a) and *gauche* conformation (b) which is present in 'amorphous' regions (after Grime and Ward [5])

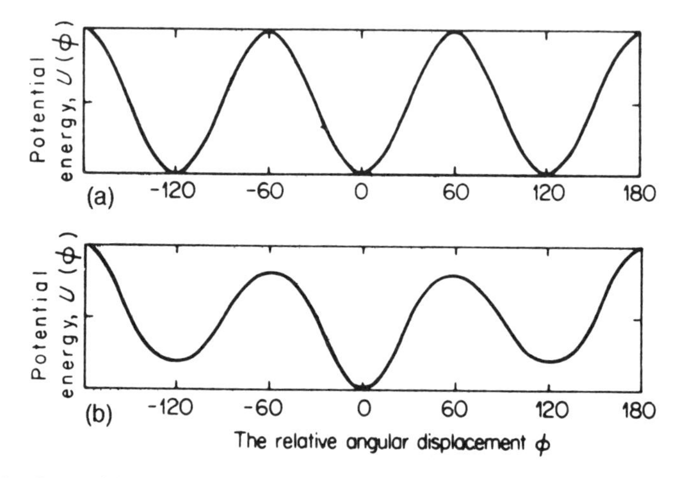

Figure 1.8 Potential energy for rotation (a) around the C—C bond in ethane and (b) around the central C—C bond in n-butane. (Reproduced with permission from McCrum, Read and Williams, *Anelastic and Dielectric Effects in Polymeric Solids*, Wiley, London, 1967)

chain molecules changing their conformations depends on the relative magnitude of the energy barrier compared with thermal energies and the perturbing effects of applied stress. Hence arises the possibility of linking molecular flexibility to deformation mechanisms, a theme to which we will return on several occasions.

1.2.2 Orientation and Crystallinity

When we consider the arrangement of molecular chains with respect to each other there are again two largely separate aspects, those of molecular orientation and crystallinity. In semicrystalline polymers this distinction may at times be an artificial one.

When cooled from the melt many polymers form a disordered structure called the amorphous state. Some of these materials, such as polymethyl methacrylate, polystyrene and rapidly cooled (melt-quenched) polyethylene terephthalate, have a comparatively high modulus at room temperature, but others, such as natural rubber and atactic polypropylene, have a low modulus. These two types of polymer are often termed *glassy* and *rubber-like* respectively, and we shall see that the form of behaviour exhibited depends on the temperature relative to a glass–rubber transition temperature (T_g) that is dependent on the material and the test method employed. Although an amorphous polymer may be modelled as a random tangle of molecules (Figure 1.9(a)), features such as the comparatively high density [6] show that the packing cannot be completely random. X-ray diffraction techniques indicate no distinct structure, rather a broad diffuse maximum (the amorphous halo) that indicates a preferred distance of separation between the molecular chains.

When an amorphous polymer is stretched the molecules may be preferentially aligned along the stretch direction. In polymethyl methacrylate and polystyrene such molecular orientation may be detected by optical methods, which measure the small difference between the refractive index in the stretch direction and that in the perpendicular direction. X-ray diffraction methods still reveal no evidence of three-dimensional order, so the structure may be regarded as a somewhat oriented tangled skein (Figure 1.9(b)), which is oriented amorphous but not crystalline.

In polyethylene terephthalate, however, stretching produces both molecular orientation and small regions of three-dimensional order, termed crystallites, because the orientation processes have brought the molecules into adequate juxtaposition for regions of three-dimensional order to form.

Many polymers, including polyethylene terephthalate, also crystallize if they are cooled slowly from the melt. In this case we may say that they are crystalline but unoriented. Although such specimens are unoriented in the macroscopic sense, i.e. they possess isotropic bulk mechanical properties,

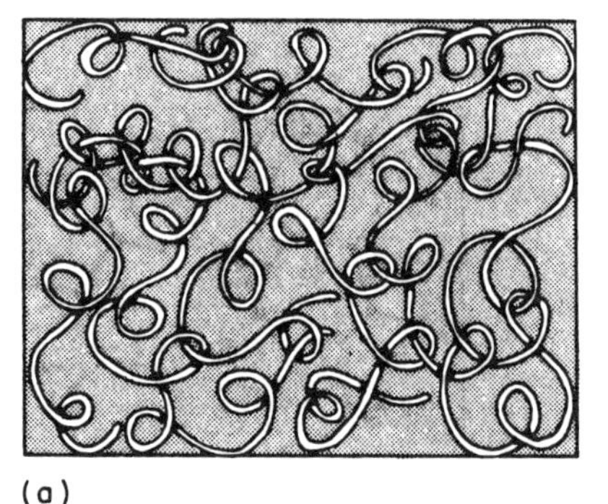

(a)

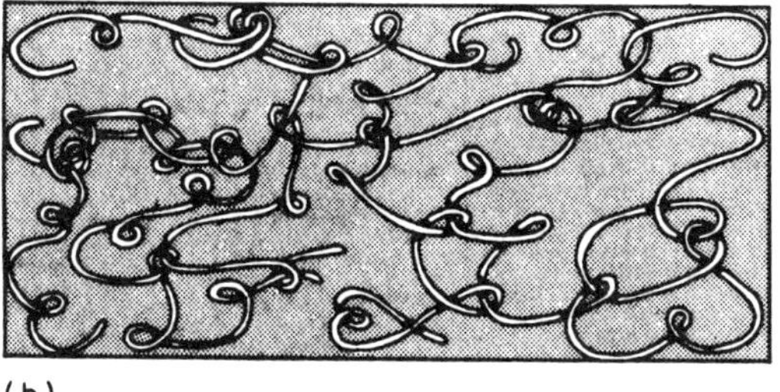

(b)

Figure 1.9 Schematic diagrams of (a) unoriented amorphous polymer and (b) oriented amorphous polymer.

they are not homogeneous in the microscopic sense and often show a spherulitic structure under a polarizing microscope.

In summary, it may be said that for a polymer to crystallize the molecule must have a regular structure, the temperature must be below the crystal melting-point, and sufficient time must be available for the long molecules to become ordered in the solid state.

The structure of the crystalline regions of polymers can be deduced from wide angle X-ray diffraction patterns of highly stretched specimens. When the stretching is uniaxial the patterns are related to those obtained from fully oriented single crystals. The crystal structure of polyethylene was determined by Bunn [7] as long ago as 1939 (Figure 1.10).

In addition to the discrete reflections from the crystallites, the diffraction pattern of a polymer also shows diffuse scattering which is attributed to amorphous regions. Such polymers are said to be *semicrystalline*, with the crystalline fraction being controlled by molecular regularity. By comparing the relative amounts of crystalline and amorphous scattering of X-rays

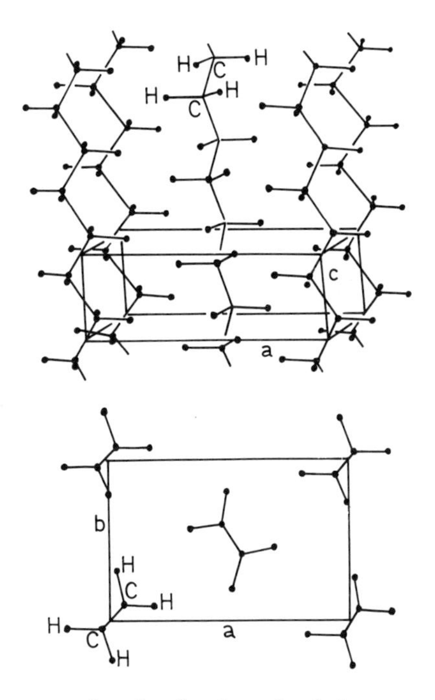

Figure 1.10 Arrangement of molecules in polyethylene crystallites. (From *Fibres from Synthetic Polymers* (ed. R. Hill), Elsevier, Amsterdam, 1953)

the crystallinity has been found to vary from more than 90% for linear polyethylene to about 30% for oriented polyethylene terephthalate.

Single crystals of many polymers may be grown from dilute solution [8]. Linear polyethylene, for example, forms single crystal lamellae with lateral dimensions of the order of 10–20 μm and a thickness about 10 nm. Electron diffraction shows that the molecular chains are oriented approximately normal to the lamellar surface, and as the molecules are typically about 1 μm in length they must be folded back and forth within the crystals. A diagrammatic representation, in which the folds are sharp and regular, with adjacent re-entry, is given in Figure 1.11, but neutron-scattering experiments suggest that such a picture may be oversimplified [9].

The crystallization of polymers from the melt is a more controversial process, as a single molecule is unlikely to be laid down on a crystalline substrate without interference from its neighbours, and it might be expected that the highly entangled topology of the chains which exists in the melt must be substantially retained in the semicrystalline state. There is, however, much evidence to support the existence of a lamellar morphology,

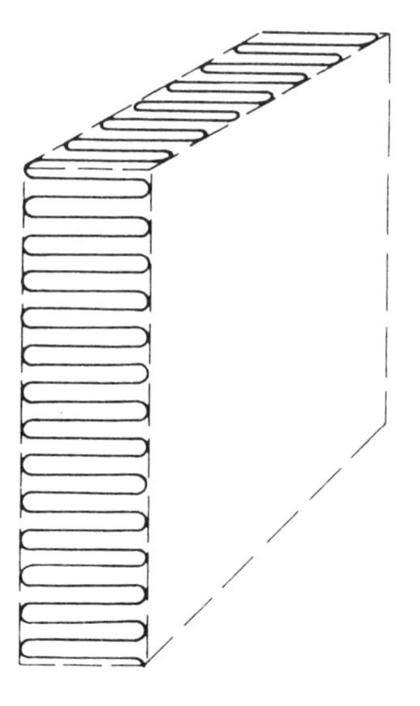

Figure 1.11 Diagrammatic representation of chain folding in polymer crystals with the folds drawn sharp and regular (Reproduced with permission from Keller, *Rep. Progr. Phys.*, **31**, 623 (1969))

with the separation between chain folds being greatest for material formed during the first stage of crystallisation. Folded-chain lamellae grow outwards from initial nucleation centres, with the spaces between lamellae filled by material which crystallized later. Typically, spherulites 1–10 μm in diameter are formed, which grow outwards until they impinge upon neighbouring spherulites (Figure 1.12). Although chain folding is predominant in the crystallization process there are still many chains which thread their way through the structure and provide continuity. Early models of spherulites are now considered to be oversimplified. For a good review, itself overtaken in some aspects, see the text by Bassett [10], and also more recent work directed by the same author.

Orientation through plastic deformation (*drawing*) destroys the spherulitic structure. What remains is determined to a large extent by the degree of crystallinity. Mechanical testing, described in the subsequent chapters, has helped to establish several models. At one extreme, some highly oriented, highly crystalline specimens of linear polyethylene behave as blocks or lamellae of crystalline material, connected together by tie molecules or crystalline bridges, and separated by the amorphous component. Such materials can in some respects be treated as microscopic composites. At the

Figure 1.12 A photograph of typical spherulitic structure under a polarizing microscope

other extreme one has materials like polyethylene terephthalate in which the crystalline and amorphous components are so intermixed that a single phase model appears to be more appropriate.

The current state of knowledge suggests that chain folding and the threading of molecules through the crystalline region both occur in typical polymers.

A schematic attempt to illustrate this situation, and other types of irregularity, is given in Figure 1.13.

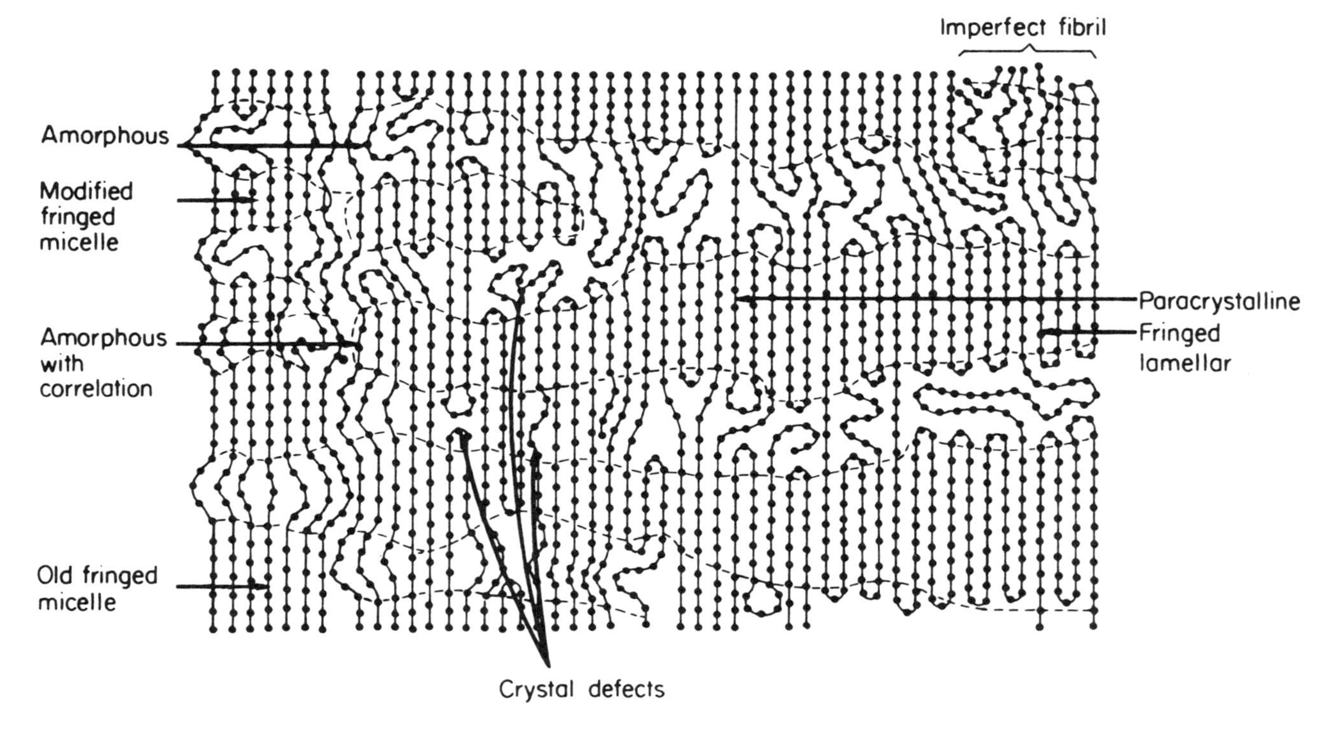

Figure 1.13 Schematic composite diagram of different types of order and disorder in oriented polymers. (Reproduced with permission from Hosemann, Polymer, **3**, 349 (1962)

REFERENCES

1. M. F. Vaughan, *Nature*, **188**, 55 (1960).
2. J. C. Moore, *J. Polymer Sci., A,* **2**, 835 (1964).
3. C. Noël and P. Navard, *Prog. Polym. Sci.,* **16**, 55 (1991).
4. S. I. Mizushima, *Structure of Molecules and Internal Rotation*, Academic Press, New York, 1954.
5. D. Grime and I. M. Ward, *Trans. Faraday Soc.,* **54**, 959 (1958).
6. R. E. Robertson, *J. Phys. Chem.,* **69**, 1575 (1965).
7. C. W. Bunn, *Trans. Faraday Soc.,* **35**, 482 (1939).
8. E. W. Fischer, *Naturforschung*, **12a**, 753 (1957); A. Keller, *Phil. Mag.,* **2**, 1171 (1957); P. H. Till, *J. Polymer Sci.,* **24**, 301 (1957).
9. A. Keller, *Disc. Faraday Soc.,* **68**, 145 (1979).
10. D. C. Bassett, *Principles of Polymer Morphology*, Cambridge University Press, 1981.

FURTHER READING

F. W. Billmeyer, *Textbook of Polymer Science,* Wiley, New York.
H. Tadokoro, *Structure of Crystalline Polymers,* Wiley, New York, 1979.
I. M. Ward (ed.), *Structure and Properties of Oriented Polymers,* Applied Science Publishers, London, 1975.
B. Wunderlich, *Macromolecular Physics*, Vols 1 and 2, Academic Press, New York, 1973, 1976.

Chapter 2

The Deformation of an Elastic Solid

In several of the following chapters we consider the behaviour of solid polymers subject to large deformations and show also that in general these materials are viscoelastic, which means that stress (or strain) varies with time. As a starting-point, however, we need to consider a polymer as a linear elastic solid: when a load is applied the deformation is instantaneous, after which it remains constant until the load is removed, when the recovery is instantaneous and complete; linearity means that stress and strain are always proportional to one another.

2.1 THE STATE OF STRESS

The convention used in Figure 2.1 shows stresses designated as positive in the direction of the outward-facing normal. As a consequence of this definition inward-acting stresses, such as hydrostatic pressure above that of the surrounding atmosphere, are defined as negative quantities.

It is, however, customary when considering yield behaviour to envisage that hydrostatic pressure causes an increase in the yield stress. For this reason the hydrostatic pressure p in Chapters 10 and 11 is defined as $p = -(\sigma_{xx} + \sigma_{yy} + \sigma_{zz})$.

The components of stress in a body are defined by considering the forces acting on an infinitesimal cubical volume element (Figure 2.1) whose edges are parallel with coordinate axes x, y, z. In equilibrium the forces per unit area acting on the cube faces are P_1 on the yz plane, P_2 on the zx plane, P_3 on the xy plane. Equilibrium implies that similar forces must act on the directly opposite hidden faces of the cube in Figure 2.1.

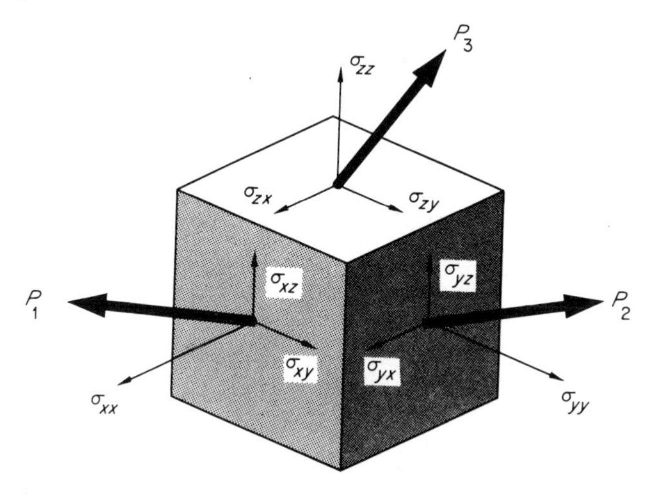

Figure 2.1 The components of stress

The forces are then resolved into their nine components in the x, y and z directions as follows:

$$P_1\text{: } \sigma_{xx}, \sigma_{xy}, \sigma_{xz},$$
$$P_2\text{: } \sigma_{yx}, \sigma_{yy}, \sigma_{yz},$$
$$P_3\text{: } \sigma_{zx}, \sigma_{zy}, \sigma_{zz},$$

where the first subscript refers to the direction of the *normal* to the plane on which the stress acts, and the second subscript to the *direction* of the stress. As the cube is in equilibrium the net torque acting on it is zero, which implies the equalities:

$$\sigma_{xy} = \sigma_{yx}, \sigma_{xz} = \sigma_{zx}, \sigma_{yz} = \sigma_{zy}.$$

The components of stress are therefore defined by six independent quantities: the normal stresses σ_{xx}, σ_{yy} and σ_{zz}, together with the shear stresses σ_{xy}, σ_{yz} and σ_{zx}. It is usual to write these components as the elements of a matrix, which is called the stress tensor σ_{ij} (for an explanation of tensors see Appendix 1).

$$\sigma_{ij} = \begin{bmatrix} \sigma_{xx} & \sigma_{xy} & \sigma_{xz} \\ \sigma_{xy} & \sigma_{yy} & \sigma_{yz} \\ \sigma_{xz} & \sigma_{yz} & \sigma_{zz} \end{bmatrix}$$

The state of stress at a point in a body is determined when we can specify the normal components and the shear components of stress acting on a plane drawn in any direction through the point. If we know these six components at a given point the stresses acting on any plane through the point can be calculated (see [1], section 67 and [2], section 47).

2.2 THE STATE OF STRAIN

In elementary elasticity, which is concerned with the elastic behaviour of isotropic materials, it is usual to consider two types of strain only. First, there is extensional strain which is defined as the fractional increase in length in the stretching direction (Figure 2.2(a)), and this definition of strain is an essential ingredient of the very familiar Hooke's law of elasticity. Secondly, there is a simple shear strain which is defined by the displacement of parallel planes, as shown in Figure 2.2(b). The lateral displacement divided by the perpendicular distance between the planes defines the 'engineering' shear strain which is the angle θ in figure. (For further discussion of shear strain see Appendix 1.)

For deformation of a material which involves extensions and shears superimposed in quite general directions, we require a more general starting-point to define extensional and shear strains, i.e. components of extensional and shear strain, analogous to the components of tensile and shear stress.

2.2.1 The Engineering Components of Strain

The displacement of any point P_1 (see Figure 2.3) in the body may be resolved into its components u, v and w parallel to x, y and z (Cartesian coordinate axes chosen in the undeformed state), so that if the coordinates of the point in the undisplaced position were (x, y, z) they become $(x + u, y + v, z + w)$ on deformation. In defining the strains we are not interested in the absolute displacement or in rotation but in the deformation. The

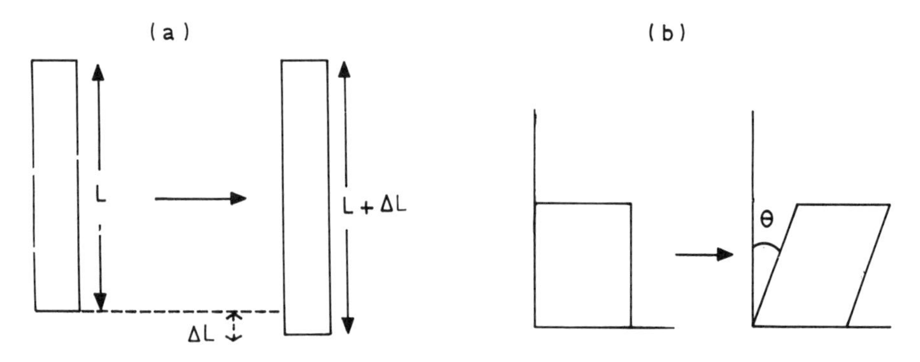

Figure 2.2 Illustration of (a) extensional strain, (b) simple shear strain

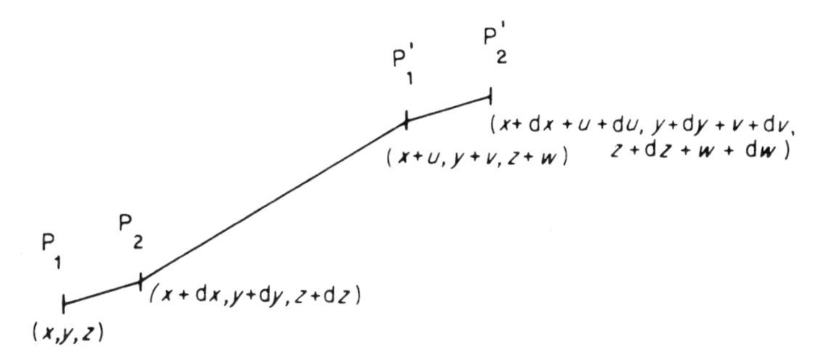

Figure 2.3 The displacement produced by deformation

latter is the displacement of a point relative to adjacent points. Consider a point P_2, very close to P_1, which in the undisplaced position had coordinates $(x + \mathrm{d}x,\ y + \mathrm{d}y,\ z + \mathrm{d}z)$ and let the displacement which it has undergone have components $(u + \mathrm{d}u,\ v + \mathrm{d}v,\ w + \mathrm{d}w)$. The quantities required are then $\mathrm{d}u$, $\mathrm{d}v$ and $\mathrm{d}w$, the relative displacements.

If $\mathrm{d}x$, $\mathrm{d}y$ and $\mathrm{d}z$ are sufficiently small i.e. infinitesimal,

$$\mathrm{d}u = \frac{\partial u}{\partial x}\mathrm{d}x + \frac{\partial u}{\partial y}\mathrm{d}y + \frac{\partial u}{\partial z}\mathrm{d}z,$$

$$\mathrm{d}v = \frac{\partial v}{\partial x}\mathrm{d}x + \frac{\partial v}{\partial y}\mathrm{d}y + \frac{\partial v}{\partial z}\mathrm{d}z,$$

$$\mathrm{d}w = \frac{\partial w}{\partial x}\mathrm{d}x + \frac{\partial w}{\partial y}\mathrm{d}y + \frac{\partial w}{\partial z}\mathrm{d}z.$$

Thus we require to define the nine quantities

$$\frac{\partial u}{\partial x},\ \frac{\partial u}{\partial y},\ \ldots,\ \text{etc.}$$

For convenience these nine quantities are regrouped and denoted as follows:

$$e_{xx} = \frac{\partial u}{\partial x} \qquad e_{yy} = \frac{\partial v}{\partial y}, \qquad e_{zz} = \frac{\partial w}{\partial z},$$

$$e_{yz} = \frac{\partial w}{\partial y} + \frac{\partial v}{\partial z}, \qquad e_{zx} = \frac{\partial u}{\partial z} + \frac{\partial w}{\partial x}, \qquad e_{xy} = \frac{\partial v}{\partial x} + \frac{\partial u}{\partial y},$$

$$2\bar{\omega}_x = \frac{\partial w}{\partial y} - \frac{\partial v}{\partial z}, \qquad 2\bar{\omega}_y = \frac{\partial u}{\partial z} - \frac{\partial w}{\partial x}, \qquad 2\bar{\omega}_z = \frac{\partial v}{\partial x} - \frac{\partial u}{\partial y}.$$

The first three quantities e_{xx}, e_{yy} and e_{zz} correspond to the fractional expansions or contractions along the x, y and z axes of an infinitesimal element at P_1. The second three quantities e_{yz}, e_{zx} and e_{xy} correspond to the components of shear strain in the yz, zx and xy planes respectively. The last three quantities $\bar{\omega}_x$, $\bar{\omega}_y$ and $\bar{\omega}_z$ do not correspond to a deformation of the element at P_1, but are the components of its rotation as a rigid body.

The concept of shear strain can be conveniently illustrated by a diagram showing the two-dimensional situation of shear in the *yz* plane (Figure 2.4). All strains are considered small, but must be depicted as large for clarity.

ABCD is an infinitesimal square which has been displaced and deformed into the rhombus A′B′C′D′, θ_1 and θ_2 being the (small) angles which A′D′ and A′B′ make with the *y* and *z* axes respectively. Now

$$\tan\theta_1 = \theta_1 = \frac{\mathrm{d}w}{\mathrm{d}y} \rightarrow \frac{\partial w}{\partial y},$$

$$\tan\theta_2 = \theta_2 = \frac{\mathrm{d}v}{\mathrm{d}z} \rightarrow \frac{\partial v}{\partial z}.$$

The shear strain in the *yz* plane is given by

$$e_{yz} = \frac{\partial w}{\partial y} + \frac{\partial v}{\partial z} = \theta_1 + \theta_2.$$

Here $2\bar{\omega}_x = \theta_1 - \theta_2$ does not correspond to a deformation of ABCD but to twice the angle through which AC has been rotated. The deformation is therefore defined by the first six quantities e_{xx}, e_{yy}, e_{zz}, e_{yz}, e_{zx}, e_{xy}, which are called the components of strain. It is important to note that *engineering strains* have been defined (see below).

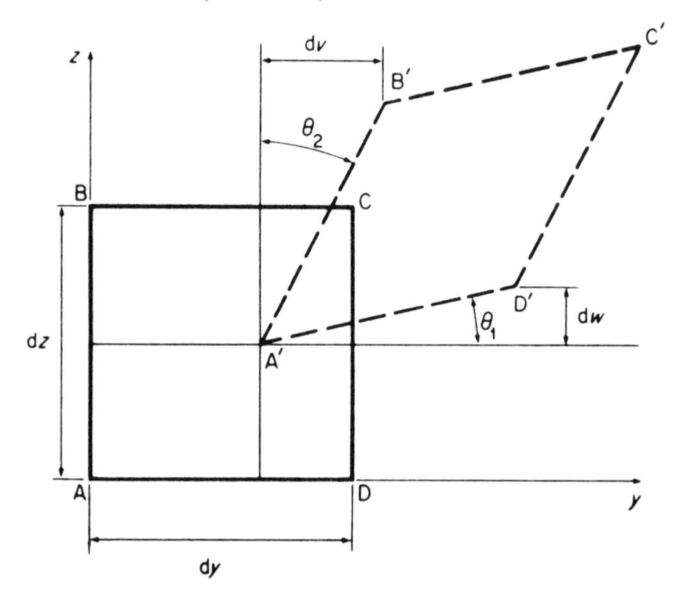

Figure 2.4 Shear strains. (Reproduced with permission from Kolsky, *Stress Waves in Solids*, Dover, New York, (1963))

2.3 THE GENERALIZED HOOKE'S LAW

The most general *linear* relationship between stress and strain is obtained by assuming that each of the six independent components of stress is linearly related to each of the six independent components of strain. Thus

$$\sigma_{xx} = ae_{xx} + be_{yy} + ce_{zz} + de_{xz} + \ldots \text{etc.}$$

and

$$e_{xx} = a'\sigma_{xx} + b'\sigma_{yy} + c'\sigma_{zz} + d'\sigma_{xz} + \ldots \text{etc.}$$

where a, $b \ldots$, a', b', are constants. This expression is the generalized Hooke's law for both isotropic and anisotropic solids. Most of this textbook is concerned with isotropic solids where there is no coupling between tensile stresses and shear stresses, or between shear stresses and extensional strains, so that these equations reduce to forms such as

$$\sigma_{xx} = ae_{xx} + be_{yy} + ce_{zz} \qquad \sigma_{xz} = fe_{xz}$$

and

$$e_{xx} = a'\sigma_{xx} + b'\sigma_{yy} + c'\sigma_{zz}$$
$$e_{xz} = f'\sigma_{xz}.$$

To develop the generalized Hooke's law for isotropic materials it is convenient to construct equations for the strains e_{xx}, e_{yy}, etc. in terms of the applied stresses σ_{xx}, σ_{yy}, etc, and so define Young's modulus E and Poisson's ratio ν. An applied stress σ_{xx} will produce a strain

$$e_{xx} = \frac{\sigma_{xx}}{E}$$

in the x direction and strains

$$e_{yy} = \frac{-\nu}{E}\sigma_{xx} \quad \text{and} \quad e_{zz} = \frac{-\nu}{E}\sigma_{xx}$$

in the y and z directions respectively. (Note that Poisson's ratio ν which defines the ratio of the contraction strain e_{yy} to the extensional strain e_{xx}, is conventionally positive, whereas the contraction e_{yy} is negative.)

A shear strain e_{xz} is related to the corresponding shear stress σ_{xz} by the relationship $e_{xz} = \sigma_{xz}/G$, where G is the shear modulus.

Thus we obtain the stress–strain relationships which are the starting-point in many elementary textbooks of elasticity ([1], pp. 7–9):

$$e_{xx} = \frac{1}{E}\sigma_{xx} - \frac{\nu}{E}(\sigma_{yy} + \sigma_{zz}),$$

$$e_{yy} = \frac{1}{E}\sigma_{yy} - \frac{\nu}{E}(\sigma_{xx} + \sigma_{zz}),$$

$$e_{zz} = \frac{1}{E}\sigma_{zz} - \frac{\nu}{E}(\sigma_{xx} + \sigma_{yy})$$

$$e_{xz} = \frac{1}{G}\sigma_{xz}, \qquad e_{yz} = \frac{1}{G}\sigma_{yz}, \qquad e_{xy} = \frac{1}{G}\sigma_{xy}.$$

A bulk modulus K, related to the fractional change in volume, can also be defined, but only two of the quantities E, ν, G, K are independent. For example

$$G = \frac{E}{2(1+\nu)} \quad \text{and} \quad K = \frac{E}{3(1-2\nu)}$$

2.4 FINITE STRAIN ELASTICITY: THE BEHAVIOUR OF POLYMERS IN THE RUBBER-LIKE STATE

In the rubber-like state a polymer may be subjected to large deformation and still show complete recovery. The behaviour of a rubber band stretching to two or three times its original length and, when released, recovering instantaneously to its original shape, is a matter of common experience. This is *elastic* behaviour at large strains. The first stage in developing an understanding of this behaviour is to consider how the fact that the strains are large affects the definition of stress and strain.

2.4.1 The Definition of Components of Stress

In small strain elasticity theory, the components of stress in the deformed body are defined by considering the equilibrium of an elemental cube of material (section 2.1 above). When the strains are small the areas of the cube faces are to a first approximation unaffected by the strain. It is therefore of no consequence whether the components of stress are referred to an elemental cube in the deformed body or to an elemental cube in the undeformed body. For finite strains this is no longer true, but for simplicity we will *choose* to define the components of stress with reference to the equilibrium of a cube in the *deformed* body. The components of stress can then be defined in exactly the same manner as for small strain elasticity.

In practice, the significance of this decision is that we must be careful to distinguish between true stresses σ which relate to the stress tensor and the convenience of using nominal stresses f which are defined as the force per unit area of unstrained cross-section. This distinction will be further exemplified when the theory is developed for simple tension (section 2.4.3 below).

2.4.2 The Generalized Definition of Strain

In this introduction to finite elasticity it is only necessary to develop the most elementary definition of finite strains (for a more comprehensive discussion, see [3], Chapter 3).

Consider a system of rectangular coordinates axes with an origin O (Figure 2.5) in which the point P has the co-ordinates x, y, z. When the body is deformed, consider that the origin of these coordinates is fixed at a point in the body, and that the point P moves to a point P′ with coordinates x', y', z' where

$$x' = \lambda_1 x \qquad y' = \lambda_2 y \qquad z' = \lambda_3 z.$$

A useful schematic way of representing the deformation, is to consider the change in the dimensions of a cube of unit dimensions (Figure 2.6). On deformation the cube deforms to a parallelepiped with sides of length λ_1, λ_2 and λ_3 in the x, y and z directions respectively.

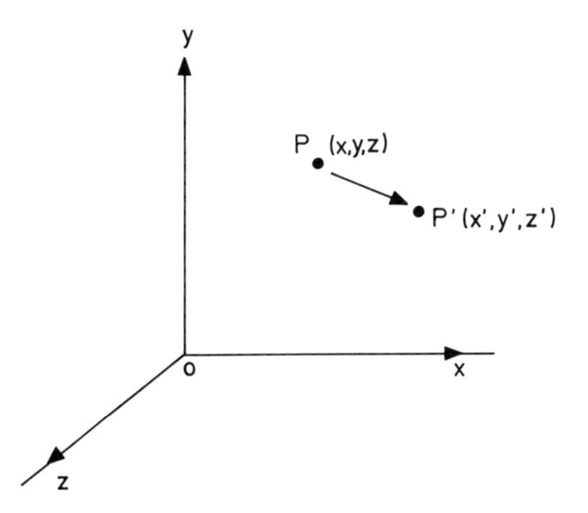

Figure 2.5 Displacement of point P to P′

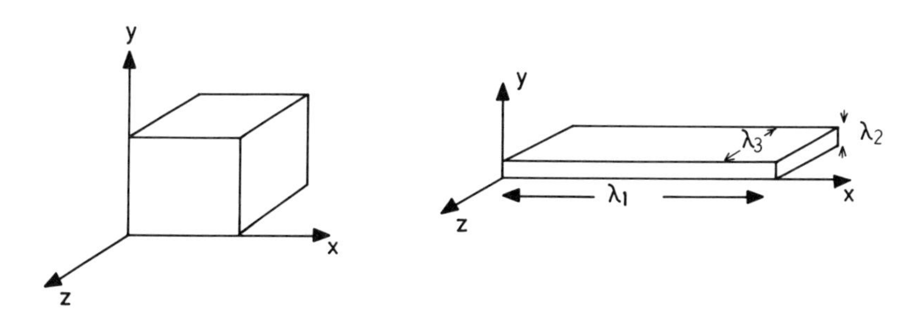

Figure 2.6 Deformation of a cube to a parallelepiped defines deformation ratios λ_1, λ_2, λ_3.

The quantities λ_1, λ_2 and λ_3 which define the deformation are called the deformation ratios, because they define the ratio of the length of lines in the x, y and z directions in the *deformed* body to their length in the *undeformed* body. Note that with the exception of bodies for which Poisson's ratio is negative, if one of the ratios is greater than unity at least one of the ratios must be less than unity.

Finite strain is most conveniently defined by these three deformation ratios, although we can equally well define the three components of extensional finite strain $\boldsymbol{\varepsilon}_{xx}$, $\boldsymbol{\varepsilon}_{yy}$ and $\boldsymbol{\varepsilon}_{zz}$ (written bold to distinguish them from the small strain components defined in section 2.4.1 above) as

$$\boldsymbol{\varepsilon}_{xx} = \tfrac{1}{2}(\lambda_1^2 - 1), \qquad \boldsymbol{\varepsilon}_{yy} = \tfrac{1}{2}(\lambda_2^2 - 1), \qquad \boldsymbol{\varepsilon}_{zz} = \tfrac{1}{2}(\lambda_3^2 - 1).$$

This generalized definition of strain, which is not limited to small strains, is compatible with our definition of strain in section 2.2 above. For example, for small strain

$$\lambda_1^2 = (1 + e_{xx})^2 \to 1 + 2e_{xx} \quad \text{and} \quad \boldsymbol{\varepsilon}_{xx} = e_{xx}.$$

A deformation in which the three coordinate axes x, y and z in the undeformed state are not rotated by the deformation is called pure strain (often pure homogeneous strain to include the idea that it is also uniform throughout the body) because the shear strain components are zero. We will always develop the theories of rubber elasticity with this simplification, because it involves no loss of generality, as can be appreciated from the following.

Consider a sphere drawn in the undeformed solid (Figure 2.7). For a quite general deformation this sphere is transformed into an ellipsoid (called the strain ellipsoid). It can be seen that the principal axes of this ellipsoid do not coincide with the x, y, z axes which were chosen to refer to the undeformed state. However, without any loss of generality, we could have *chosen* to define the x, y, z axes in the undeformed state to coincide with the principal axes of the strain ellipsoid. We can therefore define the deformation in terms of pure strain (only extensional strain components defined by λ_1, λ_2

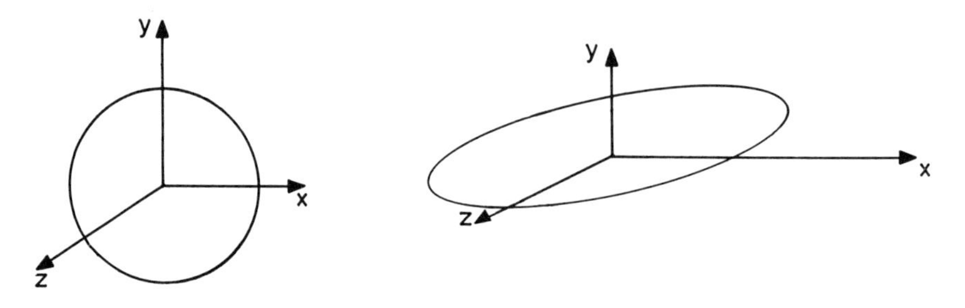

Figure 2.7 The strain ellipsoid.

and λ_3) provided that we choose our axes carefully by reference to the deformed state. In practice this is the most convenient way of proceeding because, as explained in section 2.4.1 above, stress components also are defined in the deformed state.

2.4.3 The Strain Energy Function

It has been shown that the finite strain deformation can be readily described by the deformation of a cube of unit dimensions in the undeformed state to the rectangular parallelepiped shown in Figure 2.6 which has edges λ_1, λ_2, λ_3 in the x, y, z directions respectively. In the deformed state the forces acting on the faces are f_1, f_2, f_3 with f = force per unit of undeformed cross-section, i.e. the forces are calculated in terms of the applied loads per unit cross-section in the undeformed state.

The corresponding stress components as defined in the deformed state are $\boldsymbol{\sigma}_{xx}$, $\boldsymbol{\sigma}_{yy}$, $\boldsymbol{\sigma}_{zz}$ where

$$\boldsymbol{\sigma}_{xx} = \frac{f_1}{\lambda_2\lambda_3} = \lambda_1 f_1 \qquad \boldsymbol{\sigma}_{yy} = \frac{f_2}{\lambda_3\lambda_1} = \lambda_2 f_2 \qquad \boldsymbol{\sigma}_{zz} = \frac{f_3}{\lambda_1\lambda_2} = \lambda_3 f_3$$

where the bold type indicates the distinction between these components of stress and those defined for small strain elasticity.

The work done (per unit of initial undeformed volume) in an infinitesimal displacement from the deformation state where λ_1, λ_2, λ_3 change to $\lambda_1 + \mathrm{d}\lambda_1$, $\lambda_2 + \mathrm{d}\lambda_2$, $\lambda_3 + \mathrm{d}\lambda_3$, is

$$\mathrm{d}W = f_1\,\mathrm{d}\lambda_1 + f_2\,\mathrm{d}\lambda_2 + f_3\,\mathrm{d}\lambda_3 \tag{2.1}$$

$$= \frac{\boldsymbol{\sigma}_{xx}}{\lambda_1}\,\mathrm{d}\lambda_1 + \frac{\boldsymbol{\sigma}_{yy}}{\lambda_2}\,\mathrm{d}\lambda_2 + \frac{\boldsymbol{\sigma}_{zz}}{\lambda_3}\,\mathrm{d}\lambda_3. \tag{2.1a}$$

For an elastic material the work done can be equated to a change in the stored elastic energy U. In the case of rubbers, it is usual to consider a reversible isothermal change of state at constant volume, so that the work done can be equated to the change in the Helmholtz free energy A, i.e. $\Delta U = \Delta A$. Here U is often called the strain energy function because it defines the energy stored as result of the strain, i.e.

$$U = f(\lambda_1, \lambda_2, \lambda_3). \tag{2.2}$$

Because rubber is an isotropic material the form of this function f must be independent of the choice of coordinate axes. For simplicity it should also become zero when $\lambda_1 = \lambda_2 = \lambda_3 = 1$, i.e. for zero strain. A further requirement is that for small strains, we should obtain Hooke's law for simple tension and the equivalent equation for simple shear.

An equation which satisfies these requirements is

$$U = C_1(\lambda_1^2 + \lambda_2^2 + \lambda_3^2 - 3). \tag{2.3}$$

To obtain a stress–strain relationship from this equation we invoke equation (2.1), together with the assumption that rubber is incompressible, i.e. there is no change in volume on deformation, which is true to a good approximation. For example, consider extension under a tensile force f in the x direction. This gives

$$\lambda_1 = \lambda \quad \text{and} \quad \lambda_2 = \lambda_3 = \lambda^{-1/2}, \tag{2.4}$$

where we have used the incompressibility assumption $\lambda_1\lambda_2\lambda_3 = 1$.

Equation (2.3) becomes

$$U = C_1\left(\lambda^2 + \frac{2}{\lambda} - 3\right) \tag{2.5}$$

and from (2.1) we have

$$f = \frac{\partial U}{\partial \lambda} = 2C_1\left(\lambda - \frac{1}{\lambda^2}\right) \tag{2.6}$$

This familiar equation is more usually represented as a consequence of the molecular theories of a rubber network. Here we see that it follows from purely phenomenological considerations as a simple constitutive equation for the finite deformation of an isotropic, incompressible solid. Materials which obey this relationship are sometimes called neo-Hookeian.

Note that for small strain, where $\lambda = 1 + e$, equation (2.6) reduces to

$$f = 2C_1\{1 + e - (1 - 2e)\} = 6C_1 e, \tag{2.7}$$

i.e. Hooke's law.

REFERENCES

1. S. Timoshenko and J. N. Goodier, *Theory of Elasticity*, McGraw-Hill, New York, 1951.
2. A. E. H. Love, *A Treatise on the Mathematical Theory of Elasticity*, 4th edn, Macmillan, New York, 1944.
3. I. M. Ward, *Mechanical Properties of Solid Polymers*, 2nd edn, Wiley, Chichester, 1983.

Chapter 3

Rubber-like Elasticity

3.1 GENERAL FEATURES OF RUBBER-LIKE BEHAVIOUR

The most noticeable feature of natural rubber and other elastomers is the ability to undergo large and reversible elastic deformation. It is not unexpected that stress can cause polymeric molecules to adopt an extended configuration, but at first sight it may seem surprising that on removal of the stress the molecules retract, on average, to their initial coiled form. Simple theories of rubber-like elasticity assume, as an approximation, that both extension and retraction occur instantaneously, and neglect any permanent deformation. Natural rubber [*cis*-polyisoprene] in its native state does not satisfy this last criterion, as molecules in extended configurations tend to slide past one another and do not recover completely. Molecules need to be chemically cross-linked by sulphur bonds (vulcanization) to prevent any permanent flow, and we shall show that the degree of cross-linking determines the extensibility of the rubber for a given stress.

The application of stress is considered to cause molecules to change from a coiled to an extended configuration instantaneously. For this reason it is possible to apply equilibrium thermodynamics to determine how the stress is related to changes in both internal energy and entropy. The general nature of thermodynamics implies that this type of approach can give no direct information on molecular rearrangements but, when augmented by molecular theories of a statistical nature, it is possible to derive an equation of state that relates the force causing extension to molecular parameters. We will show that this equation of state is identical in form to equation (2.6) above, which was derived as the equivalent of Hooke's law for finite deformations. It will be shown that the reason for this direct link between the behaviour at a molecular level and the mechanics of finite elasticity arises because of the applicability of the so-called 'affine deformation' assumption. Affine deformation in the molecular theory of rubber elasticity means that we can assume that the changes in the length and orientation of lines joining adjacent cross-links in the molecular network are identical to the changes in lines marked on the macroscopic rubber. It is also assumed that there is no change in volume on deformation. This assumption can be justified as a good approximation because the bulk modulus (K) is some 10^4 times greater than the shear modulus (G): typical values being 10^{10} Pa and 10^6 Pa. As a consequence, at low strains, Poisson's ratio, given by

$$\nu = \frac{3K - 2G}{2(3K + 2G)}$$

is effectively 0.500, and deformation occurs essentially at constant volume.

A schematic force–extension curve for a typical rubber is shown in Figure 3.1, with the maximum extensibility varying between 500 and 1000%, depending on the extent of cross-linking. The behaviour is Hookean, with a linear relationship between stress and strain, only at strains of the order of 1% of so. At larger strains the force–extension relation is non-linear, and we will show that its form is determined essentially by changes in configurational entropy rather than internal energy.

High extension results in a greatly reduced entropy, so that retraction is a consequence of the necessity for entropy to be maximized. A fully extended chain is a state of zero entropy because there is only one possible conformation of bonds through which it can occur. In contrast there are a very large number of ways of obtaining a given end-to-end distance for a contracted configuration of the chain. As all configurations have approximately the same internal energy, in the absence of external stress an extended chain will return to a more probable state. For this reason rubber is sometimes referred to as a 'probability or entropy spring', in contrast with

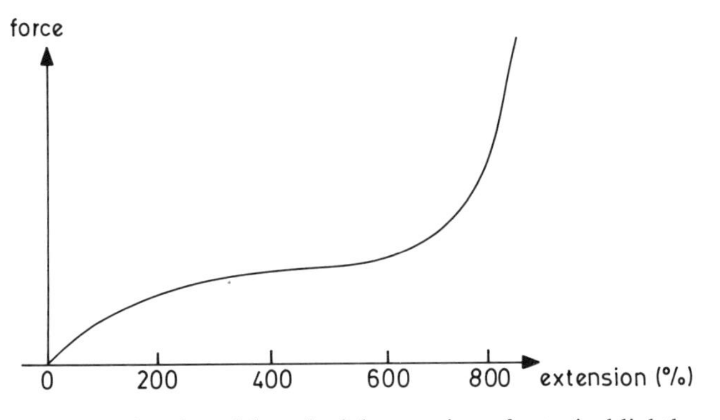

Figure 3.1 The force developed in uniaxial extension of a typical lightly cross-linked rubber

the 'energy spring' characteristics of the elasticity of materials of low molecular mass, where extension causes an increase in internal energy. For a fuller discussion see Treloar [1].

3.2 THE THERMODYNAMICS OF DEFORMATION

For a reversible isothermal change of state at constant volume, the work done can be equated to the change in the Helmholtz function (or free energy) A. When an elastic solid, initial length l, is extended uniaxially under a tensile force f, the work done on the solid in an infinitesimal displacement $\mathrm{d}l$ is

$$\mathrm{d}W = f\,\mathrm{d}l = \mathrm{d}A = \mathrm{d}U - T\,\mathrm{d}S - S\,\mathrm{d}T, \tag{3.1}$$

Where U represents internal energy and S entropy. Under isothermal conditions

$$f = \left(\frac{\partial A}{\partial l}\right)_T = \left(\frac{\partial U}{\partial l}\right)_T - T\left(\frac{\partial S}{\partial l}\right)_T \tag{3.2}$$

To obtain the change in internal energy with extension in terms of observable quantities it is necessary to replace

$$\left(\frac{\partial S}{\partial l}\right)_T$$

by an equivalent thermodynamic relation. Now

$$\mathrm{d}U = T\,\mathrm{d}S + \mathrm{d}W = T\,\mathrm{d}S + f\,\mathrm{d}l. \tag{3.3}$$

Substituting (3.3) in (3.1)

$$\mathrm{d}A = f\,\mathrm{d}l - S\,\mathrm{d}T. \tag{3.4}$$

Then

$$\left(\frac{\partial A}{\partial l}\right)_T = \mathrm{f} \quad \text{and} \quad \left(\frac{\partial A}{\partial T}\right)_l = -S. \tag{3.5}$$

But

$$\frac{\partial}{\partial l}\left(\frac{\partial A}{\partial T}\right)_l = \frac{\partial}{\partial T}\left(\frac{\partial A}{\partial l}\right)_T$$

Substituting

$$\left(\frac{\partial S}{\partial l}\right)_T = -\left(\frac{\partial f}{\partial T}\right)_l \tag{3.6}$$

Hence (3.2) becomes

$$\left(\frac{\partial U}{\partial l}\right)_T = f - T\left(\frac{\partial f}{\partial T}\right)_l \tag{3.7}$$

As long ago as 1935 Meyer and Ferri [2] showed that the tensile force at constant length was very nearly proportional to the absolute temperature, i.e. $f = \alpha T$. Differentiating this relationship we obtain

$$\left(\frac{\partial f}{\partial T}\right)_l = \alpha, \text{ a constant.}$$

By substitution (3.7) gives

$$\left(\frac{\partial U}{\partial l}\right)_T = 0,$$

which demonstrates that elasticity arises entirely from changes in entropy to this good first approximation.

3.3 THE STATISTICAL THEORY

The kinetic or statistical theory of rubber elasticity, originally proposed by Meyer, Susich and Valko [3], assumes that the very long molecular chains are each capable of assuming a wide variety of configurations in response to the thermal vibrations of their constituent atoms. Although the molecular chains are interlinked to form a coherent network, the number of cross-links is assumed to be small enough not to interfere markedly with the motion of the chains. In the absence of external forces the chain molecules will adopt configurations corresponding to a state of maximum entropy. When forces are applied the chains will tend to extend in the direction of the force, so reducing the entropy, and produce a state of strain.

Quantitative evaluation of the stress–strain characteristics of the rubber network then involves the calculation of the configurational entropy of the whole assembly of chains as a function of the state of strain. This calculation is considered in two stages, first the calculation of the entropy of a single chain and, secondly, the change in entropy of a network of chains as a function of strain.

3.3.1 Simplifying Assumptions

In reality, atoms are tightly packed along the length of a molecular chain. It is, however, convenient to represent molecular chains in terms of ‘ball and stick’ models, such as that shown for polyethylene in Figure 3.2. Here we show the fully extended molecule which takes the form of a planar zigzag. If

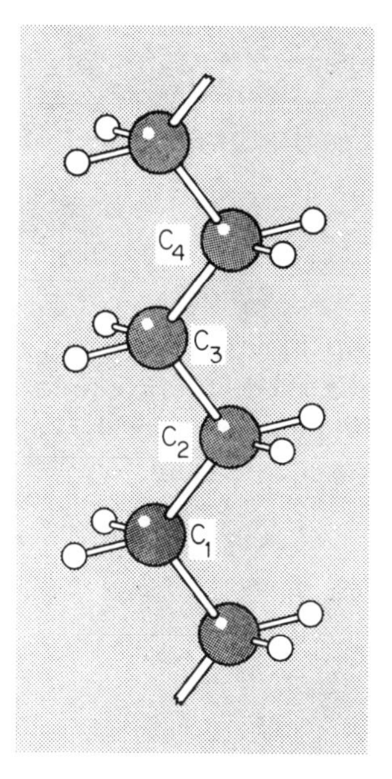

Figure 3.2 The polyethylene chain.

essentially free rotation from one conformation to another occurs, subject only to the limitation that the valence bond angle between carbon atoms must remain at 109.5°, the local situation $C_1C_2C_3C_4$ can change from the planar zigzag to a variety of conformations. In principle it is possible to calculate the number of molecular configurations that correspond to any chosen end-to-end length of the molecule: there will be only one fully extended configuration, but for a molecule containing possibly hundreds of backbone atoms the number of alternative contracted configurations will be very large. In practice it is more convenient to consider the 'freely jointed' chain, a mathematical abstraction in which the atoms are reduced to mere points, joined by one-dimensional equal links, with no restriction on the angle between adjacent links. It is assumed in this simple model that there is no difference in internal energy between the different molecular conformations along the chain.[†]

3.3.2 Average Length of a Molecule Between Cross-links

Consider a freely jointed chain with n links, each of length l. The length of a single chain is then given by

---

[†]Conformation is used to denote differences in the immediate situation of a bond, e.g. *trans* and *gauche* conformations. Configuration is retained to refer to the arrangement of the whole molecular chain.

$$r = \sum_{i=1}^{n} \boldsymbol{l}_i. \tag{3.8}$$

For a large number of chains (q), or one chain considered at many different times, the mean length

$$r(q) = \frac{l}{q} \sum_{j=1}^{q} r_j = 0, \tag{3.9}$$

as the vector length is equally likely to be positive or negative.

We follow the procedure used, for example, with sinusoidally varying quantities such as alternating current and voltage, where the mean value is zero, and calculate the mean square chain length

$$\overline{r^2} = \frac{l}{q} \sum_{1}^{q} r_j^2 = \frac{l}{q} \sum_{1}^{q} \left(\sum_{1}^{n} l_i \right)_j^2 \tag{3.10}$$

Expand, giving

$$\overline{r^2} = \frac{l}{q} \sum_{1}^{q} \left(l_1^2 + l_2^2 + \ldots l_n^2 + \boldsymbol{l}_1 \cdot \boldsymbol{l}_2 + \boldsymbol{l}_1 \cdot \boldsymbol{l}_3 + \ldots + \boldsymbol{l}_{n-1} \cdot \mathrm{l}_n \right)$$

but $l_1^2 = l_2^2 = l_n^2$ and $\boldsymbol{l}_\mathrm{m} \cdot \boldsymbol{l}_\mathrm{n} = l_\mathrm{m} l_\mathrm{n} \cos\theta$. In a freely jointed chain θ can have any value; hence $\Sigma \boldsymbol{l}_m \cdot \boldsymbol{l}_n = 0$, $m \neq n$. Therefore

$$\overline{r^2} = \frac{l}{q} \sum_{1}^{q} nl^2 = nl^2.$$

The root mean square chain length is therefore

$$\surd(\overline{r^2}) = l\surd(n) = r_\mathrm{rms} \tag{3.11}$$

compared with the fully extended chain length of ln; i.e. for a chain of 100 bonds between cross-links the maximum extensibility is 10.

3.3.3 The Entropy of a Single Chain

The expression just derived indicates the reason for the high extensibility of lightly cross-linked rubbers and serves to introduce the important concept of a mean square length, but yields no information on the probability of a chain having a particular end-to-end length. This latter problem was first analysed mathematically by Kuhn [4] and by Guth and Mark [5].

Consider a chain of n links each of length l, which has a configuration such that one end P is at the origin (Figure 3.3). The probability distribution for the position of the end Q is derived using approximations which are valid providing that the distance between the chain ends P and Q is much less than the extended chain length nl. The probability that Q lies within the elemental volume $\mathrm{d}x\,\mathrm{d}y\,\mathrm{d}z$ at the point (x, y, z) can be shown to be

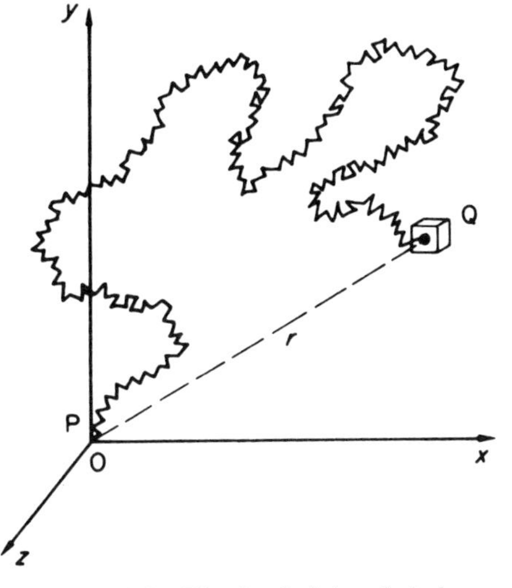

Figure 3.3 The freely jointed chain

$$p(x, y, z)\,\mathrm{d}x\,\mathrm{d}y\,\mathrm{d}z = \frac{b^3}{\pi^{3/2}}\exp(-b^2r^2)\,\mathrm{d}x\,\mathrm{d}y\,\mathrm{d}z. \tag{3.12}$$

This distribution has the form of the Gaussian error function, and is spherically symmetrical about the origin, where the value is a maximum (Figure 3.4). The most probable end-to-end length is not, however, zero, as the probability that Q falls within an elemental volume situated between r and $(r + \mathrm{d}r)$ from the origin, irrespective of direction, is the product of the probability distribution $p(r)$ and the volume of the concentric shell, $4\pi r^2\,\mathrm{d}r$. The overall probability is then

$$\begin{aligned} P(r)\,\mathrm{d}r &= p(r)4\pi r^2\,\mathrm{d}r = (b^3/\pi^{3/2})\exp(-b^2r^2)\cdot 4\pi^2\,\mathrm{d}r \\ &= (4b^3/\pi^{1/2})r^2\exp(-b^2r^2)\,\mathrm{d}r, \end{aligned} \tag{3.13}$$

which is illustrated in Figure 3.5.

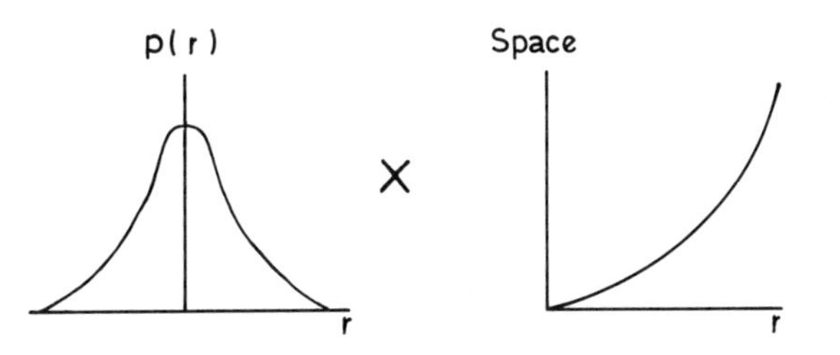

Figure 3.4 The Gaussian probability distribution for the free end of a chain must be multiplied by the volume in which that end can reside: $P(r) = p(r) \times 4\pi r^2\,\mathrm{d}r$

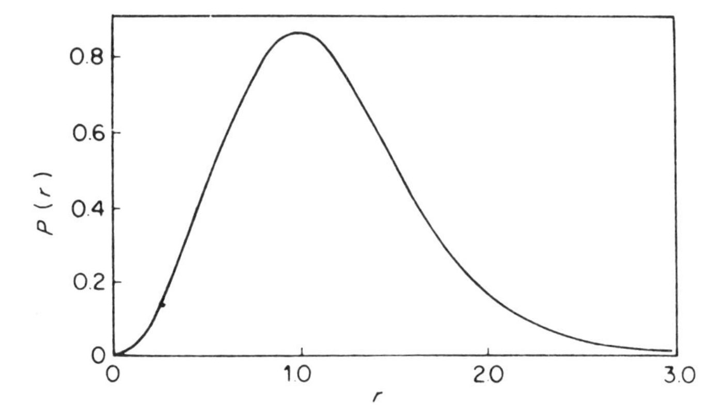

Figure 3.5 The distribution function $P(r) = \text{const}\, r^2 \exp(-b^2 r^2)$

It is seen that the most probable end-to-end distance, irrespective of direction, is not zero, but it is a function of b, i.e. of the length l of the links and the number n of links in the chain, as shown in section 3.3.2 above.

Another important quantity is the root mean square chain length $(\overline{r^2})^{1/2}$

$$\overline{r^2} = \int_0^\infty r^2 P(r)\,\mathrm{d}r.$$

Substitution of the above expression for P (r) gives

$$\overline{r^2} = 3/2b^2 = nl^2 \tag{3.14}$$

so that the root mean square length $(\overline{r^2})^{1/2} = l\sqrt{n}$, i.e. it is proportional to the square root of the number of links in the chain, as shown in section 3.3.2 above.

The entropy of the freely jointed chain s is proportional to the logarithm of the number of configurations Ω so that

$$s = k \ln \Omega,$$

where k is Boltzmann's constant. If $\mathrm{d}x\,\mathrm{d}y\,\mathrm{d}z$ is constant, the number of configurations available to the chain is proportional to the probability per unit volume $p(x, y, z)$. The entropy of the chain is thus given by

$$s = c - kb^2 r^2 = c - kb^2(x^2 + y^2 + z^2), \tag{3.15}$$

where c is an arbitrary constant.

3.3.4 The Elasticity of a Molecular Network

We wish to calculate the strain-energy function for a molecular network, assuming that this is given by the change in entropy of a network of chains as a function of strain.

The actual network is replaced by an ideal network in which each segment of a molecule between successive points of cross-linkage is considered to be a Gaussian chain.

Three additional assumptions are introduced:

1. In either the strained or unstrained state, each junction point may be regarded as fixed at its mean position.
2. The effect of the deformation is to change the components of the vector length of each chain in the same ratio as the corresponding dimensions of the bulk material (the 'affine' deformation assumption).
3. The mean square end-to end distance for the whole assembly of chains in the unstrained state is the same as for a corresponding set of free chains and is therefore given by equation (3.14).

In effect it is necessary to calculate the difference in probability between a spherical distribution of chain end-to-end vectors in the unstrained state, and an ellipsoidal distribution for uniaxial extension (Figure 3.6). This difference is related to changes in entropy, and so to tensile force.

As discussed in section 2.2, we can restrict our discussion to the case of pure strain without loss of generality. We choose principal extension ratios λ_1, λ_2, λ_3 parallel to the three rectangular coordinate axes x, y, z. The affine deformation assumption implies that the relative displacement of the chain ends is defined by the macroscopic deformation. Thus, in Figure 3.7, we take a system of coordinates x, y, z in the undeformed body.

In this coordinate system a representative chain PQ has one end P at the origin. We refer any point in the deformed body to this system of coordinates. Thus the origin, i.e. the end of the chain P, could be moved bodily during the deformation. The other end Q (x, y, z) is displaced to the point $Q'(x', y', z')$ and from the affine deformation assumption we have

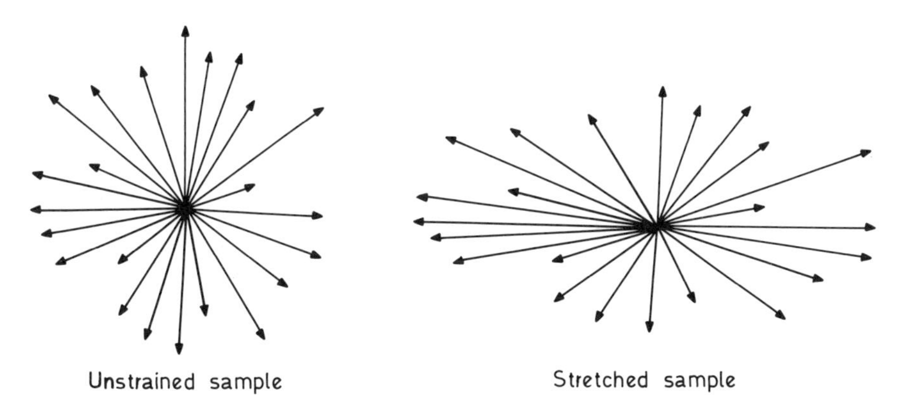

Figure 3.6 Schematic representation of chain end-to-end vectors in the initial and final states

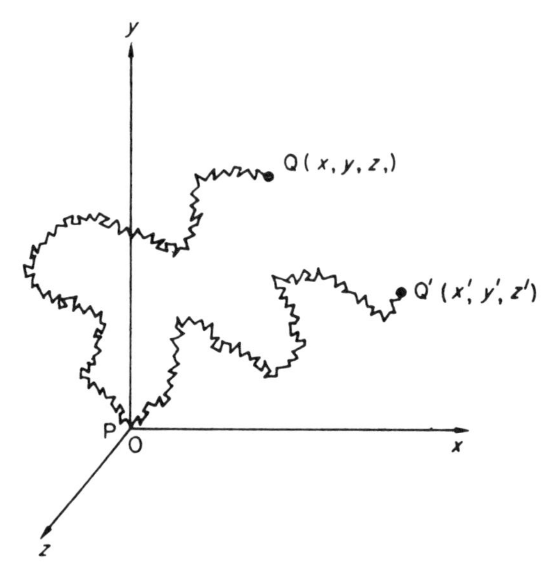

Figure 3.7 The end of the chain Q (x, y, z) is displaced to Q (x', y', z')

$$x' = \lambda_1 x, \qquad y' = \lambda_2 y, \qquad z' = \lambda_3 z.$$

The entropy of the chain in the undeformed state is given by equation (3.15) as

$$s = c - kb^2(x^2 + y^2 + z^2).$$

After deformation the entropy becomes

$$s' = c - kb^2(\lambda_1^2 x^2 + \lambda_2^2 y^2 + \lambda_2^3 z^2) \tag{3.16}$$

giving the change in entropy

$$\Delta s = s' - s = -kb^2\{(\lambda_1^2 - 1)x^2 + (\lambda_2^2 - 1)y^2 + (\lambda_3^2 - 1)z^2\}. \tag{3.17}$$

Let there be N chains per unit volume in the network, with m of these having a given value of b (say b_p). The total entropy change for this particular group of chains is

$$\Delta s_b = \sum_1^m \Delta s = -kb_p^2\left\{(\lambda_1^2 - 1)\sum_1^m x^2 + (\lambda_2^2 - 1)\sum_1^m y^2 + (\lambda_3^2 - 1)\sum_1^m z^2\right\} \tag{3.18}$$

where $\Sigma_1^m x^2$ is the sum of the squares of the x components for these m chains in the underformed network. As there is no preferred direction for the chain vectors in the underformed (isotropic) state, there is no preference for the x, y or z directions, so that

$$\sum_1^m x^2 = \sum_1^m y^2 = \sum_1^m z^2,$$

but

$$\sum_1^m x^2 + \sum_1^m y^2 + \sum_1^m z^2 = \sum_1^m r^2,$$

giving

$$\sum_1^m x^2 = \sum_1^m y^2 = \sum_1^m z^2 = \tfrac{1}{3}\sum_1^m r^2. \tag{3.19}$$

From (3.14)

$$\sum_1^m r^2 = m\overline{r^2} = m\left(\frac{3}{2b_p^2}\right) \tag{3.20}$$

Combining (3.19) and (3.20) with (3.18)

$$\Delta s_b = -\tfrac{1}{2}mk\{\lambda_1^2 + \lambda_2^2 + \lambda_3^2 - 3\}. \tag{3.21}$$

We can now add the contribution of all the chains in the network (N per unit volume), and obtain the entropy change of the network ΔS where

$$\Delta S = \sum_1^N \Delta s = -\tfrac{1}{2}Nk\{(\lambda_1^2 + \lambda_2^2 + \lambda_3^2) - 3\}. \tag{3.22}$$

Assuming no change in internal energy on deformation this gives the change in the Helmholtz free energy.

$$\Delta A = -T\,\Delta S = \tfrac{1}{2}NkT(\lambda_1^2 + \lambda_2^2 + \lambda_3^2 - 3).$$

If we assume that the strain-energy function U is zero in the undeformed state this gives

$$U = \Delta A = \tfrac{1}{2}NkT(\lambda_1^2 + \lambda_2^2 + \lambda_3^2 - 3). \tag{3.23a}$$

Consider simple elongation λ in the x direction. The incompressibility relationship gives $\lambda_1\lambda_2\lambda_3 = 1$. Hence, by symmetry, $\lambda_2 = \lambda_3 = \lambda^{-1/2}$ and

$$U = \tfrac{1}{2}NkT\left(\lambda^2 + \frac{2}{\lambda} - 3\right) \tag{3.23b}$$

From equation (2.6) above

$$f = \frac{\partial U}{\partial \lambda} = NkT\left(\lambda - \frac{1}{\lambda^2}\right) \tag{3.24}$$

We have therefore obtained the neo-Hookean relationship of section 2.4.3 above with a constant NkT. For small strain we can put $\lambda = 1 + e_{xx}$ and it follows from equation (3.24) that

$$f = \sigma_{xx} = 3NkTe_{xx} = Ee_{xx},$$

where E is Young's modulus. Since for an incompressible material $E \equiv 3G$, we see that the quantity NkT in equation (3.23) is equivalent to the shear modulus of the rubber, G. This term is sometimes written in terms of the mean molecular mass M_c of the chains, i.e. between successive points of cross-linkage. Then

$$G = NkT = \rho RT/M_c,$$

where ρ is the density of the rubber and R is the gas constant.

3.4 MODIFICATIONS OF THE SIMPLE MOLECULAR THEORY

The equation of state deduced above rests on many simplifying assumptions: freely jointed chains of negligible volume and with a Gaussian distribution of end-to-end lengths, that give rise to no changes in internal energy on deformation. We mention briefly some of the investigations regarding those approximations, but direct the reader who requires further information to Treloar's monograph [1].

Even N, the number of chains per unit volume, is not a simple concept, for junction points can be either (permanent) chemical cross-links or (temporary) physical entanglements. Not all cross-links will be effective, as seen in Figure 3.8, which shows (a) 'loose loops', where a chain folds back on itself and (b) 'loose ends', where a chain does not contribute to the network [6].

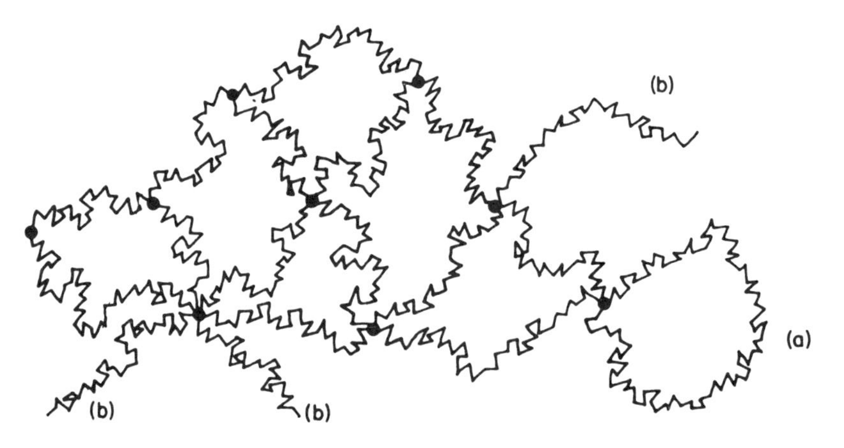

Figure 3.8 Types of network defect: (a) loose loop; (b) loose end

Real polymer chains have fixed bond lengths and possibly hindered rotation. These effects are taken into account by the concept of 'the equivalent freely joined chain' [7]; for example, a paraffin-type chain with unhindered rotation will have r_{rms} $\sqrt{2}$ times greater than the simple model. More sophisticated treatments are possible in terms of random walk statistics.

Kuhn and Grün showed that removing the restriction of a Gaussian distribution of chain end-to-end distances (but retaining other features of the simple model) gave a probability distribution $p(r)$ of the form

$$\ln p(r) = \text{const} - n\left[\frac{r}{nl}\beta + \ln(\beta/\sinh\beta)\right] \tag{3.25}$$

where β is defined by

$$\frac{r}{nl} = \coth\beta - 1/\beta = \mathcal{L}(\beta),$$

where $\mathcal{L}$ is the Langevin function and $\beta = \mathcal{L}^{-1}\,(r/nl)$ the inverse Langevin function.

The expression for probability can be expanded to give

$$\ln p(r) = \text{const} - n\left[\frac{3}{2}\left(\frac{r}{nl}\right)^2 + \frac{9}{20}\left(\frac{r}{nl}\right)^4 + \frac{99}{350}\left(\frac{r}{nl}\right)^6 + \ldots\right] \tag{3.26}$$

The Gaussian distribution is the first term of this series, and so is adequate when $r \ll nl$. James and Guth [8] subsequently used the inverse Langevin distribution function to give a revised expression for the force per unit unstrained area:

$$f = \frac{NkT}{3}n^{1/2}\left(\mathcal{L}^{-1}\left(\frac{\lambda}{n^{1/2}}\right) - \lambda^{-3/2}\mathcal{L}^{-1}\left(\frac{1}{\lambda^{1/2}n^{1/2}}\right)\right\} \tag{3.27}$$

Treloar's fit to the experimental data for natural rubber using (3.27) and a suitable choice of parameters for N and n is shown in Figure 3.9. The maximum extension of the network is primarily determined by n, the number of chain links between successive cross-links; a result that is relevant for the cold-drawing and crazing behaviour discussed later (section 11.6 and section 12.5), where the basic deformation also involves the extension of a molecular network.

Although we have seen how the extension to non-Gaussian statistics gives rise to a very large increase in tensile stress at large extensions, in the case of natural rubber it has been proposed that the observed increase in tensile stress occurs primarily because of strain-induced crystallization. The basic physical idea is that the melting-point T_{c} of the rubber is increased due to extension; $T_{\mathrm{c}} = \Delta H/\Delta S$, where ΔH and ΔS are the enthalphy and entropy

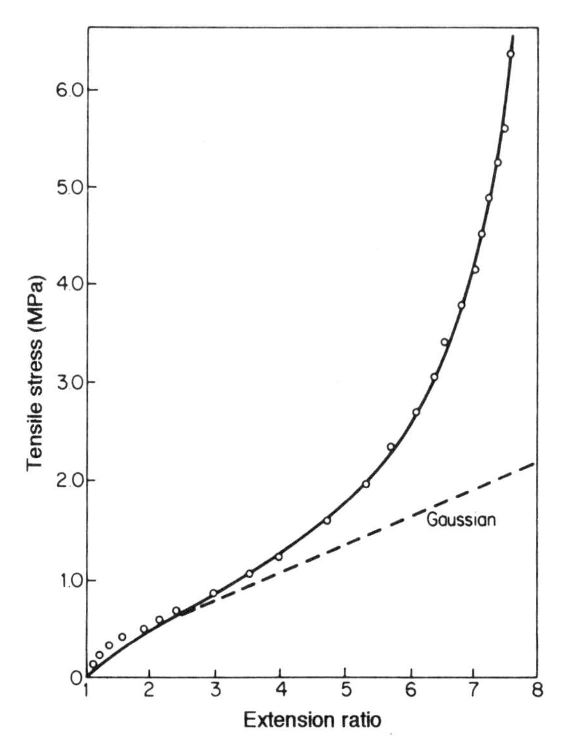

Figure 3.9 Theoretical non-Gaussian free-extension curve obtained by fitting experimental data 0 to the James and Guth theory, with $NkT = 0.273$ MPa, $n = 75$. (Reproduced with permission from Treloar, *The Physics of Rubber Elasticity*, 3rd edn, Oxford University Press, Oxford, 1975)

of fusion respectively. Because the entropy of the extended rubber is low, the change in entropy on crystallization is reduced and T_c correspondingly increased. A higher degree of supercooling then gives rise to crystallization and the crystallites act to increase the modulus by forming additional physical cross-links.

Finally, the simple treatment of rubber elasticity given above makes two assumptions which require further consideration. First, it has been assumed that the internal energy contribution is negligible, which implies that different molecular conformations of the chains have identical internal energies. Secondly, the thermodynamic formulae which have been derived are, strictly, only applicable to measurements at constant volume, whereas most experimental results are obtained at constant pressure. For comprehensive elementary accounts of these complications, the reader is referred to the textbooks by Treloar [1] and by Ward [9].

REFERENCES

1. L. R. G. Treloar, *The Physics of Rubber Elasticity*, 3rd edn, Clarendon Press, Oxford, 1975.
2. K. H. Meyer and C. Ferri, *Helv. Chim. Acta.*, **18**, 570 (1935).
3. K. H. Meyer, G. Von Susich and E. Valko, *Kolloidzeitschrift*, **59**, 208 (1932).
4. W. Kuhn, *Kolloidzeitschrift*, **68**, 2 (1934); **76**, 258 (1936).
5. E. Guth and H. Mark, *Lit. Chem.*, **65**, 93 (1934).
6. P. J. Flory, *Chem. Rev.*, **35**, 51 (1944).
7. W. Kuhn, *Kolloidzeitschrift*, **76**, 258 (1936); **87**, 3 (1939).
8. H. M. James and E. Guth, *J. Chem. Phys.*, **11**, 455 (1943).
9. I. M. Ward, *Mechanical Properties of Solid Polymers*, 2nd edn, Wiley, Chichester, 1983.

PROBLEMS FOR CHAPTERS 2 AND 3

1. The tensile stress σ in an *ideal* rubber when simply extended to a length λ times its initial length is given by

$$\sigma = NkT(\lambda^2 - \lambda^{-1}).$$

   Explain *without giving any mathematical details* the physical model which leads to this expression.

   A sample of polyisoprene (density, 1300 kg m^{-3}, monomer of relative molar mass, 68) has a shear modulus of 4×10^5 Pa at room temperature. Calculate the average number of monomers between cross-links.
2. Rubber deforms at constant volume, and so has Poisson's ratio of 0.5000 at small strains. By how much must a rod of rubber be extended before Poisson's ratio falls to 0.4900?
3. What would be the root mean square end-to-end distance of a paraffinic chain consisting of 1000 carbon atoms? The length of the C—C bond is 1.53 Å and the chain can be considered to be freely jointed (i.e. without the restriction that valence angles should remain constant).
4. A non-Gaussian rubber has the strain-energy function

$$U = C_1(\lambda_1^{1.3} + \lambda_2^{1.3} + \lambda_3^{1.3} - 3),$$

   where λ_1, λ_2, λ_3 are the principal extension ratios and $C_1 = 4 \times 10^5$ Pa.

   If a piece of such a rubber is initially 1 m long and has a cross-section area of 6×10^{-4} m^2, find the mass required to give a final extended length of 3 m.
5. The strain energy function for an ideal rubber is

$$U = C_1(\lambda_1^2 + \lambda_2^2 + \lambda_3^2 - 3),$$

   where λ_1, λ_2, λ_3 are the principal extension ratios. Derive the stress-strain relations for the following:

   (i) Simple extension $\lambda_1 = \lambda$ produced by a force applied in the 1 direction;
   (ii) an equal two-dimensional extension $\lambda_1 = \lambda_2 = \lambda$, produced by the simultaneous application of equal forces in the 1 and 2 directions.

6. A non-Gaussian rubber has a strain-energy function

$$U = C_1(\lambda_1^2 + \lambda_2^2 + \lambda_3^2 - 3) + C_2(\lambda_1^2\lambda_2^2 + \lambda_2^2\lambda_3^2 + \lambda_1^2\lambda_3^2 - 3).$$

Derive the stress–strain relation for a simple extension $\lambda_1 = \lambda$ produced by a force applied in the 1 direction, and hence show that the low strain tensile modulus for this rubber is given by $E = 6(C_1 + C_2)$.

7. The application of a mass 1 kg to a rubber strip of initial cross-sectional area 10 mm^2 causes a 100% increase in length at 300 K. How many chains are there per unit volume? (Boltzmann's constant $k = 1.38 \times 10^{-23}$ J deg^{-1}).

8. A cross-linked rubber of undeformed cross-section 15 mm × 1.5 mm is stretched at 300 K to three times its initial length by suspending a mass of 1 kg. If the density of the rubber is 900 kg m^{-3} what is the mean molecular mass of the network chains?

Chapter 4

Principles of Linear Viscoelasticity

In this chapter we describe the common forms of viscoelastic behaviour and discuss the phenomena in terms of the deformation characteristics of elastic solids and viscous fluids. The discussion is confined to linear viscoelasticity, for which the Boltzmann superposition principle enables the response to multistep loading processes to be determined from simpler creep and relaxation experiments. Phenomenological mechanical models are considered and used to derive retardation and relaxation spectra, which describe the time-scale of the response to an applied deformation. Finally we show that in alternating strain experiments the presence of the viscous component leads to a phase difference between stress and strain.

4.1 VISCOELASTICITY AS A PHENOMENON

The behaviour of materials of low relative molecular mass is usually discussed in terms of two particular types of ideal material: the elastic solid and the viscous liquid. The former has a definite shape and is deformed by external forces into a new equilibrium shape; on removal of these forces it reverts instantaneously to its original form. The solid stores all the energy which it obtains from the external forces during the deformation, and this energy is available to restore the original shape when the forces are removed. By contrast, a viscous liquid has no definite shape and flows irreversibly under the action of external forces.

One of the most interesting features of polymers is that a given polymer can display all the intermediate range of properties between an elastic solid and a viscous liquid depending on the temperature and the experimentally chosen time scale. Bouncing putty, a silicone product, flows over a period of hours, fractures like a ductile solid when deformed rapidly, and bounces like an elastomer when dropped. Of greater commercial importance are the rubber-like, and in extreme cases brittle, characteristics exhibited by molten polymers at high processing rates. This form of response which combines both liquid-like and solid-like features, is termed viscoelasticity.

4.1.1 Linear Viscoelastic Behaviour

Newton's law of viscosity defines viscosity η by stating that stress σ is proportional to the velocity gradient in the liquid:

$$\sigma = \eta \frac{\partial V}{\partial y}$$

where V is the velocity, and y is the direction of the velocity gradient. For a velocity gradient in the xy plane

$$\sigma_{xy} = \eta\left(\frac{\partial V_x}{\partial y} + \frac{\partial V_y}{\partial x}\right)$$

where $\partial V_x/\partial y$ and $\partial V_y/\partial x$ are the velocity gradients in the y and x directions respectively (see Figure 4.1 for the case where the velocity gradient is in the y direction).

Since $V_x = \partial u/\partial t$ and $V_y = \partial v/\partial t$, where u and v are the displacements in the x and y directions respectively, we have

$$\begin{aligned}\sigma_{xy} &= \eta\left[\frac{\partial}{\partial y}\left(\frac{\partial u}{\partial t}\right) + \frac{\partial}{\partial x}\left(\frac{\partial v}{\partial t}\right)\right] \\ &= \eta \frac{\partial}{\partial t}\left(\frac{\partial u}{\partial y} + \frac{\partial v}{\partial x}\right) \\ &= \eta \frac{\partial e_{xy}}{\partial t}\end{aligned}$$

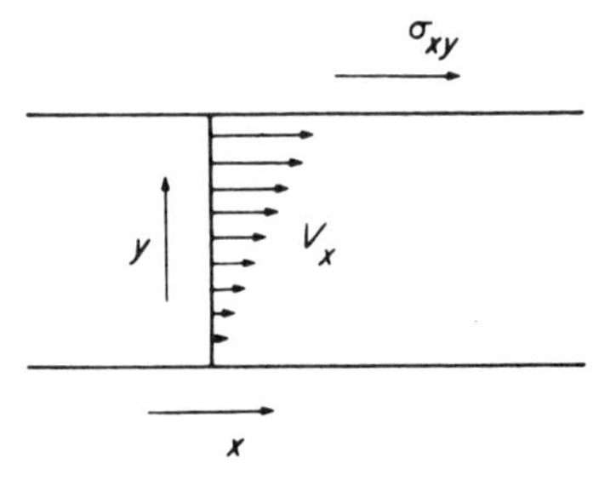

Figure 4.1 The velocity gradient

It can be seen that the shear stress σ_{xy} is directly proportional to the rate of change of shear strain with time. This formulation brings out the analogy between Hooke's law for elastic solids and Newton's law for viscous liquids. In the former the stress is linearly related to the strain, in the latter the stress is linearly related to the rate of change of strain or strain rate.

Hooke's law describes the behaviour of a linear elastic solid and Newton's law that of a linear viscous liquid. A simple constitutive relation for the behaviour of a linear viscoelastic solid is obtained by combining these two laws:

1. For elastic behaviour $(\sigma_{xy})_E = Ge_{xy}$, where G is the shear modulus.
2. For viscous behaviour $(\sigma_{xy})_V = \eta(\partial e_{xy}/\partial t)$.

A simple possible formulation of linear viscoelastic behaviour combines these equations, making the assumption that the shear stresses related to strain and strain rate are additive:

$$\sigma_{xy} = (\sigma_{xy})_E + (\sigma_{xy})_V = Ge_{xy} + \eta\frac{\partial e_{xy}}{\partial t}$$

The equation represents one of the simple models for linear viscoelastic behaviour, the Kelvin or Voigt model and is discussed in detail in section 4.2.3.1 below.

As most of the experiments on linear viscoelasticity examine a single mode of deformation, usually corresponding to a measurement of Young's modulus or the shear modulus, our initial discussion of linear viscoelasticity will be confined to the one-dimensional situation.

For elastic solids Hooke's law is valid only at small strains, and Newton's law of viscosity is restricted to relatively low flow rates, as only when the stress is proportional either to the strain or the strain rate is analysis of the deformation feasible in simple form. A comparable limitation holds for viscoelastic materials: general quantitative predictions are possible only in the case of linear viscoelasticity, for which the results of changing stresses or strains are simply additive, but the time at which the change is made must be taken into account. For a single loading process there will be a linear relation between stress and strain at a given time. Multistep loading can be analysed in terms of the Boltzmann superposition principle (section 4.2.1) because each increment of stress can be assumed to make an independent contribution to the overall strain.

In practice most useful plastics show some non-linearity at strains around 1%, and so the following discussion may not be relevant for particular practical applications: for instance the fibres in a carpet pile suffer bending strains as high as 4%.

4.1.2 Creep

Creep is the time-dependent change in strain following a step change in stress. Unlike the creep discussed by metallurgists, creep in polymers at low strains (1%) is essentially recoverable after unloading, without the need for annealing at a raised temperature. The responses to two levels of stress for linear elastic and linear viscoelastic materials are compared in Figure 4.2. In the former case the strain follows the pattern of the loading programme exactly and in exact proportionality to the magnitude of the stresses applied. For the most general case of a linear viscoelastic solid the total strain e is the sum of three essentially separate parts: e_1 the immediate elastic deformation, e_2 the delayed elastic deformation and e_3 the Newtonian flow, which is identical with the deformation of a viscous liquid obeying Newton's law of viscosity.

Because the material shows linear behaviour the magnitudes of e_1, e_2 and e_3 are exactly proportional to the magnitude of the applied stress, so that a creep compliance $J(t)$ can be defined, which is a function of time only:

$$J(t) = \frac{e(t)}{\sigma} = J_1 + J_2 + J_3,$$

where J_1, J_2 and J_3 correspond to e_1, e_2 and e_3. Linear amorphous

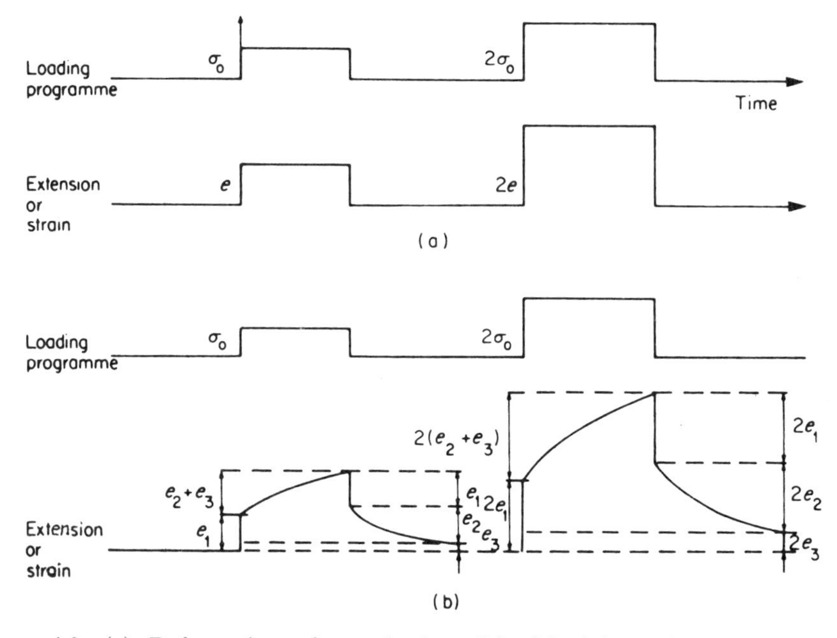

Figure 4.2 (a) Deformation of an elastic solid; (b) deformation of a linear viscoelastic solid

polymers show a significant J_3 above their glass transition temperatures, when creep may continue until the specimen ruptures, but at lower temperatures J_1 and J_2 dominate. Cross-linked polymers do not show a J_3 term, and to a very good approximation neither do highly crystalline polymers.

The separation of compliance into immediate and delayed terms is somewhat arbitrary, as J_1 is generally the limiting compliance observed at the shortest experimentally accessible times; for the purposes of analysis, however, we shall assume that there is a real distinction between the two phases of response. The initial elastic deformation is sometimes called the unrelaxed response to distinguish it from the relaxed response observed at times long enough for the various relaxation mechanisms to have occurred.

The maximum insight into the nature of creep is obtained by plotting the logarithm of creep compliance against the logarithm of time over a very wide time-scale (Figure 4.3). We shall show in section 5.3.1 that an extended time-scale can be achieved through a series of short-term experiments at a range of temperatures. This diagram shows that at very short times the compliance (typically 10^{-9} Pa^{-1}) is that for a glassy solid and is independent of time; in contrast, at very long times the compliance (typically 10^{-5} Pa^{-1}) is that for a rubber-like solid and is again time independent. At intermediate times the compliance lies between these extremes and is time dependent, so that the behaviour is viscoelastic.

It is convenient to define a retardation time τ' in the middle of the viscoelastic region to characterize the time-scale for creep. The distinction between a rubber and a glassy plastic is then seen to be somewhat artificial, because it depends only on the value of τ' at room temperature. Compared with typical experimental response times, which can rarely be less than 1 s,

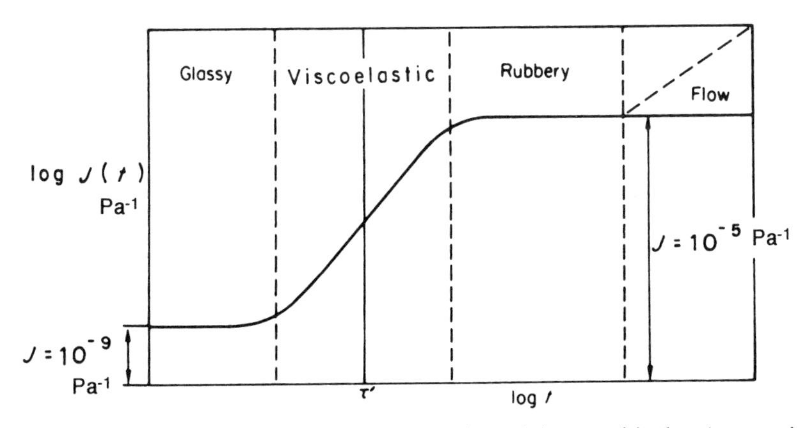

Figure 4.3 The creep compliance $J(t)$ as a function of time t; τ' is the characteristic time (the retardation time)

τ' for a rubber is very small at room temperature, whereas the opposite is true for a glassy polymer. As the temperature is raised the frequency of molecular rearrangements increases, so reducing the value of τ'. Thus at sufficiently low temperatures a rubber behaves like a glassy plastic, and will shatter under impact conditions; correspondingly a glassy plastic will become rubber-like at a sufficiently high temperature.

Recovery curves from creep under constant stress are also illustrated in Figure 4.2, which indicates that at any selected time the extent of recovery (and so the unrecovered strain) is directly proportional to the stress that had formerly been applied. The relation between recovery and creep will be derived in section 4.2.1.

4.1.3 Stress Relaxation

When an instantaneous strain is applied to an ideal elastic solid a finite and constant stress will be recorded. For a linear viscoelastic solid subjected to a nominally instantaneous strain the initial stress will be proportional to the applied strain and will decrease with time (Figure 4.4), at a rate characterized by the relaxation time τ. This behaviour is called stress relaxation. For amorphous linear polymers at high temperatures the stress may eventually decay to zero. In the following discussion we shall ignore transient behaviour.

Making the assumption of linear viscoelastic behaviour we can define the stress relaxation modulus $G(t) = \sigma(t)/e$. Where there is no viscous flow the stress decays to a finite value (Figure 4.5), to give an equilibrium or relaxed modulus G_r at finite time. As with creep we see that there are regions of glassy, viscoelastic, rubber-like and flow behaviour; and similarly changing temperature is equivalent to changing the time-scale. The relaxation time τ is of the same general magnitude as the retardation time τ', but the two are identical only for the simpler models.

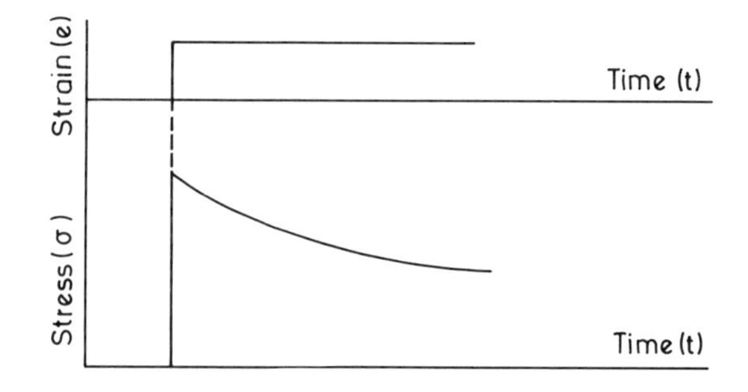

Figure 4.4 Stress relaxation (idealized)

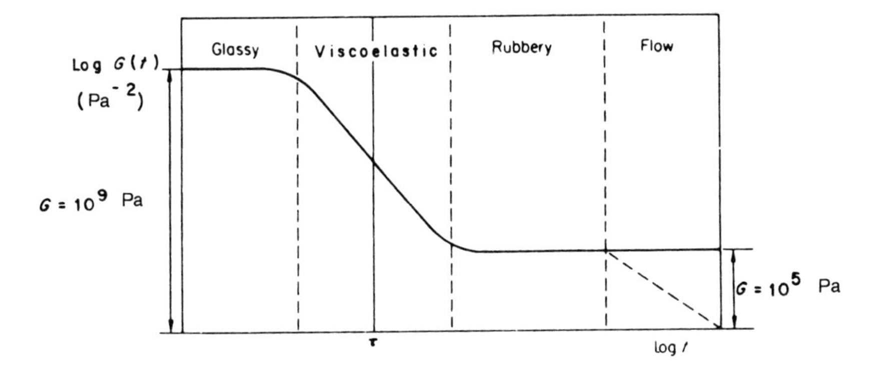

Figure 4.5 The stress relaxation modulus $G(t)$ as a function of time t; τ is the characteristic time (the relaxation time)

4.2 MATHEMATICAL REPRESENTATION OF LINEAR VISCOELASTICITY

The models discussed here which are phenomenological and have no direct relation with chemical composition or molecular structure enable, in principle, the response to a complicated loading pattern to be deduced from a single creep (or stress-relaxation) plot extending over a long time interval. Interpretation depends on the assumption in linear viscoelasticity that the total deformation can be considered as the sum of *independent* elastic (Hookean) and viscous (Newtonian) components. In essence, the simple behaviour is modelled by a set of either integral or differential equations, which are then applicable in other situations.

4.2.1 The Boltzmann Superposition Principle

Boltzmann proposed, as long ago as 1876 [1], that:

1. The creep is a function of the entire past loading history of the specimen;
2. Each loading step makes an independent contribution to the final deformation, so that the total deformation can be obtained by the addition of all the contributions.

Figure 4.6 illustrates the creep response to a multistep loading programme, in which incremental stresses $\Delta\sigma_1$, $\Delta\sigma_2$, . . ., are added at times τ_1, τ_2, . . . respectively. The total creep at time t is then given by

$$e(t) = \Delta\sigma_1 J(t - \tau_1) + \Delta\sigma_2 J(t - \tau_2) + \Delta\sigma_3 J(t - \tau_3) + \ldots, \quad (4.1)$$

where $J(t - \tau)$ is the creep compliance function. For a particular loading step the form of the function that is relevant is that for the time interval

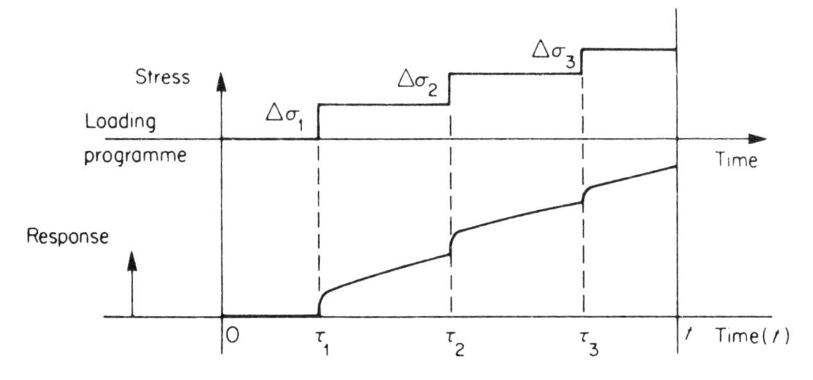

Figure 4.6 The creep behaviour of a linear viscoelastic solid

between the present instant and that at which the load increment was applied.

The summation of (4.1) can be generalized in integral form as

$$e(t) = \int_{-\infty}^{t} J(t - \tau)\, \mathrm{d}\sigma(t). \tag{4.2}$$

It is usual to separate out the instantaneous elastic response in terms of the unrelaxed modulus G_u, giving

$$e(t) = \frac{\sigma}{G_u} + \int_{-\infty}^{t} J(t - \tau)\, \frac{\mathrm{d}\sigma(\tau)}{\mathrm{d}\tau}\, \mathrm{d}\tau, \tag{4.3}$$

where σ represents the total stress at the end of the experiment. Note that the integral extends from $-\infty$ to t, which implies that all previous elements of loading history must be taken into account and, in principle the user must know the history of each specimen since its manufacture. In fact, when creep levels are low enough for linearity to apply, the deformation effectively levels off at sufficiently long times, so that only comparatively recent history is relevant, and this can be standardized by a conditioning treatment (see section 5.1.1). For this reason viscoelastic solids are sometimes said to be materials with 'fading memory'.

The integral in equation (4.3) is called a Duhamel integral, and it is a useful illustration of the consequences of the Boltzmann superposition principle to evaluate the response for a number of simple loading programmes. Recalling the development which leads to equation (5.2) it can be seen that the Duhamel integral is most simply evaluated by treating it as the summation of a number of response terms. Consider two specific cases:

1. *Single-step loading of a stress σ_0 at time $\tau = 0$* (Figure 4.7). For this case

$$J(t - \tau) = J(t) \quad \text{and} \quad e(t) = \sigma_0 J(t).$$

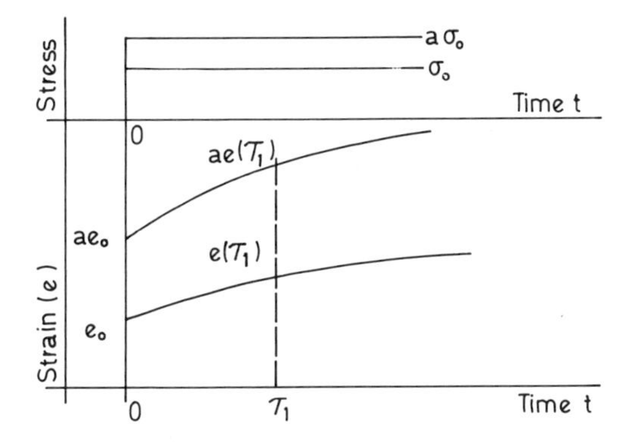

Figure 4.7 Creep for single step loading. For stress $a\sigma_0$ the strain at any time ($\tau \geqslant 0$) is a times greater than that for stress σ_0

2. *Two-step loading of a stress* σ_0 *at time* $\tau = 0$, *followed by an additional stress* σ_0 *at time* $\tau = t_1$ (Figure 4.8). For this case the creep deformation produced by the two loading steps are

$$e_1 = \sigma_0 J(t) \quad \text{and} \quad e_2 = \sigma_0 J(t - t_1),$$

so that

$$e(t) = e_1 + e_2 = \sigma_0 J(t) + \sigma_0 J(t - t_1).$$

The 'additional creep' $e'_c(t - t_1)$ produced by the second loading step is given by

$$e'_c(t - t_1) = \sigma_0 J(t) + \sigma_0 J(t - t_1) - \sigma_0 J(t) = \sigma_0 J(t - t_1).$$

The above illustrates one consequence of the Boltzmann principle, viz. that the additional creep $e'_c(t - t_1)$ produced by adding the stress σ_0 is

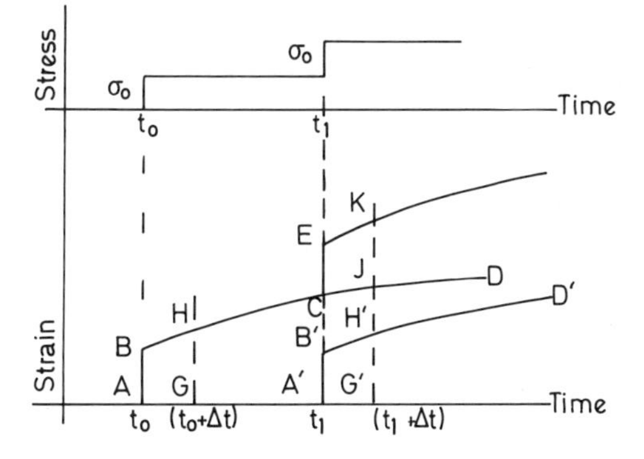

Figure 4.8 Creep for two equal loading steps. Additional instantaneous strain $CE = AB$. Additional total strain at $(t_1 + \Delta t)$: $JK = GH$.

identical with the creep which would have occurred had this stress σ_0 been applied without any previous loading at the same instant in time t_1.

The principle is illustrated in Figure 4.8 where ABD represents creep under σ_0 alone. The response to an additional stress σ_0 at t_1 is found by sliding ABD along the time axis by t_1 to give curve A′B′D′, and at any time adding together the individual strains due to ABD and A′B′D′; e.g. at Δt after the initial loading step the deformation is GH; at Δt after t_1 the deformation due to the initial σ_0 is G′J. The total deformation at $(t_1 + \Delta t)$ is found by adding to G′J the strain JK = G′H′ = GH.

If the second load had been $a\sigma_0$, where a is a constant, then CE = A′B′ = aAB; JK = aG′H′ = aGH, etc.

3. *Creep and recovery*. In this case (Figure 4.9) the stress σ_0 is applied at time $\tau = 0$ and removed at time $\tau = t_1$. The deformation $e(t)$ at a time $t > t_1$ is given by the addition of two terms $e_1 = \sigma_0 J(t)$ and $e_2 = -\sigma_0 J(t - t_1)$, which express the application and removal of the stress σ_0 respectively. Thus $e(t) = \sigma_0 J(t) - \sigma_0 J(t - t_1)$.

The recovery $e_r(t - t_1)$ will be defined as the difference between the anticipated creep under the initial stress and the actual measured response. Thus

$$e_r(t - t_1) = \sigma_0 J(t) - [\sigma_0 J(t) - \sigma_0 J(t - t_1)] = \sigma_0 J(t - t_1),$$

which is identical with the creep response to a stress σ_0 applied at a time t_1. This procedure demonstrates a second consequence of the Boltzmann superposition principle, that the creep and recovery responses are identical in magnitude.

The initial creep curve, ABD in Figure 4.9, is again moved along the time axis by t_1 to give A′B′D′. At any subsequent time the overall deformation is given by the difference between the two curves (CFG). It

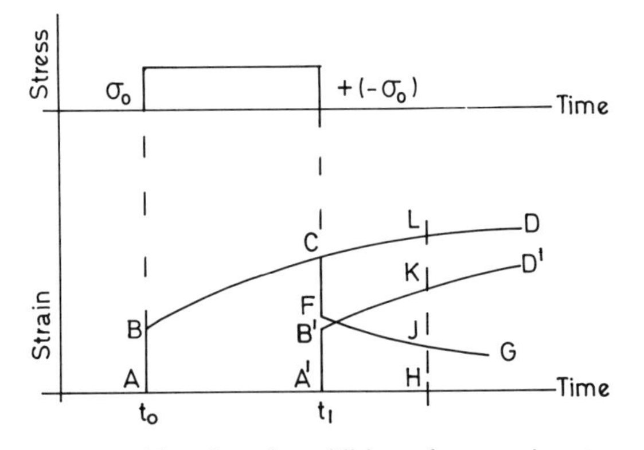

Figure 4.9 Recovery considered as the addition of a negative stress increment, i.e. subtract *B′D′* from *CD*

is important to realize that at a time $t_2(> t_1)$ the residual deformation is obtained by subtracting A′B′D′ from the deformation which would have occurred had unloading not taken place; e.g. residual deformation HJ = HL − HK, and not A′C − HK, where A′C is the maximum strain attained before unloading

4.2.2 The Stress Relaxation Modulus

Stress relaxation behaviour can be represented in an exactly complementary fashion using the Boltzmann superposition principle. Consider a stress relaxation programme in which incremental strains Δe_1, Δe_2, Δe_3, etc. are added at times τ_1, τ_2, τ_3, etc. respectively. The total stress at time t is then given by

$$\sigma(t) = \Delta e_1 G(t - \tau_1) + \Delta e_2 G(t - \tau_2) + \Delta e_3 G(t - \tau_3) + \ldots, \quad (4.4)$$

where $G(t - \tau)$ is the stress relaxation modulus. Equation (4.4) may be generalized in an identical manner in which (4.1) leads to (4.2) and (4.3) to give

$$\sigma(t) = [G_r e] + \int_{-\infty}^{t} G(t - \tau) \frac{\mathrm{d}e(\tau)}{\Delta \tau} \mathrm{d}t, \quad (4.5)$$

where G_r is the equilibrium or relaxed modulus.

4.2.3 Mechanical Models, Retardation and Relaxation Time Spectra

Linear viscoelasticity may be represented pictorially by models comprising massless Hookean springs and Newtonian dashpots, the latter being considered as oil-filled cylinders in which a loosely fitting piston moves at a rate proportional to the viscosity of the oil and to the applied stress. The models are used to establish differential equations which describe the deformation of the polymer under investigation. We start by considering the two possible combinations of a single spring and a single dashpot, and then discuss more realistic models. In the outcome the differential equation becomes the basis for further development, and the model itself is discarded.

4.2.3.1 The Kelvin or Voigt model

This model (Figure 4.10(a)) consists of a spring of modulus E_K, in parallel with a dashpot of viscosity η_K. If a constant stress σ is applied at time $t = 0$ there can be no instantaneous extension of the spring, as it is retarded by the dashpot. Deformation then occurs at a varying rate, with the stress shared between the two components until, after a time dependent on the dashpot viscosity, the spring approaches a finite maximum extension. When the

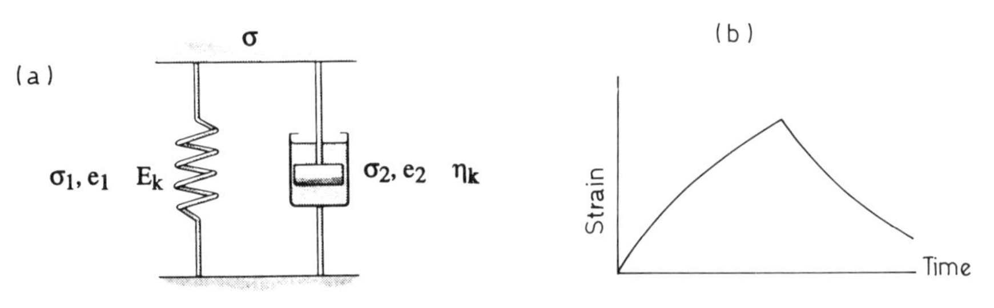

Figure 4.10 (a) The Kelvin, or Voigt, unit; (b) creep and recovery behaviour

stress is removed the reverse process occurs: there is no instantaneous retraction, but the initial unstretched length is eventually recovered (Figure 4.10(b)). The model does represent the time-dependent component of creep to a first approximation.

The stress–strain relations are, for the spring, $\sigma_1 = E_K e_1$ and for the dashpot

$$\sigma_2 = \eta_K \frac{de_2}{dt}$$

The total stress σ is shared between spring and dashpot: $\sigma = \sigma_1 + \sigma_2$; but the strain in each component is the total strain: $e = e_1 = e_2$.

$$\therefore \quad \sigma = E_K e + \eta_K \frac{de}{dt} \tag{4.6}$$

Solving for $0 < t < t_1$, when the stress is σ.

$$\frac{E_K}{\eta_K} \int_0^t dt = \int_0^e \frac{de}{(\sigma/E_K) - e}$$

where η_K/E_K has the dimensions of time, and represents the rate at which the deformation occurs: it is the retardation time τ'. Hence by integration

$$\frac{t}{\tau'} = \ln\left(\frac{\sigma/E_K}{(\sigma/E_K) - e}\right)$$

giving

$$\frac{\sigma}{E_K} = \left(\frac{\sigma}{E_K} - e\right) \exp(t/\tau').$$

Rearranging, we obtain

$$e = \frac{\sigma}{E_K}[1 - \exp(-t/\tau')]. \tag{4.7}$$

It is convenient in creep experiments to replace E_K by $1/J$, where J is the spring compliance, to give

$$e = J\sigma[1 - \exp(-t/\tau')]. \tag{4.8}$$

For $t > t_1$ after unloading, the solution becomes

$$e = e_{t_1} \exp\left(\frac{t_1 - t}{\tau'}\right), \text{ where } e_{t_1} = J\sigma[1 - \exp(-t_1/\tau')]. \tag{4.9}$$

The retardation time τ' is the time after loading for the strain to reach

$$\left(1 - \frac{1}{\exp}\right)$$

of its equilibrium value; after stress removal the strain decays to (1/exp) of its maximum value in time τ'.

The Kelvin model is unable to describe stress relaxation, as at constant strain the dashpot cannot relax. In mathematical terms $(\mathrm{d}e/\mathrm{d}t) = 0$, giving $\sigma = E_K e$.

4.2.3.2 *The Maxwell model*

The Maxwell model consists of a spring and dashpot in series as shown in Figure 4.11(a).

The equations for the stress–strain relations are

$$\sigma_1 = E_m e_1, \tag{4.10a}$$

relating the stress σ_1 and the strain e_1 in the spring and

$$\sigma_2 = \eta_m \frac{\mathrm{d}e_2}{\mathrm{d}t}, \tag{4.10b}$$

relating the stress σ_2 and the strain e_2 in the dashpot. Because the stress is identical for the spring and the dashpot the total stress $\sigma = \sigma_1 = \sigma_2$. The total strain e is the sum of the strain in the spring and the dashpot, i.e. $e = e_1 + e_2$.

To find the relationship between total stress and total strain equation (4.10a) can be written as

$$\frac{\mathrm{d}\sigma}{\mathrm{d}t} = E_m \frac{\mathrm{d}e_1}{\mathrm{d}t}$$

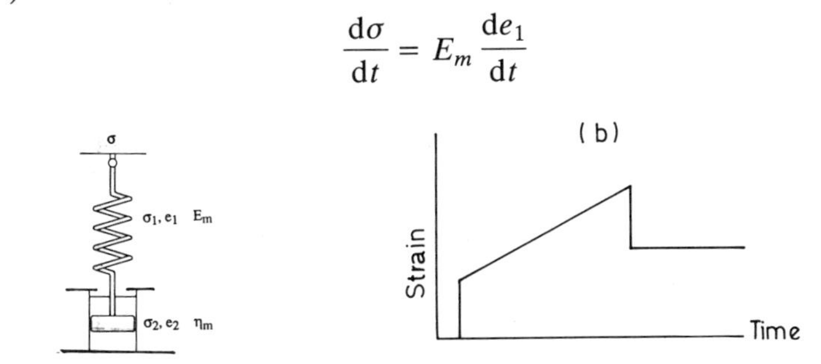

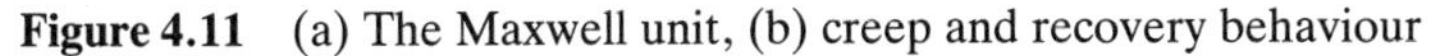
Figure 4.11 (a) The Maxwell unit, (b) creep and recovery behaviour

and added to (4.10b), giving

$$\frac{de}{dt} = \frac{1}{E_m}\frac{d\sigma}{dt} + \frac{\sigma}{\eta_m} \tag{4.11}$$

The Maxwell model is of particular value in considering a stress relaxation experiment. In this case

$$\frac{de}{dt} = 0 \quad \text{and} \quad \frac{1}{E_m}\frac{d\sigma}{dt} + \frac{\sigma}{\eta_m} = 0.$$

Thus

$$\frac{d\sigma}{\sigma} = -\frac{E_m}{\eta_m}dt.$$

At time $t = 0$, $\sigma = \sigma_0$, the initial stress, and integrating we have

$$\sigma = \sigma_0 \exp\left(\frac{-E_m}{\eta_m}\right)t. \tag{4.12}$$

This equation shows that the stress decays exponentially with a characteristic time constant $\tau = \eta_m/E_m$:

$$\sigma = \sigma_0 \exp\left(\frac{-t}{\tau}\right)$$

where τ is called the 'relaxation time'. There are two inadequacies of this simple model which can be understood immediately.

First, under conditions of constant stress, i.e

$$\frac{d\sigma}{dt} = 0, \qquad \frac{de}{dt} = \frac{\sigma}{\eta_m}$$

and Newtonian flow is observed. This is clearly not generally true for viscoelastic materials where the creep behaviour is more complex.

Secondly, the stress relaxation behaviour cannot usually be represented by a single exponential decay term, nor does it necessarily decay to zero at infinite time.

4.2.3.3 The standard linear solid

We have seen that the Maxwell model describes the stress relaxation of a viscoelastic solid to a first approximation, and the Kelvin model the creep behaviour, but that neither model is adequate for the general behaviour of a viscoelastic solid where it is necessary to describe both stress relaxation and creep.

A response closer to that of a real polymer is obtained by adding a second

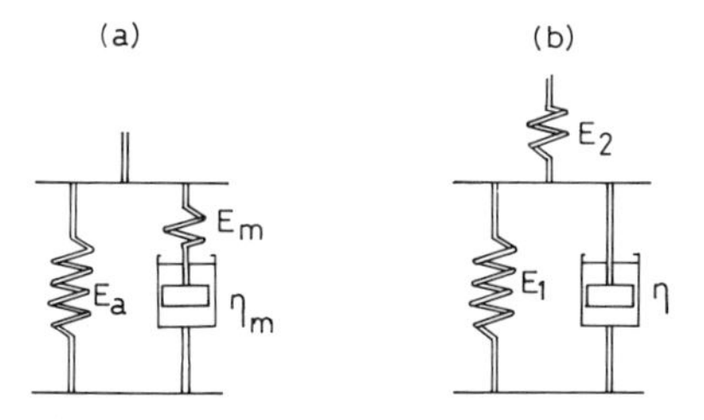

Figure 4.12 Three-element models. (a) is known as the standard linear solid (SLS)

spring of modulus E_a in parallel with a Maxwell unit (Figure 4.12). This model is known as the 'standard linear solid' and is usually attributed to Zener [2]. It provides an approximate representation to the observed behaviour of polymers in their viscoelastic range. In creep, both springs extend, so that

$$\tau' = \eta_m \left[\frac{1}{E_a} + \frac{1}{E_m} \right]$$

but in stress relaxation E_a is unaffected, giving

$$\tau = \frac{\eta_m}{E_m}$$

The stress–strain relationship is

$$\sigma + \tau \frac{d\sigma}{dt} = E_a e + (E_a + E_m)\tau \frac{de}{dt} \tag{4.13}$$

This model describes both creep and stress relaxation and the transition from the glassy modulus E_m (for $E_m \gg E_a$) at short times, where the viscous dashpot has an infinite viscosity, to the rubbery modulus E_a at long times.

4.2.3.4 Multi-element models

For real materials a simple exponential response in creep or stress relaxation is not an adequate description of the time dependence. A good representation can be obtained by simulating creep by an array of Kelvin models in series and stress relaxation by an array of Maxwell models in parallel (Figure 4.13). The dashpot in the first unit can be considered to be filled with a low-viscosity fluid, so that its response is effectively instantaneous. Figure 4.14 illustrates how such a system approaches reality, with successive units responding as time elapses.

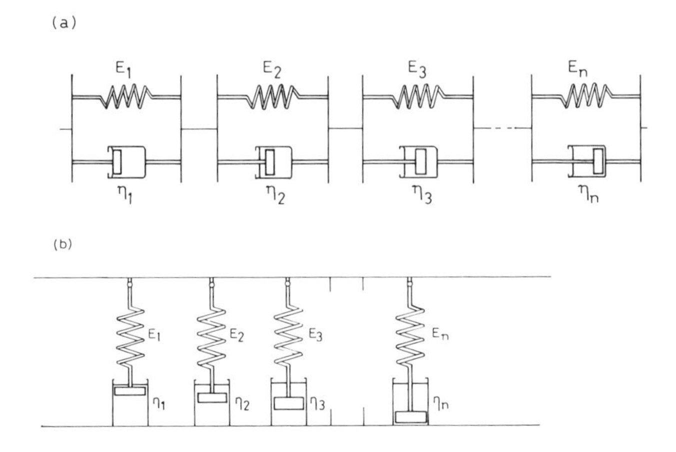

Figure 4.13 (a) Kelvin units in series for creep simulation; (b) Maxwell units in parallel for stress relaxation simulation

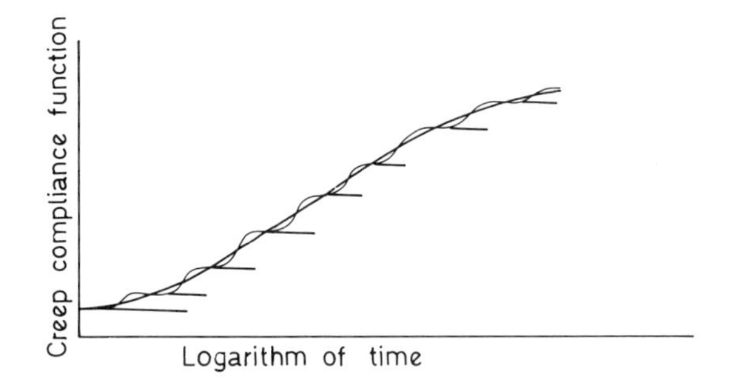

Figure 4.14 Simulation of creep function through combining Kelvin units in series (schematic)

4.2.3.5 Retardation and relaxation time spectra

Let the number of individual units in multi-element models tend to infinity. For creep, an infinite number of Kelvin units gives an infinite number of retardation times: this is called the spectrum of retardation times. The analagous development for stress relaxation leads to the spectrum of relaxation times.

For stress relaxation at constant strain e the Maxwell model gives

$$\sigma(t) = eE \exp\left(\frac{-t}{\tau}\right) \tag{4.14}$$

so that the stress relaxation modulus

$$G(t) = E\exp\left(\frac{-t}{\tau}\right)$$

For a series of Maxwell units all at strain e

$$\sigma(t) = e\sum^{n} E_n \exp\left(\frac{-t}{\tau_n}\right) \tag{4.15}$$

where E_n and τ_n refer to the nth Maxwell unit.

The summation can be written as an integral, giving

$$\sigma(t) = [G_{\mathrm{r}}e] + e\int_0^\infty f(\tau)\exp\left(\frac{-t}{\tau}\right)\mathrm{d}\tau, \tag{4.16}$$

where the first term defines the eventual 'relaxed' stress, and the spring constant E_n is replaced by the weighting function $f(\tau)\,\mathrm{d}\tau$ which defines the concentration of Maxwell units with relaxation times between τ and $(\tau + \mathrm{d}\tau)$.

The stress relaxation modulus is then given by

$$G(t) = G_{\mathrm{r}} + \int_0^\infty f(\tau)\exp\left(\frac{-t}{\tau}\right)\mathrm{d}\tau. \tag{4.17}$$

As relaxation curves extend over many decades of time it is convenient to use a logarithmic time-scale, so that the stress relaxation modulus becomes

$$G(t) = G_{\mathrm{r}} + \int_{-\infty}^{\infty} H(\tau)\exp\left(\frac{-t}{\tau}\right)(\mathrm{d}\ln\tau). \tag{4.18}$$

A new relaxation time spectra $H(\tau)$ is now defined where $H(\tau)\,\mathrm{d}(\ln\tau)$, gives the contribution to stress relaxation associated with relaxation times between $\ln\tau$ and $\ln\tau + \mathrm{d}(\ln\tau)$.

For creep under constant stress σ, the Kelvin model gives (from 4.7)

$$J(t) = J\left[1 - \exp\left(\frac{-t}{\tau'}\right)\right]$$

Following a similar line of argument to that above, the creep compliance becomes

$$J(t) = J_{\mathrm{u}} + \int_{-\infty}^{\infty} L(\tau)\left[1 - \exp\left(\frac{-t}{\tau}\right)\right]\mathrm{d}(\ln\tau), \tag{4.19}$$

where J_{u} is the instantaneous (unrelaxed) elastic compliance, and $L(\tau)\,\mathrm{d}(\ln\tau)$, which gives the contributions to the creep compliance associated with retardation times between $\ln\tau$ and $\ln\tau + \mathrm{d}(\ln\tau)$, defines the retardation time spectrum $L(\tau)$. Note that when moving from a linear to logarithmic time-scale the lower limit of integration becomes minus infinity.

4.3 DYNAMIC MECHANICAL MEASUREMENTS: THE COMPLEX MODULUS AND COMPLEX COMPLIANCE

An alternative experimental procedure to creep and stress relaxation is to subject the specimen to an alternating strain and simultaneously measure the stress. For linear viscoelastic behaviour, when equilibrium is reached, the stress and strain will both vary sinusoidally, but the strain lags behind the stress. Thus we write

$$\text{strain } e = e_0 \sin \omega t,$$

$$\text{stress } \sigma = \sigma_0 \sin(\omega t + \delta),$$

where ω is the angular frequency and δ the phase lag.

Expanding: $\sigma = \sigma_0 \sin \omega t \cos \delta + \sigma_0 \cos \omega t \sin \delta$ we see that the stress can be considered to consist of two components: (1) of magnitude $(\sigma_0 \cos \delta)$ in phase with the strain; (2) of magnitude $(\sigma_0 \sin \delta)$ 90° out of phase with the strain.

The stress–strain relationship can therefore be defined by a quantity G_1 in phase with the strain and by a quantity G_2 which is 90° out of phase with the strain, i.e.

$$\sigma = e_0 G_1 \sin \omega t + e_0 G_2 \cos \omega t, \tag{4.20}$$

where

$$G_1 = \frac{\sigma_0}{e_0} \cos \delta \quad \text{and} \quad G_2 = \frac{\sigma_0}{e_0} \sin \delta.$$

A phasor diagram (Figure 4.15) then indicates that G_1 and G_2 define a complex modulus G^*. If $e = e_0 \exp(\mathrm{i}\omega t)$, then $\sigma = \sigma_0 \exp[\mathrm{i}(\omega t + \delta)]$, so that

$$\begin{aligned} G^* &= \frac{\sigma}{e} = \frac{\sigma_0}{e_0} \exp(\mathrm{i}\delta) \\ &= \frac{\sigma_0}{e_0} (\cos \delta + \mathrm{i} \sin \delta) \\ &= G_1 + G_2, \end{aligned} \tag{4.21}$$

where G_1, which is in phase with the strain, is called the storage modulus because it defines the energy stored in the specimen due to the applied strain, and G_2, which is $\pi/2$ out of phase with the strain, defines the dissipation of energy, and is called the loss modulus, for a reason which becomes evident through calculating the energy (ΔE) dissipated per cycle:

$$\Delta E = \oint \sigma \, \mathrm{d}e = \int_0^{2\pi/\omega} \sigma \frac{\mathrm{d}e}{\mathrm{d}t} \mathrm{d}t.$$

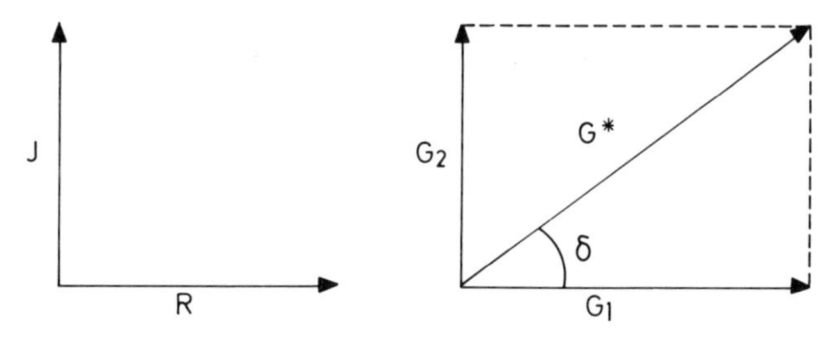

Figure 4.15 Phasor diagram for complex modulus $G^* = G_1 + iG_2$ and phase lag, $\tan\delta = G_2/G_1$

Substituting for σ and e,

$$\Delta E = \omega e_0^2 \int_0^{2\pi/\omega} (G_1 \sin\omega t \cos\omega t + G_2 \cos^2\omega t)\, dt. \tag{4.22}$$

The integral is solved by using $\sin\omega t \cos\omega t = \frac{1}{2}\sin 2\omega t$ and $\cos^2\omega t = \frac{1}{2}(1 + \cos 2\omega t)$, to give

$$\Delta E = \pi G_2 e_0^2. \tag{4.23}$$

If the integral for ΔE is evaluated over a quarter-cycle rather than over the complete period, the first term

$$G_1 \omega e_0^2 \int_0^{\pi/2\omega} \sin\omega t \cos\omega t\, dt \tag{4.24}$$

gives the maximum stored elastic energy (E).

Evaluating as before, we obtain

$$E = \tfrac{1}{2} G_1 e_0^2 \tag{4.25}$$

which, as expected, is independent of frequency. Rewriting (4.23) and (4.25)

$$G_1 = \frac{2E}{e_0^2} \qquad G_2 = \frac{\Delta E}{\pi e_0^2}$$

Hence

$$\frac{G_2}{G_1} = \tan\delta = \frac{\Delta E}{2\pi E} \tag{4.26}$$

The ratio $\Delta E/E$ is called the specific loss

$$\frac{\Delta E}{E} = 2\pi \tan\delta. \tag{4.27}$$

Typical values of G_1, G_2 and $\tan\delta$ for a polymer are 10^9 Pa, 10^7 Pa and 0.01 respectively. In such cases $|G^*|$ is approximately equal to G_1, and it is customary to define the dynamic mechanical behaviour in terms of the 'modulus' $G \approx G_1$, and the phase angle δ, or $\tan\delta = G_2/G_1$.

A complementary treatment can be developed to define a complex compliance $J^* = J_1 - \mathrm{i}J_2$, which is directly related to the complex modulus, as $G^* = 1/J^*$.

4.3.1 Experimental Patterns for G_1, G_2, etc. as a Function of Frequency

Consider the variation of G_1, G_2 and $\tan\delta$ with frequency, for a viscoelastic solid which shows no flow (Figure 4.16). At very low frequencies the polymer is rubber-like and has a low modulus (G_1 probably about 10^5 Pa) which is independent of frequency. At the highest frequencies the rubber is glassy, with a modulus around 10^9 Pa, which is again independent of frequency. In the intermediate region, where the material behaves viscoelastically, the modulus will increase with increasing frequency. The loss modulus will be zero at low and high frequencies, where stress and strain are in phase for the rubbery and glassy states. In the intermediate viscoelastic region G_2 rises to a maximum value, close to the frequency at which G_1 is changing most rapidly with frequency. The loss factor $\tan\delta$ also has a maximum in the viscoelastic region, but this occurs at a slightly lower frequency than that in G_2, since $\tan\delta = G_2/G_1$, and G_1 is also changing rapidly in that frequency region.

An analogous diagram (Figure 4.17) shows the variation of the compliances J_1 and J_2 with frequency.

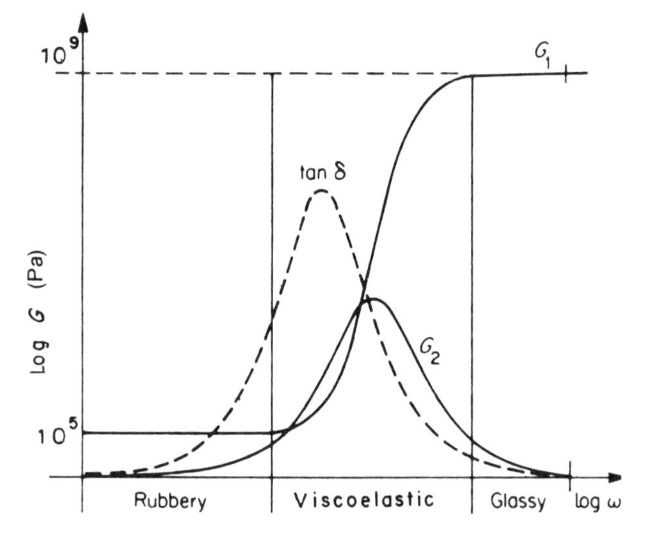

Figure 4.16 The complex modulus $G_1 + \mathrm{i}G_2$ as a function of frequency ω

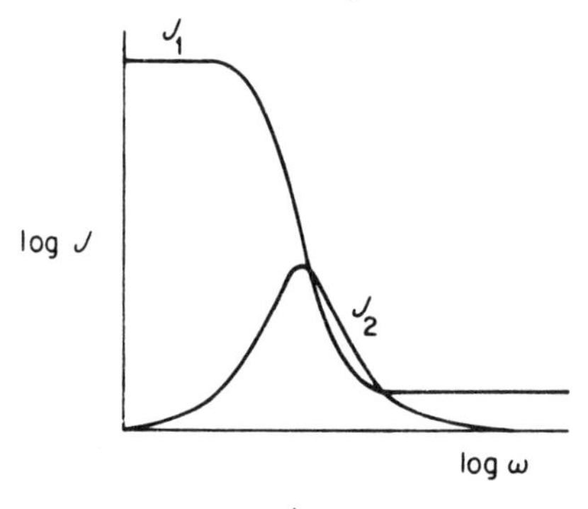

Figure 4.17 The complex compliance $J^* = J_1 - \mathrm{i}J_2$ as a function of frequency ω

4.3.2 The Alfrey Approximation

Relaxation and retardation time spectra can be calculated exactly from stress relaxation, creep and dynamic mechanical measurements using Fourier or Laplace transform methods. However, it is often adequate to use simple approximations due to Alfrey in which the exponential term for a single Kelvin or Maxwell unit is replaced by a step function, as shown schematically in Figure 4.18.

Consider

$$G(t) = [G_r] + \int_{-\infty}^{\infty} H(\tau) \exp\left(\frac{-t}{\tau}\right) \mathrm{d}(\ln \tau). \tag{4.28}$$

If we assume that $e^{-t/\tau} = 0$ up to the time $\tau = t$, and $e^{-t/\tau} = 1$ for $\tau > t$. We can write

$$G(t) = [G_r] + \int_{\ln \tau}^{\infty} H(\tau) \mathrm{d}(\ln \tau).$$

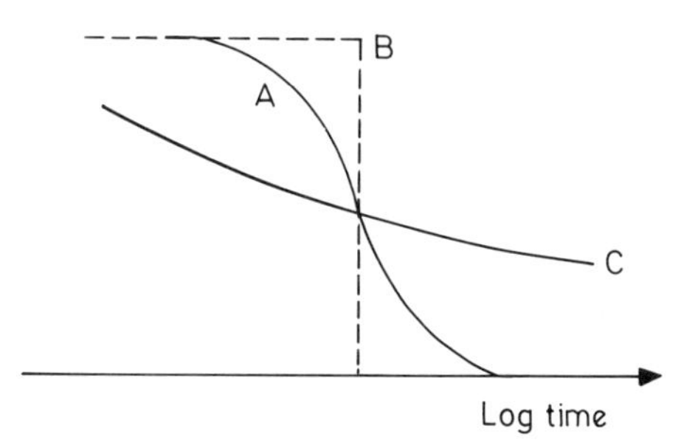

Figure 4.18 The Alfrey approximation: the stress relaxation of a Maxwell unit A is replaced by a step function B. The curve C represents relaxation of a typical viscoelastic polymer

This gives the relaxation time spectrum

$$H(\tau) = \left[\frac{\mathrm{d}G(t)}{\mathrm{d}\ln t}\right]_{t=\tau} \tag{4.29}$$

which is known as the 'Alfrey approximation' [3].

The relaxation time spectrum can be expressed to a similar degree of approximation in terms of the real and imaginary parts of the complex modulus:

$$H(\tau) = \left[\frac{\mathrm{d}G_1(\omega)}{\mathrm{d}\ln \omega}\right]_{1/\omega=\tau} = \frac{2}{\pi}\,[G_2(\omega)]_{1/\omega=\tau}. \tag{4.30}$$

These relationships are illustrated diagrammatically for the case of a single relaxation transition in Figure 4.19(a) and (b). To obtain the complete relaxation time spectrum the longer time part of $H(\tau)$ will be found from the stress relaxation modulus data of Figure 4.19(a) and the shorter time part from the dynamic mechanical data of Figure 4.19(b).

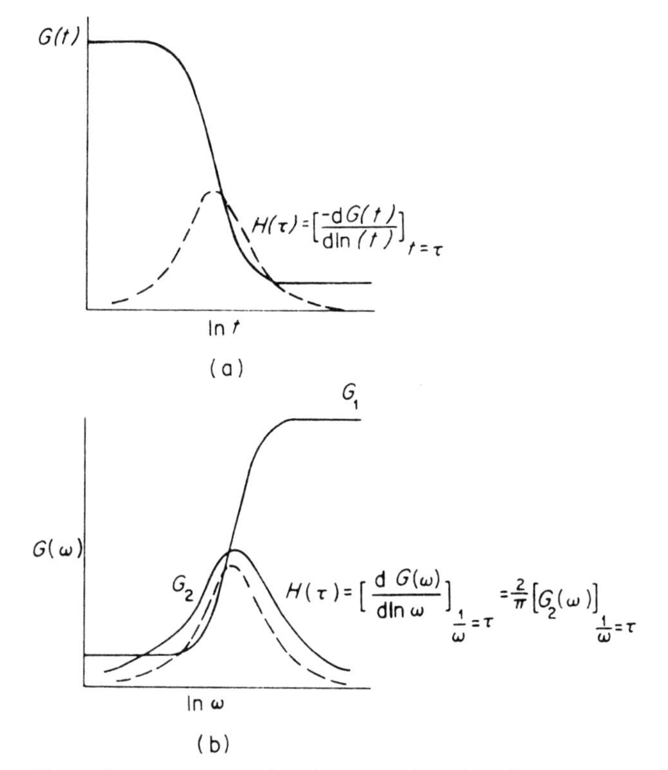

Figure 4.19 The Alfrey approximation for the relaxation time spectrum $H(\tau)$: (a) from the stress relaxation modulus $G(t)$; (b) from the real and imaginary parts G_1 and G_2 respectively of the complex modulus $G(\omega)$

Figure 4.19 shows that the relaxation time spectrum can be determined directly from the gradient of plots of the relaxation modulus or the dynamic modulus G_1 versus logarithm of time, or from G_2 even more directly.

Complementary relationships can be used to obtain the retardation time spectrum in terms of the complex compliances and the creep compliance.

REFERENCES

1. L. Boltzmann, *Pogg. Ann. Phys. Chem.* **7**, 624 (1876).
2. C. Zener, *Elasticity and Anelasticity of Metals*, Chicago University Press, Chicago, 1948.
3. T. Alfrey, *Mechanical Behaviour of High Polymers*, Interscience, New York, 1948.

PROBLEMS FOR CHAPTER 4

1. The stress relaxation modulus of a certain polymer can be described approximately by

$$G(t) = G_0 e^{-t/\tau}$$

and has the values 2.0 GPa and 1.0 GPa at $t = 0$ and 10^4 s respectively. Calculate the form of the creep compliance and so evaluate the strain 1000 s after the rapid application of a stress of 100 MPa.

2. The creep deformation of a linear viscoelastic solid under constant stress can be represented by the model shown, in which a spring is in series with a Kelvin unit.

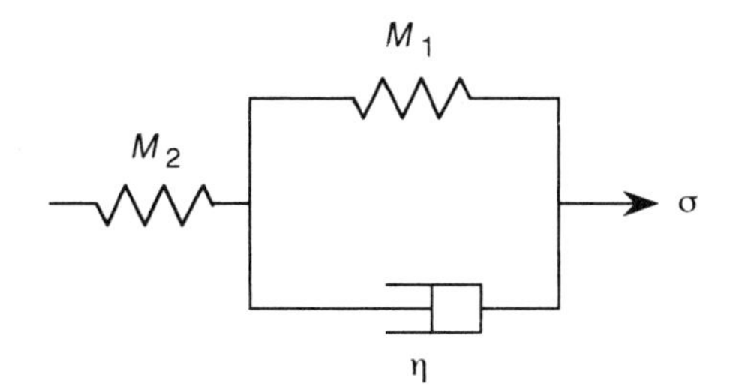

Here M_1 and M_2 represent Young's moduli, η the viscosity and σ the total constant stress. Show that the total strain is dependent on the relaxation time, $\tau_2 = \eta/M_1$, and is given by

$$\varepsilon = \frac{\sigma}{M_2} + \frac{\sigma}{M_1}[1 - \exp(-t/\tau_2)].$$

Immediately after applying the stress the strain is 0.002; after 1000 s the strain is 0.004; after a very long time the strain tends to 0.006. What is the retardation time τ_2?

NB It is not necessary to know the stress or the values of M_1, M_2 and η.

3. A strip of a linear viscoelastic polymer 200 mm long, 10 mm wide and 1 mm thick, and with an extensional Young's modulus of 2 GPa, is mounted in a dynamic testing apparatus. The specimen is initially extended by 1 mm, and then subjected to a sinusoidally varying strain with an amplitude of $\pm$ 1 mm.

   At 20 °C and 5 Hz the phase lag between stress and strain is 0.1 rad. Calculate the maximum stress developed, the elastic energy stored during the positive quarter-cycle and the work dissipated per cycle.

   Another specimen of the same polymer is tested in a simple torsion pendulum at 20 °C. The period of vibration is 2 s and the logarithmic decrement is 0.20. What would you expect the phase lag in a dynamic tester to be at 20 °C and 0.5 Hz? Comment on the result.
4. A long fibre is loaded by a mass of 0.1 kg attached to its lower end. The extension (%) is measured at various times after loading

Ext (%)	t (min)
0.300	0
0.328	10
0.350	20
0.390	40
0.428	60
0.462	80
0.490	100
0.514	120
0.535	140
0.555	160
0.572	180
0.585	200
0.593	220
0.600	240

(The reading at $t = 0$ gives the immediate elastic extension.)

Assuming the behaviour to be linear viscoelastic, calculate the extension under the given conditions.

(i) Load with 0.1 kg at $t = 0$; remove load at $t = 40$; reload at $t = 80$; remove load at $t = 120$. Calculate the net residual extension at $t = 240$ min.

(ii) Load with 0.1 kg at $t = 0$; add a further 0.2 kg at $t = 40$; unload completely at $t = 200$. What is the extension at $t = 80$; what is the immediate elastic recovery; what is the net residual extension at $t = 240$? NB A graphical solution is not necessary.

5. It is found that the stress relaxation behaviour of a certain polymer can be represented by a Maxwell model, where an elastic spring of modulus 10^8 Pa is in series with a viscous dashpot of viscosity 10^{10} Pa. State what is meant by the relaxation time and calculate its value for this polymer. If a strain of 1% is applied at time $t = 0$, followed by a further additional strain of 2% at time $t = 25$ s, calculate the stress at time $t = 50$ s.
6. For a polymer whose creep behaviour can be described by the Kelvin model, the creep compliance $J(t) = J_0\ (1 - e^{-t/\tau'})$. If this polymer is subjected to an instantaneous stress σ_1 at time $t = 0$, which is increased to a value σ_2 at time $t = t_1$, find the strain at a time $t > t_1$.

7. A Maxwell element consists of an elastic spring of modulus $E_m = 10^9$ Pa and a dashpot of viscosity 10^{11} Pa. Calculate the stress at time $t = 100$ s in the following loading programme.

   (i) At time $t = 0$ an instantaneous strain of 1% is applied.
   (ii) At time $t = 30$ s the strain is increased instantaneously from 1% to 2%.

8. The stress relaxation behaviour of a certain polymer can be represented by a Maxwell model of a spring and dashpot in series. The spring has the properties of an ideal rubber with a room temperature (293 K) modulus of 10^6 Pa. The dashpot has a viscosity η_{293} at room temperature of 10^8 Pa, with a temperature dependence such that the viscosity at a temperature T is given by

$$\eta_T = \eta_{293} \exp\left(\frac{-\Delta H}{RT}\right)$$

   where $\Delta H = 4\ \mathrm{kJ\,mol^{-1}}$. Calculate the stress relaxation time of the polymer at 100 °C. You may neglect the change in the density of the rubber with temperature. (The gas constant R may be taken as $8.3\ \mathrm{JK^{-1}\,mol^{-1}}$.)

Chapter 5

The Measurement of Viscoelastic Behaviour

For a satisfactory understanding of the viscoelastic behaviour of polymers data are required over a wide range of frequency (or time) and temperature. The number of experiments required can sometimes be reduced by using either the equivalence of creep, stress relaxation and dynamic mechanical data (described in Chapter 4) or the equivalence of time and temperature as variables (to be discussed in Chapter 6). Nevertheless a variety of techniques need to be combined to cover a wide range of both time and temperature.

There are five main classes of experiment, which will be discussed in turn.

1. Transient measurements: creep and stress relaxation;
2. Low-frequency vibrations: free oscillation methods;
3. High-frequency vibrations: resonance methods;
4. Forced vibration non-resonance methods;
5. Wave propagation methods.

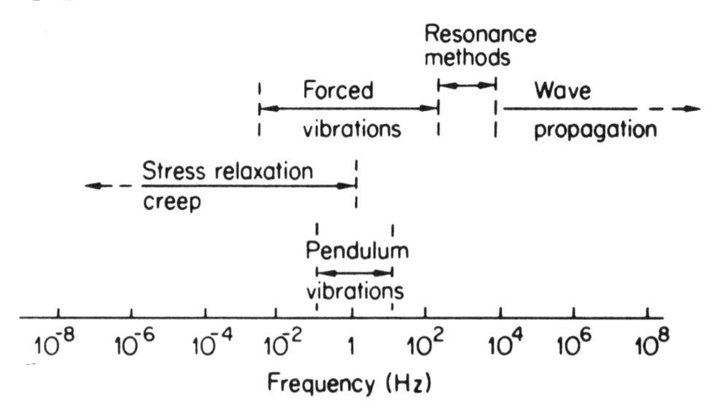

Figure 5.1 Approximate frequency scales for different experimental techniques. (Reproduced with permission from Becker, *Mater. Plast. Elast.*, **35**, 1387 (1969))

The approximate frequency scale for each technique is indicated in Figure 5.1.

5.1 CREEP AND STRESS RELAXATION

Reliable creep and stress relaxation data are obtainable only if the specimens are well defined and strictly comparable. As deformations and deformation rates are usually quite small if linearity is to hold, precision measurements are required: conditions that may be difficult to attain throughout the highly significant short-time regime.

5.1.1 Creep conditioning

Leaderman [1] was the first worker to emphasize that specimens must be cyclically conditioned at the highest temperature of measurement in order to obtain reproducible measurements in creep and recovery. Each cycle consists of application of the maximum load for the maximum period of loading, followed by a recovery period after unloading of about 10 times the loading period; cycling must be continued until reproducibility is obtained.

The conditioning procedure has two major effects on the creep and recovery behaviour. First, subsequent creep and recovery responses under a given load are then identical, i.e. the sample has lost its 'long-term' memory and now only remembers loads applied in its immediate past history. Secondly, after the conditioning procedure the deformation produced by any loading programme is almost completely recoverable provided that the recovery period is about 10 times the period during which loads are applied. For tensile creep measurements over a wide range of temperature greater elaboration is required.

5.1.2 Specimen characterization

Many early experiments on the viscoelastic behaviour of polymers were unsatisfactory because the specimens were inadequately characterized, so that like was not compared with like.

Average molecular mass and its distribution are both critical parameters, and all polymers contain processing and stabilizing additives, which can sometimes have a significant effect on the response to stress and strain, particularly at temperatures well above that of the laboratory.

The physical structure of the polymer, in terms of morphology, crystallinity and molecular orientation, will also be important and should be well characterized.

5.1.3 Experimental Precautions

Measurements made in the vicinity of a major relaxation region, for instance creep tests on isotactic polypropylene close to room temperature, are sensitive to small changes in temperature, so that a controlled temperature environment is essential. For some polymers, such as nylon, it is also essential to control the humidity, because the presence of moisture in the polymer has a dramatic effect on the mechanical behaviour [2] (in nylon by reducing the effect of interchain hydrogen bonding). Whenever possible, several nominally identical specimens should be examined, to confirm that inter-specimen variability is small compared with the effects under examination.

In the linear viscoelastic region strains are unlikely to exceed 1%, so that the change of cross-section, and hence stress, with strain will be small. At larger strains the effective load should be reduced in a manner proportional to the decrease in cross-sectional area, to maintain a constant stress. Leaderman [3] used the device illustrated in Figure 5.2, in which a flexible tape (B) attached to the specimen (A) is wound round the periphery of a cylindrical drum C. A similar tape F, attached to a profiled cam D, supports a fixed mass E. As the specimen extends under load, the moment of E decreases according to the cam profile.

The creep and relaxation plots in Figures 4.2 and 4.4 are idealized for, as already mentioned, the immediate response is no more than that occurring before the first measurement after loading, and it is evident that the relevant

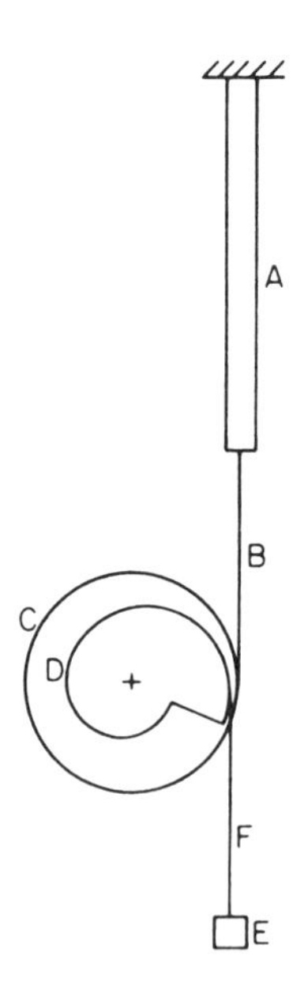

Figure 5.2 Cam arrangement for creep under constant stress. (Reproduced with permission from Leaderman, *Tran. Soc. Rheol.*, **6**, 361 (1962))

time interval should be no longer than necessary. For instance, if the load is applied very slowly during a stress relaxation experiment, a low value of maximum stress will be obtained (Figure 5.3). Conversely, too rapid a stress application will result in the complications of dynamic loading (Figure 5.4). This point is of particular importance, because some attempts to relate viscoelasticity to molecular parameters are particularly dependent on the short-time response. As explained in section 6.2 below, some of the difficulties concerned with the early stages of deformation can be removed by employing time–temperature equivalence.

Specimens in extensional and torsional tests must be firmly clamped at their ends. However, the stresses in the clamp region will differ greatly from those in the bulk of the specimen, and if the complete length of the specimen is measured, end effects can be ignored only where the length is at least ten times the diameter. For oriented samples end effects are even more important. As an approximate guide it is reasonable to consider that the ratio of length to diameter should be greater than $10\sqrt{E/G}$, where E and

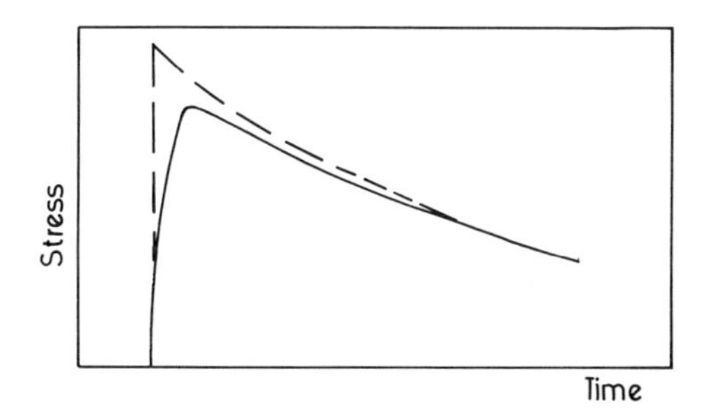

Figure 5.3 Effect on stress relaxation of the strain being applied slowly. Ideal --- actual ——

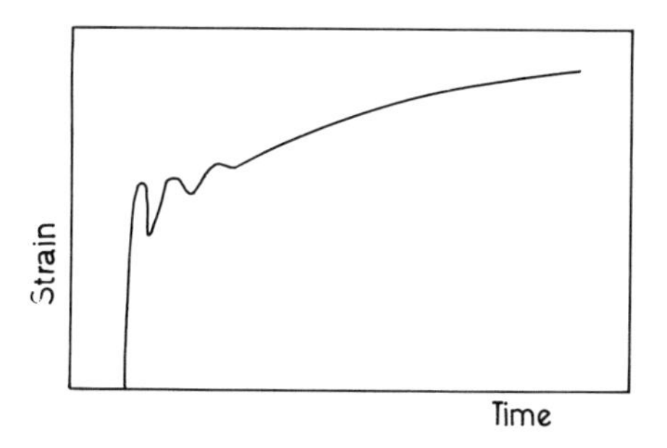

Figure 5.4 Damped vibrations resulting from rapid loading in a creep experiment

G are the Young's modulus (in the fibre direction) and the torsional modulus respectively. A more satisfactory technique for high accuracy with robust specimens is to use an extensometer attached to the specimen away from its ends, strain being converted either into rotation of light-reflecting mirrors [4] or into an electrical signal by a displacement transducer [5].

In stress relaxation measurements, as in standard mechanical testing devices, changing stress may be monitored through the changing deformation of a stiff spring in series with the sample. It must be confirmed that the spring stiffness is indeed much higher than that of the specimen, otherwise corrections must be applied.

A range of measurement equipment is described and illustrated in the books by Turner [6] and Ward [7].

Where materials are being examined for their suitability for specific applications it is essential that creep measurements are performed over an extended time-scale. It is possible that a material which shows good short-term creep shows accelerated creep at longer times (Figure 5.5). As discussed later comparable problems may occur for materials that are viscoelastically non-linear, so that the recovery response is very different from that during the early stages of creep.

5.2 DYNAMIC MECHANICAL MEASUREMENTS

A variety of techniques is required, capable of covering wide ranges of both time and temperature (Figure 5.1). Free oscillation pendulum methods, which have the advantage of simplicity, are confined to the frequency range 10^{-1}–10 Hz. Forced vibration techniques, though more complicated, may yield higher reproducibility, and can extend the frequency range by a further decade on either side, linking up with creep and stress relaxation at the

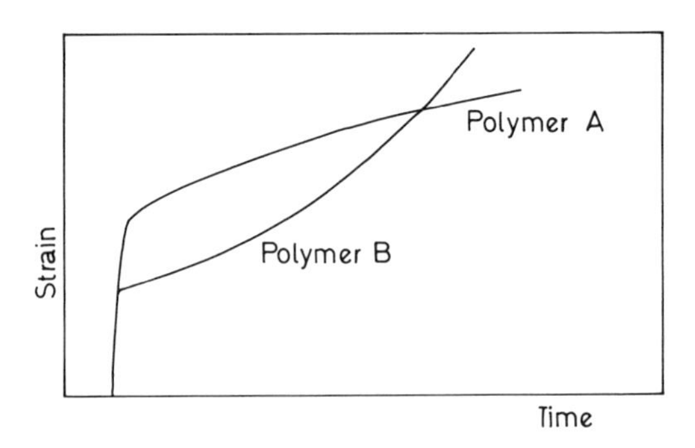

Figure 5.5 Short-term creep behaviour at a single temperature implies nothing about the eventual deformation

lowest frequencies and resonance methods at the higher end. These last, which are very sensitive to inter-specimen variability, are important above 10 kHz.

5.2.1 The Torsion Pendulum

The simplest device consists of a specimen of circular cross-section suspended vertically with its upper end rigidly clamped [8]. Its lower end supports a disc, or preferably a bar, fitted with adjustable weights (Figure 5.6), whose distance from the axis can be altered, so changing the moment of inertia (I) and the period of oscillation. When the bar is twisted and released the oscillations gradually decrease in amplitude, and the logarithmic decrement Λ, the natural logarithm of the ratio of amplitude of successive oscillations, is recorded.

Because the specimen is loaded by the inertia bar, the specimen is subjected to tensile as well as torsional stresses, which perturb the nominally free vibrations. For more precise work the specimen can be mounted as in Figure 5.7, with the inertia bar clamped at its upper end. The assembly is then suspended by an elastic wire or ribbon, which has a negligible effect on damping.

For an *elastic* rod the equation of motion is $I\ddot{\theta} + \tau\theta = 0$, where τ is the torsional rigidity of the rod, which is related to the shear modulus G, through

$$\tau = \frac{\pi r^4 G}{2l}$$

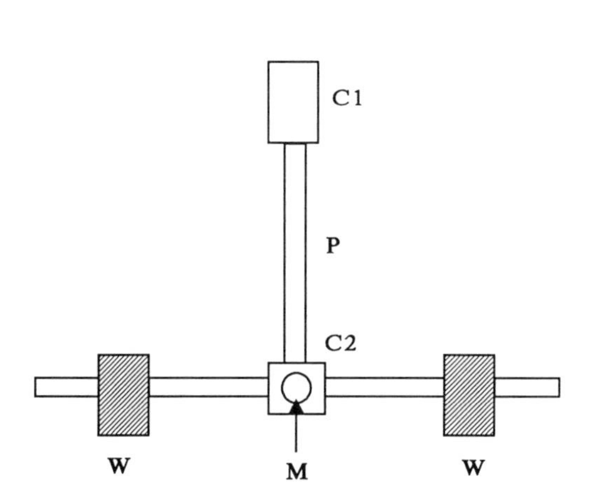

Figure 5.6 Free vibration torsion pendulum. P, polymer specimen; C1 fixed upper clamp; C2, lower clamp, fixed to inertia bar; W, W, sliding masses to change moment of inertia; M, mirror to reflect light beam

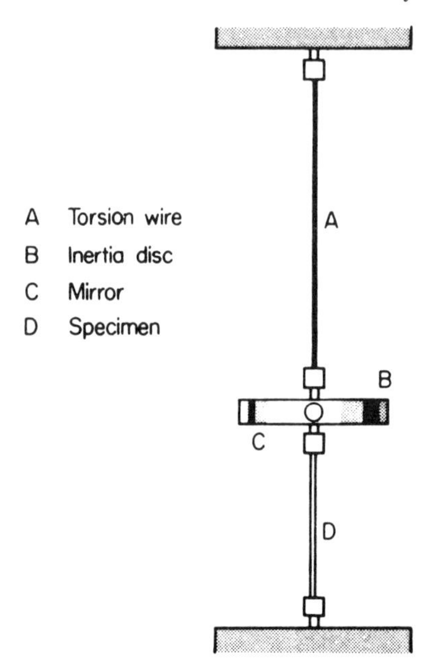

Figure 5.7 Apparatus for measuring torsional rigidity at low frequencies

where l is the length and r the radius of the rod. The elastic system executes simple harmonic motion with an angular frequency

$$\omega = \sqrt{\frac{\tau}{I}} = \sqrt{\frac{\pi r^4 G}{2lI}} \tag{5.1}$$

The high sensitivity to sample dimensions implies that inter-specimen comparisons can be subject to large uncertainties.

When the vibrations are damped the amplitude decreases with time, but with light damping there is only a small effect on the period

$$\left(\frac{2\pi}{\omega}\right)$$

For linear viscoelastic solids the torsional modulus is complex, and may be written as $G^* = G_1 + iG_2$. When the damping is small, it is justified to replace G_1 for G in (5.1), hence

$$\omega^2 = \frac{\pi r^4 G_1}{2lI} \tag{5.2}$$

The logarithmic decrement can then be related to the specific loss (4.33) and hence tan δ.

$$\Lambda = \ln\left(\frac{A_n}{A_{n+1}}\right)$$

where A_n denotes the amplitude of the nth oscillation. For small damping

$$\Lambda = \ln\left(1 + \frac{\Delta A}{A_n}\right) = \frac{\Delta A}{A_n} - \frac{1}{2}\frac{\Delta A^2}{A_n^2}$$

Hence

$$\Lambda = \frac{1}{2}\left(\frac{A_n^2 - A_{n+1}^2}{A_n^2}\right)$$

But (amplitude)^2 is proportional to stored energy, giving

$$\Lambda = \frac{1}{2}\frac{\Delta E}{E} = \pi \tan \delta \qquad (5.3)$$

from which $G_2 = G_1 \tan \delta$ can be obtained.

5.2.2 Forced Vibration Methods

Free vibration methods suffer from the disadvantage that the frequency of vibration depends on the stiffness of the specimen, which varies with temperature, so that forced vibration methods are to be preferred when the frequency and temperature dependence of viscoelastic behaviour are to be investigated.

As indicated in section 4 above, when a sinusoidal strain is imposed on a linear viscoelastic body the strain lags behind the stress by the phase angle δ, which determines the degree of damping. The strains must be low enough for linearity to apply, and the strain must at all times remain positive. In practice the strain amplitude is typically $\pm\frac{1}{2}\%$, superposed on an initial extension slightly in excess of $\frac{1}{2}\%$, to allow for some degree of stretch during the experiment. The specimen must be short enough for there to be no appreciable variation of stress along its length, i.e. the length must be short compared with the wavelength of the stress waves. Assuming that the lowest value of modulus is 10^7 Pa for a specimen of density 10^3 kg m^{-3}, the longitudinal wave velocity is 10^2 ms^{-1}. At 100 Hz the wavelength of the stress waves is 1 m, which suggests that at that frequency the upper limit on specimen length is about 0.1 m. As the stress must never vanish, a lower limit to frequency is set by the stress relaxation time.

Typically strain and stress are measured by unbonded strain-gauge transducers, the signals from which are then fed to a phase meter, which provides a direct reading of the relative amplitudes and the phase difference, hence giving values of modulus and tan δ [9].

5.2.3 The Vibrating Reed

This formerly popular industrial method employs a small flat specimen which is clamped in a head driven from a variable frequency oscillator, typically in the range 200–1500 Hz [10, 11]. The amplitude of vibration which can be measured optically, varies with frequency at a fixed temperature to yield a resonance curve of the form shown in Figure 5.8. The resonant frequency is usually temperature dependent.

A complete treatment of the viscoelastic behaviour of the vibrating reed has been given by Bland and Lee [12]. If losses are small the modulus E can be obtained from the solution to the *elastic* problem, and $\tan\delta$ from the resonance curve. For a cantilever of density ρ, length l and thickness h, the resonant frequency of the fundamental mode is

$$f_{\mathrm{r}} = \frac{1}{2\pi}\frac{(1.875)^2}{l^2}\sqrt{\left(\frac{E}{\rho}\right)}\frac{h}{2\sqrt{3}}$$

giving

$$E = 38.24\,\frac{l^4\rho}{h^2}\,f_{\mathrm{r}}^2. \tag{5.4}$$

As with free torsional vibrations, the sensitivity to sample dimensions can give rise to large uncertainties in inter-specimen comparisons.

An analogy between the viscoelastic reed and an equivalent electrical circuit [10] indicates that

$$\tan\delta = \Delta f/f_{\mathrm{r}}, \tag{5.5}$$

where Δf, the bandwidth, is the difference between the frequencies at which the amplitude of vibration is $1/\sqrt{2}$ the maximum value (Figure 5.8).

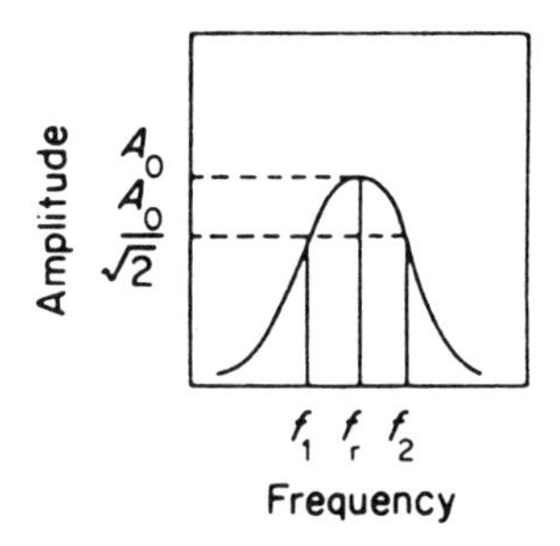

Figure 5.8 Resonance curve for the vibrating reed ($f_2 - f_1 = \Delta f$; see text)

5.3 WAVE-PROPAGATION METHODS

Wave-propagation methods are in two broad categories:

1. In the kilohertz frequency range;
2. In the megahertz frequency range: ultrasonic methods.

5.3.1 The Kilohertz Frequency Range

At frequencies in the kilohertz range the wavelength of the stress waves is of the order of the length of a viscoelastic specimen. Typically [13], a thin monofilament is stretched longitudinally, with one end attached to a stiff massive diaphragm, such as loudspeaker. A piezoelectric crystal pickup then detects the changes in signal amplitude and phase along the length of the specimen. As shown in Figure 5.9 for low-density polyethylene, a plot of the phase (θ) against distance (l) takes the form of damped oscillations about the line $\theta = kl$, where k is the propagation constant.

Where the attenuation coefficient α is small it has been shown [14] that $V_{max}/V_{min} = (\alpha l + \beta)$, where V is the signal amplitude. A plot of

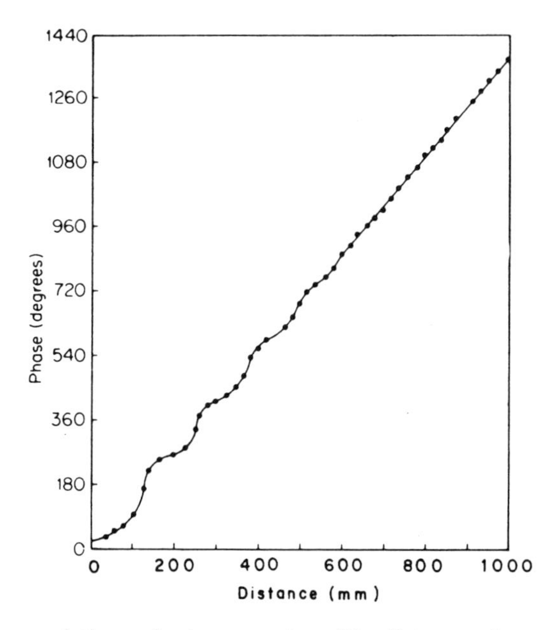

Figure 5.9 The variation of phase angle with distance along a polyethylene monofilament for transmission of sound waves at 3000 Hz. (Reproduced with permission from Hillier and Kolsky, *Proc. Phys. Soc.* **B**, **62**, 111 (1949)

$$\tanh^{-1}\left(\frac{V_{\max}}{V_{\min}}\right)$$

against l than gives a line of slope α.

It is possible to relate α and k to the storage and loss modulus E_1 and E_2, and to $\tan\delta$. For a filament of density ρ, in which c is the longitudinal wave velocity

$$E_1 = \frac{\omega^2}{k^2}\rho \qquad E_2 = \frac{2\alpha c^3 \rho}{\omega}$$

and

$$\tan\delta = \frac{2\alpha c}{\omega}$$

For further information the reader is referred to review articles by Kolsky [15].

5.3.2 The Megahertz Frequency Range: Ultrasonic Methods

Measurements of the velocity and attenuation of elastic waves at ultrasonic frequencies are important, especially for oriented polymers and composites. Compact solid specimens with dimensions of the order of 10 mm are required.

In a typical application of this technique, Chan et al. [16] measured the elastic constants of a uniaxially oriented rod, 12 mm in diameter, by cutting discs of thickness 4–8 mm parallel, perpendicular and at 45° to the axis of the rod (Figure 5.10). Quartz transducers were bonded to the discs, so that longitudinal and transverse waves were propagated along the geometrical axes of each disc. In principle nine different velocities v_{ab} can be measured, where a refers to the direction of polarization and b to the direction of wave propagation. For a specimen of density ρ we can then define $Q_{ab} = \rho v_{ab}^2$, where Q_{ab} is either an elastic stiffness constant or a linear combination of such constants. Velocities were measured using the pulse echo-overlap technique [17], and $\tan\delta$ obtained by making attenuation measurements [18].

An alternative approach [19, 20] is to immerse a specimen, thickness d, in a water-filled tank fitted with both a transmitter and a receiver of ultrasonic waves, and measure the change (τ) in transit time with and without the specimen in the beam. If υ is the velocity in the polymer and υ_w that in the water we have

$$\frac{1}{\upsilon} = \frac{1}{\upsilon_w} - \frac{\tau}{d} \qquad (5.6)$$

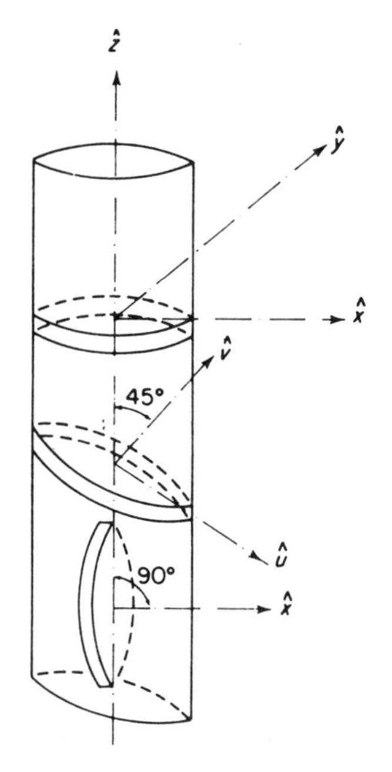

Figure 5.10 Schematic diagram illustrating the sample discs employed in ultrasonic measurements. (Reproduced with permission from Chan et al., *J. Phys. D*, **11**, 481 (1975))

The various wave velocities, which can be derived from measurements made over a range of incident angles, are related to the elastic stiffness constants.

In a variation of the above method Wright et al. [21] detected the component of the incident beam that was reflected from the immersed specimen and hence measured the critical angle of incidence.

More recently this technique has been developed to measure the anisotropy of uniaxial composites [22]. A specimen of uniform thickness, placed between the transmitting and receiving heads in a water-filled container, could be rotated about a vertical axis to change the angle of incidence and hence the direction of the beam in the sample (Figure 5.11). The velocity V and angle of refraction r of the wave are then calculated following the method of Markham [19]. Let X_1 and X_2 axes define the isotropic plane perpendicular to the fibre axis. It can then be shown [23] that for quasi-tensile waves propagating in the X_1–X_3 plane at an angle r to the X_1 axis, with the specimen axis X_2 vertical,

$$V_t^2 = \frac{B_{11} + B_{33} + [(B_{33} - B_{11})^2 + 4B^2{}_{13}]^{1/2}}{2\rho}$$

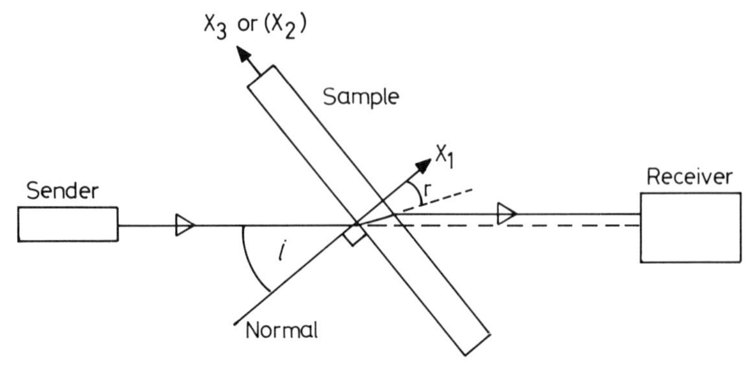

Figure 5.11 Schematic diagram showing layout of ultrasonic apparatus used for measurement of elastic constants, (From Dyer *et al.*, *J. Phys. D* (1992))

and for quasi-shear waves

$$V_s^2 = \frac{B_{11} + B_{33} - [(B_{33} + B_{11})^2 + 4B^2{}_{13}]^{1/2}}{2\rho}$$

where ρ is the specific gravity of the specimen.

The elastic stiffness constants C_{ij} are obtained from

$$B_{11} = C_{11}\cos^2 r + C_{44}\sin^2 r,$$

$$B_{33} = C_{33}\sin^2 r + C_{44}\cos^2 r,$$

$$B_{13} = (C_{44} + C_{13})\sin r\cos r.$$

REFERENCES

1. H. Leaderman, *Elastic and Creep Properties of Filamentous Materials and Other High Polymers*, Textile Foundation, Washington, DC, 1943.
2. D. W. Hadley, P. R. Pinnock and I. M. Ward, *J. Mater. Sci.*, **4**, 152 (1969).
3. H. Leaderman, *Trans. Soc. Rheol.*, **6**, 361 (1962).
4. C. M. R. Dunn, W. H. Mills and S. Turner, *Brit. Plast.*, **37**, 386 (1964).
5. I. M. Ward, *Polymer*, **5**, 59 (1964).
6. S. Turner, *Mechanical Testing of Plastic*, 2nd edn, G. Godwin, Harlow, 1983.
7. I. M. Ward, *Mechanical Properties of Solid Polymers*, Wiley, Chichester, 1983.
8. K. Schmeider and K. Wolf, *Kolloidzeitschrift*, **127**, 65 (1952).
9. P. R. Pinnock and I. M. Ward, *Proc. Phys. Soc.*, **81**, 261 (1963); M. Takayanagi, in *Proceedings of Forth International Congress on Rheology*, Part 1, Interscience, New York, 1965, p. 161; G. W. Becker, *Mater. Plast. Elast.*, **35**, 1387 (1969).
10. A. W. Noble, *J. Appl. Phys.*, **19**, 753 (1948).

11. D. W. Robinson, *J. Sci. Inst.*, **2**, 32 (1955).
12. D. R. Bland and E. H. Lee, *J. Appl. Phys.*, **26**, 1497 (1955).
13. K. W. Hillier and H. Kolsky, *Proc. Phys. Soc. B.*, **62**, 111 (1949); H. Kolsky, *Structural Mechanics*, Pergamon Press, Oxford, 1960.
14. K. W. Hillier, in, *Progress in Solid Mechanics*, (eds, I. N. Sneddon and R. Hill), North-Holland, Amsterdam, 1961.
15. H. Kolsky, *Appl. Mech. Rev.*, **11**, 9 (1958); *Structural Mechanics*, Pergamon Press, Oxford, 1960.
16. O. K. Chan, F. C. Chen, C. L. Choy and I. M. Ward, *J. Phys. D*, **11**, 481 (1975).
17. E. P. Papadakis, *J. Appl. Phys.*, **35**, 1474 (1964); S. F. Kwan, F. C. Chen and C. L. Choy, *Polymer*, **16**, 481 (1975).
18. R. L. Roderick and R. Truell, *J. Appl. Phys.*, **23**, 267 (1952).
19. M. F. Markham, *Composites*, **1**, 145 (1970).
20. F. F. Rawson and J. G. Rider, *J. Phys. D.*, **7**, 41 (1974).
21. H. Wright, C. S. N. Faraday, E. F. T. White and L. R. G. Treloar, *J. Phys. D.*, **4**, 2002 (1971).
22. S. R. A. Dyer, D. Lord, I. J. Hutchinson, I. M. Ward and R. A. Duckett, *J. Phys. D.*, **25**, 66 (1992).
23. M. J. P. Musgrove, *Proc. Roy. Soc.*, **A226**, 339 (1954).

Chapter 6

Experimental Studies of Linear Viscoelastic Behaviour as a Function of Frequency and Temperature: Time–Temperature Equivalence

6.1 GENERAL INTRODUCTION

An introduction to the extensive experimental studies of linear viscoelastic behaviour in polymers falls conveniently into three parts, in which amorphous polymers, crystalline polymers and temperature dependence are discussed in turn.

6.1.1 Amorphous Polymers

Many of the earlier investigations of linear viscoelastic behaviour in polymers were confined to amorphous polymers. During the early 1950s R. S. Marvin [1, 2] of the National Bureau of Standards, Washington, DC, coordinated the assembly of data from many laboratories who had measured the complex shear modulus and complex shear compliance of a specimen of polyisobutylene $(CH_2—CCH_3CH_3)_n$ of high relative molecular mass over a wide range of frequencies. The results, redrawn in Figure 6.1, show clearly the four regions, i.e. the glassy, the viscoelastic, rubbery and flow regions that are characteristic for amorphous high polymers. At high frequencies the

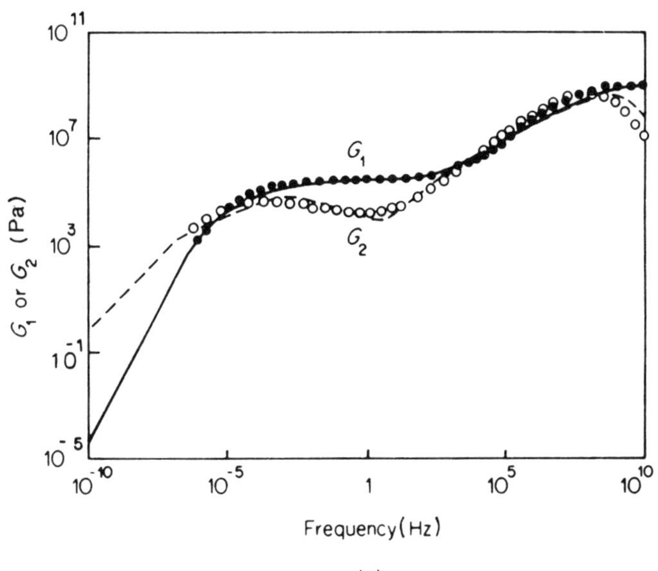

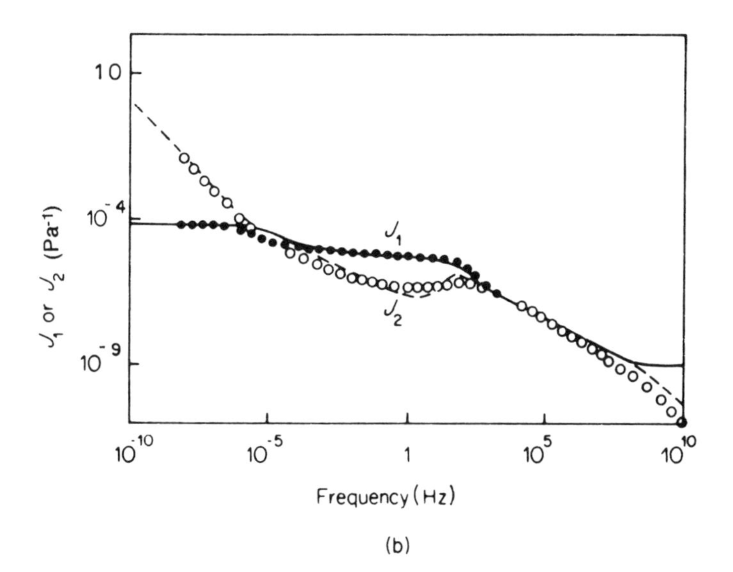

Figure 6.1 (a) Complex shear modulus, and (b) complex shear compliance for 'standard' polyisobutylene reduced to 25 °C. Points from averaged experimental measurements; curves from a theoretical model for viscoelastic behaviour. (Reproduced with permission from Marvin and Oser, *J. Res. Natl. Bur. Stand. b*, **66**, 171 (1962))

complex modulus has a value around 10^9 Pa. For material of high molecular mass a plateau in modulus occurs, and appreciable molecular flow occurs only at frequencies below 10^{-5} Hz (i.e. period >1 day).

The reader may use the Alfrey approximation to derive relaxation and retardation time spectra from the data of Figure 6.1. These spectra can be approximated by a 'wedge and box' distribution [3], shown by the dotted lines in Figure 6.2.

The observed plateau in the rubbery region is a consequence of high molecular mass, as the long molecules tend to entangle, with the formation of physical cross-links, which restrict molecular flow through the formation of temporary networks. At long times such physical entanglements are usually labile and lead to some irreversible flow, in contrast with the situation for permanent chemical cross-links, such as those introduced when rubber is vulcanized. It follows directly from the theory of rubber elasticity (Chapter 3) that the value of the modulus in the rubber-like plateau region is directly related to the number of effective cross-links per unit volume.

The influence of molecular entanglements is illustrated by Figure 6.3, which shows the stress relaxation behaviour for two samples of polymethyl methacrylate. It is seen that the lower molecular mass sample does not show a rubbery plateau region of modulus but passes directly from the viscoelastic region to the region of permanent flow.

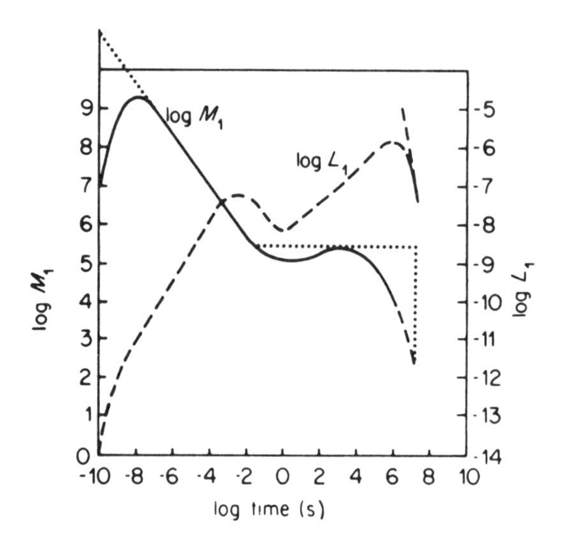

Figure 6.2 Approximate distribution functions of relaxation (M_1) and retardation (L_1) times for polyisobutylene. (Reproduced with permission from Marvin, *Proceedings of the 2nd International Congress of Rheology*, Butterworths, London (1954)

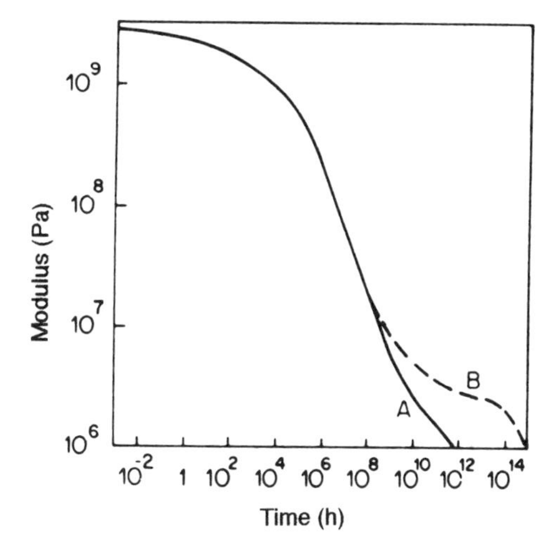

Figure 6.3 Master stress-relaxation curves for low molecular mass (molecular mass 1.5×10^5 daltons), curve A, and high molecular mass (molecular mass 3.6×10^6 daltons), curve B, polymethyl methacrylate. (Reproduced with permission from McLoughlin and Tobolsky, *J. Colloid Sci.*, **7**, 555 (1952)

6.1.2 Temperature Dependence of Viscoelastic Behaviour

Previously we have referred only indirectly to the effect of temperature on viscoelastic behaviour. From a practical viewpoint, however, the temperature dependence of polymer properties is of paramount importance because plastics and rubbers show very large changes in properties with changing temperature.

In purely scientific terms, the temperature dependence has two primary points of interest. In the first place, as we have seen in Chapter 5, it is not possible to obtain from a single experimental technique a complete range of measuring frequencies to evaluate the relaxation spectrum at a single temperature. It is therefore a matter of considerable experimental convenience to change the temperature of the experiment, and so bring the relaxation processes of interest within a time-scale which is readily available. This procedure, of course, assumes that a simple interrelation exists between time-scale and temperature, and we will discuss shortly the extent to which this assumption is justified.

Secondly, there is the question of obtaining a molecular interpretation of the viscoelastic behaviour. In most general terms polymers change from glass-like to rubber-like behaviour as either the temperature is raised or the time-scale of the experiment is increased. In the glassy state at low temperatures we would expect the stiffness to relate to changes in the stored elastic energy on deformation which are associated with small displacements of the molecules from their equilibrium positions. In the rubbery state at high temperatures, on the other hand, the molecular chains have considerable flexibility, so that in the undeformed state they can adopt conformations which lead to maximum entropy (or more strictly, minimum free energy). The rubber-like elastic deformations are then associated with changes in the molecular conformations.

The molecular physicist is interested in understanding how this conformational freedom is achieved in terms of molecular motions, for example to establish which bonds in the structure become able to rotate as the temperature is raised. One approach, which has proved successful to some degree, has been to compare the viscoelastic behaviour with dielectric relaxation behaviour and more particularly with nuclear magnetic resonance behaviour.

We have tacitly assumed that there is only one viscoelastic transition, corresponding to the change from the glassy low-temperature state to the rubbery state. In practice there are several relaxation transitions. For a typical amorphous polymer the situation is summarized in Figure 6.4. At low temperatures there are usually several secondary transitions involving comparatively small changes in modulus. These transitions are attributable to such features as side-group motions, e.g. methyl(—CH_3) groups in polypropylene

$$\left[CH_2 - \underset{|}{\overset{CH_3}{CH}} - \right]_n$$

In addition, there is one primary transition, usually called the 'glass transition', which involves a large change in modulus. The temperature at which it occurs is commonly denoted by T_g.

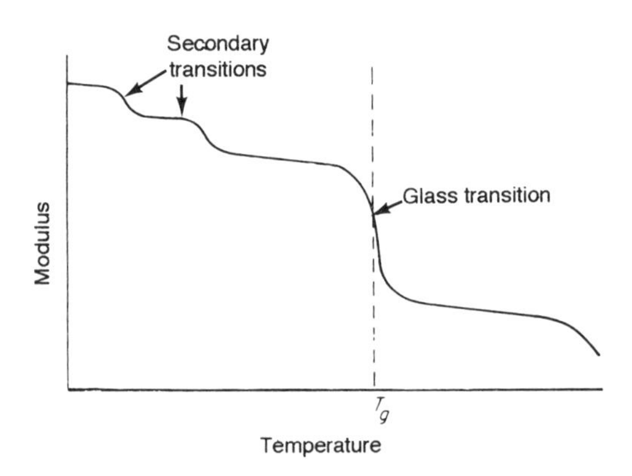

Figure 6.4 Temperature dependence of modulus in a typical polymer

6.1.3 Crystalline Polymers

Although the viscoelastic behaviour of semicrystalline polymers gives some indication of the four characteristic regions which can be identified for amorphous polymers, they are much less clearly defined, as is illustrated in Figure 6.5, which shows data for polychlorotrifluoroethylene $(CClF—CF_2)_n$ and polyvinyl fluoride $(CH_2CHF)_n$ obtained by Schmeider and Wolf [4]. The fall in modulus over the glass transition region for semicrystalline materials is, at between one and two orders of magnitude, much less than for amorphous polymers, and the change in modulus or loss factor with temperature or frequency is much more gradual, indicating a broader relaxation time spectrum. At high temperatures (or low frequencies) molecular mobility is severely curtailed by the crystalline regions, so it is no longer correct to regard the polymer as rubber-like. These differences are clearly illustrated by the data of Thompson and Woods [5] for polyethylene terephthalate

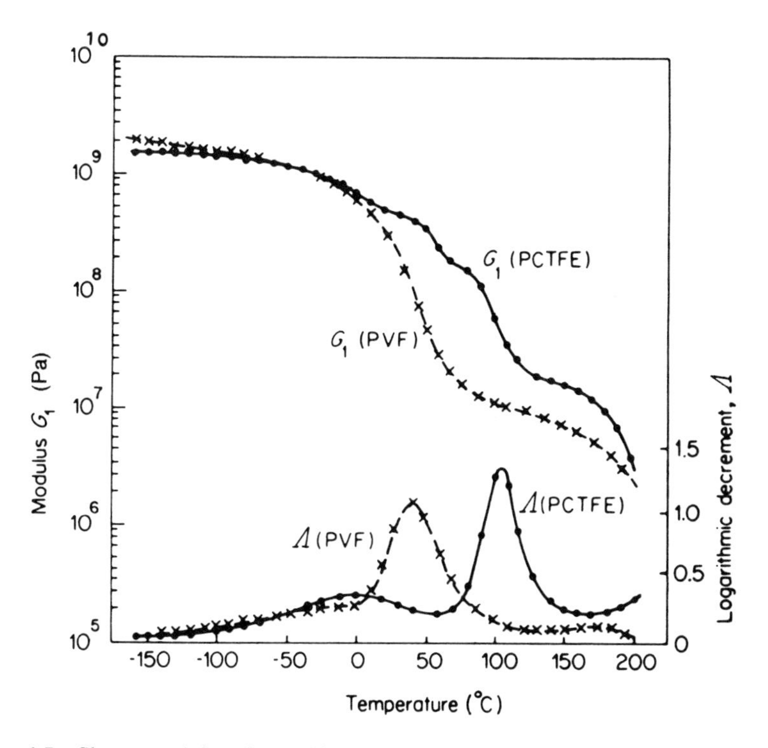

Figure 6.5 Shear modulus G_1 and logarithmic decrement Λ of polychlorotrifluoroethylene (PCTFE) and polyvinyl fluoride (PVF), as a function of temperature at ~3 Hz. (Reproduced with permission from Schmieder and Wolf, *Kolloidzeitschrift*, **134**, 149 (1953))

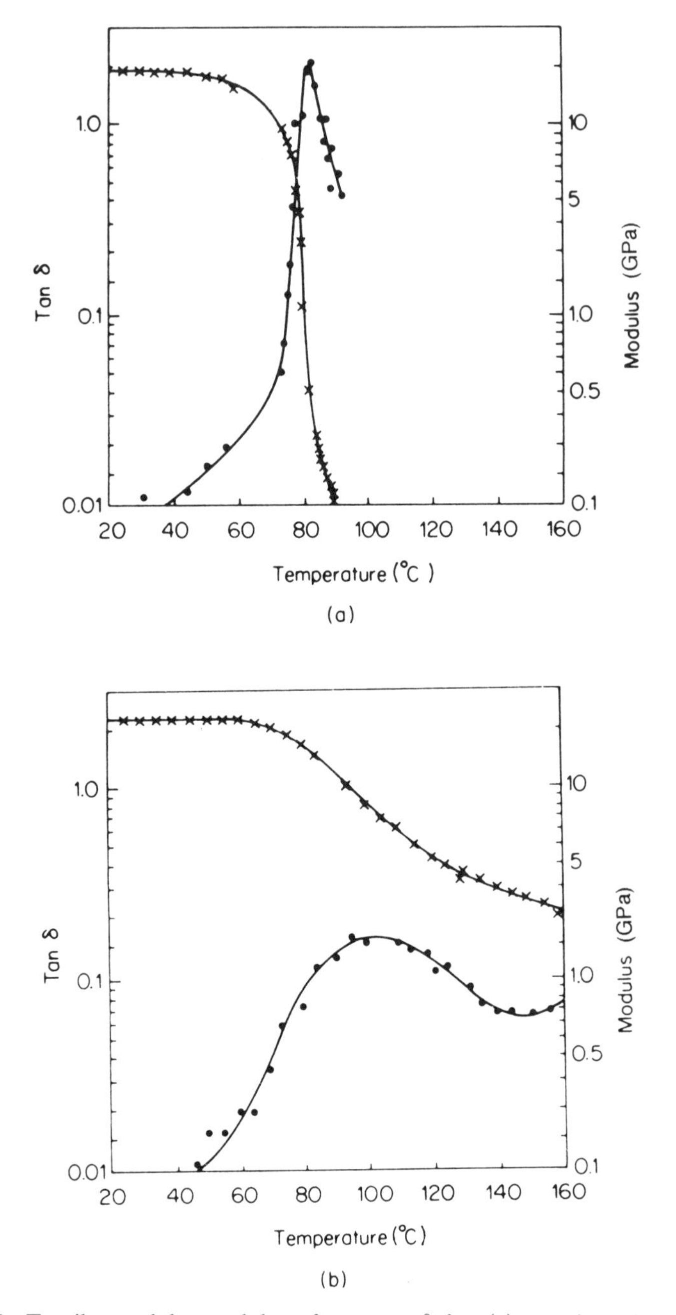

Figure 6.6 Tensile modulus and loss factor $\tan\delta$ for (a) unoriented amorphous polyethylene terephthalate, and (b) unoriented crystalline polyethylene terephthalate as a function of temperature at ~ 1.2 Hz. (Reproduced with permission from Thompson and Woods, *Trans. Faraday Soc.*, **52**, 1383 (1956)

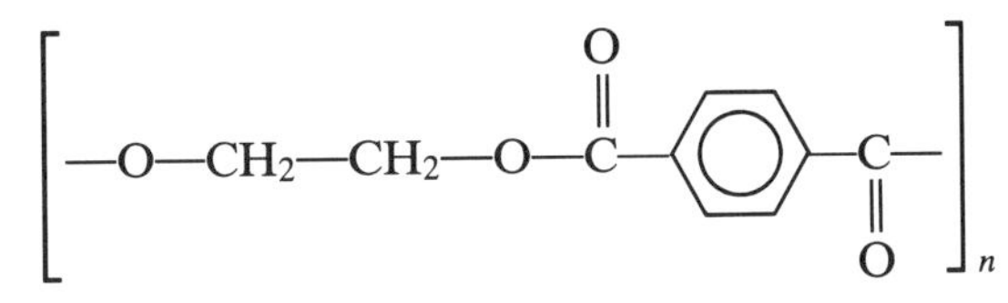

a material which is amorphous when quenched rapidly from the melt (Figure 6.6(a)), but semicrystalline when slowly cooled or subsequently heat treated (Figure 6.6(b)).

6.2 TIME–TEMPERATURE EQUIVALENCE AND SUPERPOSITION

Time–temperature equivalence in its simplest form implies that the viscoelastic behaviour at one temperature can be related to that at another temperature by a change in the time-scale only. Consider the idealized double logarithmic plots of creep compliance versus time shown in Figure 6.7(a). The compliances at temperatures T_1 and T_2 can be superimposed exactly by a horizontal displacement $\log a_t$, where a_t is called the shift factor. Similarly (Figure 6.7(b)), in dynamic mechanical experiments, double logarithmic plots of $\tan\delta$ versus frequency show an equivalent shift with temperature.

The experimental procedure is illustrated in Figures 6.8 and 6.9. A series of creep compliance curves each typically extending over 2 h, so that individual tests can be performed on successive days, is plotted using a specimen that has been mechanically conditioned at the highest temperature used. The individual plots are then transposed along the logarithmic time axis until they coincide, using any required temperature within the experimental range as the reference value. The variation of shift factor with temperature should also be recorded, for comparison with the predictions of theoretical interpretations to be discussed shortly.

Ferry and his co-workers [6], on the basis of the molecular theory of viscoelasticity, proposed that superposition should incorporate a small vertical shift factor $T_0\rho_0/T\rho$, where ρ is the density at the experimental temperature T and ρ_0 relates to the reference temperature T_0. Further corrections have been suggested by McCrum and Morris [7] to deal with the changes in unrelaxed and relaxed compliances with temperature.

The situation is illustrated schematically in Figure 6.10. When we compare the creep compliance curves at the two temperatures T_1 and T_2 we see that the relaxed and unrelaxed compliances are both changing with temperature. McCrum and Morris [7] propose a scaling procedure for obtaining a modified or 'reduced' compliance curve at the temperature T_1, to give the dashed curve $J_\rho^T(t)$ in Figure 6.10. The shift factor is now obtained by a horizontal shift of $J_\rho^{T_1}(t)$ to superimpose $J^{T_2}(t)$.

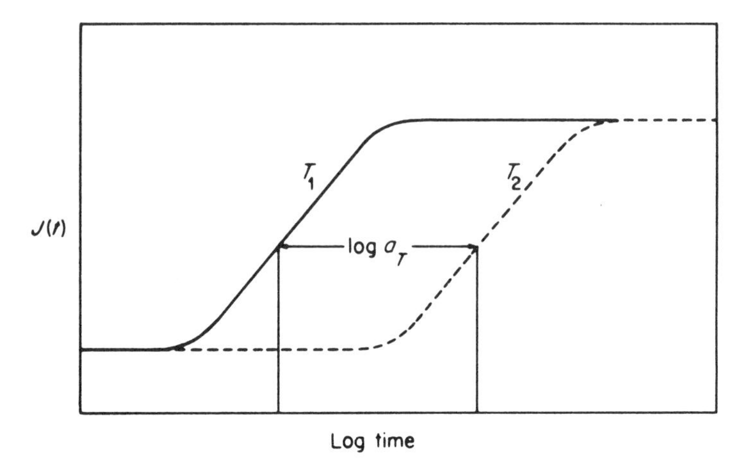

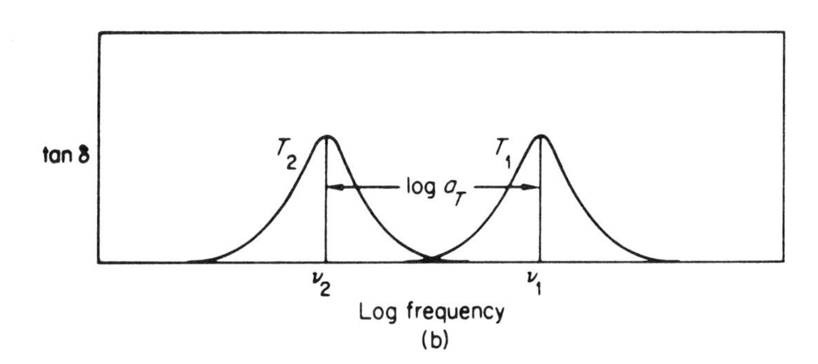

Figure 6.7 Schematic diagrams illustrating the simplest form of time–temperature equivalence for (a) compliance, $J(t)$ and (b) loss factor tan δ

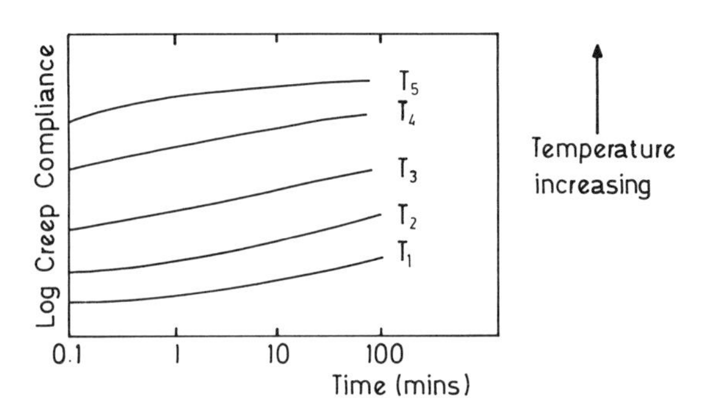

Figure 6.8 Creep plots at different temperatures (schematic)

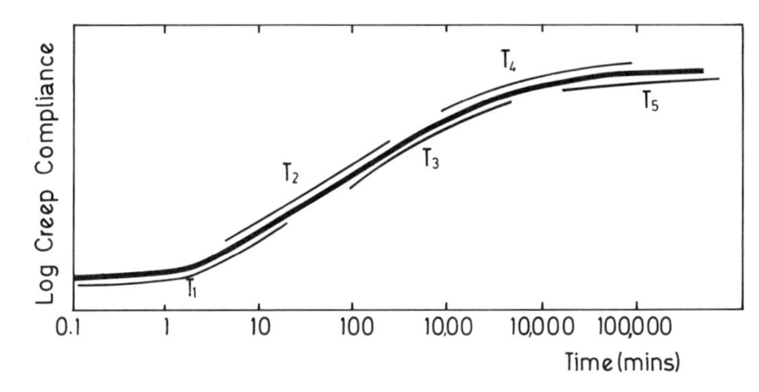

Figure 6.9 Master curve of creep from superposing plots of Figure 6.8

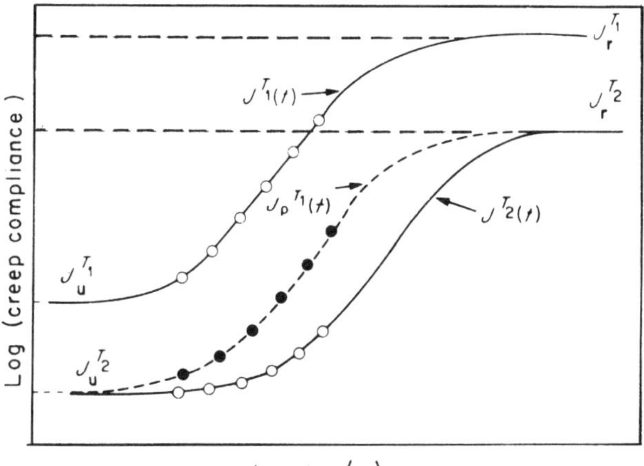

Figure 6.10 Schematic diagram illustrating McCrum's reduction procedure for superposition of creep data. $J_r^{T_1}$ and $J_u^{T_1}$ are the relaxed and unrelaxed compliances respectively at the temperature T_1; $J_r^{T_2}$; and $J_u^{T_2}$ are the corresponding quantities at the temperature T_2

6.3 MOLECULAR INTERPRETATIONS OF TIME–TEMPERATURE EQUIVALENCE

6.3.1 Molecular Rate Processes with a Constant Activation Energy: The Site Model Theory

The simplest theories which attempt to deal with the temperature dependence of viscoelastic behaviour are the transition state or barrier theories [8, 9]. The site model was originally developed to explain the dielectric behaviour of solids [10, 11], but was later applied to mechanical relaxations in polymers [12].

In its simplest form there are two sites, separated by an equilibrium free energy difference $\Delta G_1 - \Delta G_2$, the barrier heights being ΔG_1 and ΔG_2 per mole respectively (Figure 6.11).

The transition probability for a jump from site 1 to site 2 is given by

$$\omega_{12}^0 = A' \exp(-\Delta G_1/RT) \tag{6.1}$$

and for a jump from site 2 to site 1 by

$$\omega_{21}^0 = A' \exp(-\Delta G_2/RT), \tag{6.2}$$

where A' is a constant.

To give rise to a mechanical relaxation process, the energy difference between the two sites must be changed by the application of the applied stress. There is then a change in the populations of site 1 and site 2, and it is assumed that this relates directly to the strain. It is not difficult to imagine how this might arise at a molecular level if, for example, the uncoiling of a molecular chain involved internal rotations. Locally, the chain conformations could be changing from crumpled *gauche* conformations to extended *trans* conformations (see section 1.2.1).

Assume that the applied stress σ causes a small linear shift in the free energies of the sites such that

$$\delta G'_1 = \lambda_1 \sigma \tag{6.3}$$

and

$$\delta G'_2 = \lambda_2 \sigma \tag{6.4}$$

for sites 1 and 2 respectively, where λ_1 and λ_2 are constants with the dimensions of volume. The transition probabilities ω_{12} and ω_{21} in the presence of the applied stress are related to those in the absence of stress by expanding the exponential

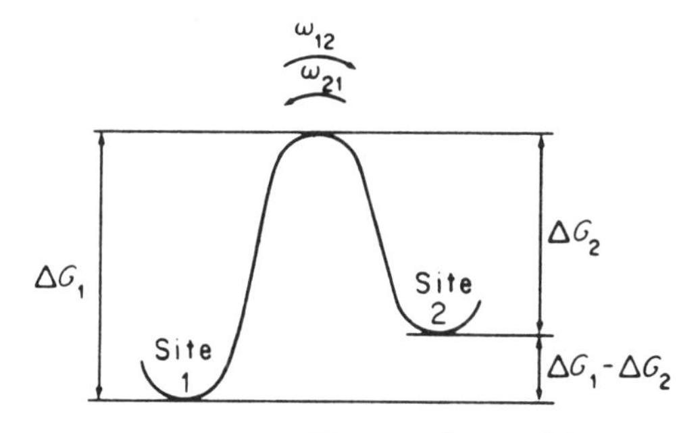

Figure 6.11 The two-site model

$$\omega_{12} \approx \omega_{12}^{0}\left[1 - \frac{\delta G'_1}{RT}\right] = \omega_{12}^{0}\left[1 - \frac{\lambda_1 \sigma}{RT}\right] \tag{6.5}$$

where ω_{12}^{0} is the transition probability in the absence of the stress. Similarly

$$\omega_{21} \approx \omega_{21}^{0}\left[1 - \frac{\lambda_2 \sigma}{RT}\right] \tag{6.6}$$

The rate equations for sites 1 and 2 are then

$$\frac{\mathrm{d}N_1}{\mathrm{d}t} = -N_1\omega_{12} + N_2\omega_{21}, \tag{6.7}$$

$$\frac{\mathrm{d}N_2}{\mathrm{d}t} = -N_2\omega_{21} + N_1\omega_{12}, \tag{6.8}$$

where we can write the occupation number N_1 of state 1 as $N_1 = N_1^0 + n$ and similarly $N_2 = N_2^0 - n$, where N_1^0 and N_2^0 are the occupation numbers at zero stress, $N_1^0 + N_2^0 = N_1 + N_2 = N$.

Combining these equations and making suitable approximations gives a rate equation

$$\frac{\mathrm{d}n}{\mathrm{d}t} + N(\omega_{12}^{0} + \omega_{21}^{0}) = N_1^0\omega_{12}^{0}\left[\frac{\lambda_1 - \lambda_2}{RT}\right]\sigma \tag{6.9}$$

which describes the change in the site population n as a function of time. Assuming that the observed strain e is a consequence of the change in site population it is given by

$$e = e_{\mathrm{u}} + n\bar{e}. \tag{6.10}$$

In this equation e_{u} is the instantaneous or unrelaxed elastic deformation and it is considered that each change in site population produces a proportionate change in strain by an amount $\bar{e}$.

Equation (6.9) can then be seen to have the form

$$\frac{\mathrm{d}e}{\mathrm{d}t} + Be = C,$$

where B and C are constants, which is formally identical to the equation of a Kelvin unit (section 4.2.3.1 above), with a characteristic retardation time given by $\tau' = 1/B$, i.e.

$$\tau' = \frac{1}{(\omega_{12}^{0} + \omega_{21}^{0})} = \frac{\exp(\Delta G_2/RT)}{A'[\exp\{-(\Delta G_1 - \Delta G_2)/RT\} + 1]} \tag{6.11}$$

Since RT is usually small compared with the equilibrium free energy difference we may approximate to

$$\tau' = \frac{1}{A'} \exp(\Delta G_2/RT). \tag{6.12}$$

Since

$$2\pi\nu = \frac{1}{\tau'}$$

where ν is the frequency of molecular jumps between the two rotational isomeric states of the chain molecule, we can write

$$\nu = \frac{A'}{2\pi} \exp(-\Delta G_2/RT). \tag{6.13}$$

This equation states that the frequency of molecular conformational changes depends on the *barrier height* ΔG_2 and not on the free energy difference between the equilibrium sites. Equation (6.13) may also be written as

$$\nu = \frac{A''}{2\pi} \exp(\Delta S/R) \exp(-\Delta H/RT) = \nu_0 \exp(-\Delta H/RT). \tag{6.14}$$

This form of equation (6.14) emphasizes the way in which temperature affects ν primarily through the activation energy ΔH. To a good approximation the activation energy for the process (actually an enthalpy) is thus given by

$$\Delta H = -R \left[\frac{\partial(\ln \nu)}{\partial(1/T)} \right]_p \tag{6.15}$$

Equation (6.14) is known as the 'Arrhenius equation', because it was first shown by Arrhenius [13] that it describes the influence of temperature on the velocity of chemical reactions.

As an example of the application of the Arrhenius equation consider the $\tan\delta$ curve of Figure 6.7(b). At temperatures T_1 and T_2 the peak value of $\tan\delta$ occurs at frequencies ν_1 and ν_2 respectively. Using

$$\nu = \nu_0 \exp(-\Delta H/RT), \tag{6.16}$$

we obtain

$$\frac{\nu_1}{\nu_2} = \frac{\exp(-\Delta H/RT_1)}{\exp(-\Delta H/RT_2)}$$

giving the shift factor as

$$\frac{\log \nu_1}{\log \nu_2} = \log a_T = \frac{\Delta H}{R} \left\{ \frac{1}{T_2} - \frac{1}{T_1} \right\} \tag{6.17}$$

The activation energy for the process can then be obtained by plotting $\log a_T$ versus the reciprocal of the absolute temperature, and for large values of

ΔH changes in temperature are equivalent to very large changes in frequency.

It can be shown [14] that the magnitude of the relaxation is proportional to

$$S\left[\frac{\exp(-\Delta G_1 - \Delta G_2)/RT}{\{\exp[-(\Delta G_1 - \Delta G_2)/RT] + 1\}^2}\right]\frac{(\lambda_1 - \lambda_2)^2}{RT} \tag{6.18}$$

where S is the number of species per unit volume. Thus the intensity of the relaxation is predicted to be low at both high and low temperatures, and passes through a maximum when the free energy difference $(\Delta G_1 - \Delta G_2)$ and RT are of the same order of magnitude.

It must be emphasized that the site model is applicable only to relaxation processes showing a constant activation energy, examples being those associated with localized motions in the crystalline regions of semicrystalline polymers. The temperature dependence of the glass transition relaxation behaviour of polymers does not fit a constant activation energy model, and where this has appeared to be true it is probably a consequence of the limited range of experimental frequencies that were available.

6.3.2 The Williams–Landel–Ferry (WLF) Equation

In considering time–temperature equivalence of the glass transition behaviour in amorphous polymers, we will follow a treatment very close to that given by Ferry [15]. To fix our ideas, consider the storage compliance J_1 of an amorphous polymer (poly-n-octyl methacrylate) as a function of temperature and frequency (Figure 6.12). It can be seen that there is an overall change in the shape of the compliance–frequency curve as the temperature changes. At high temperatures there is an approximately constant high compliance, the rubbery compliance. At low temperatures the compliance is again approximately constant but at a low value, the glassy compliance. At intermediate temperatures there is the frequency-dependent viscoelastic compliance.

Applying time–temperature equivalence (as discussed in section 6.2 above) gives the storage compliance as a function of frequency over a very wide range of frequencies, as shown in Figure 6.13. Thus it is now possible to calculate the retardation time spectrum, and compare this with any theoretical models which may be proposed.

The horizontal shift on a logarithmic time-scale is shown in Figure 6.14. Remarkably, Williams, Landel and Ferry [6] found an approximately identical shift factor–temperature relation for all amorphous polymers, which could be expressed as

$$\log a_T = \frac{C_1(T - T_S)}{C_2 + (T - T_S)}$$

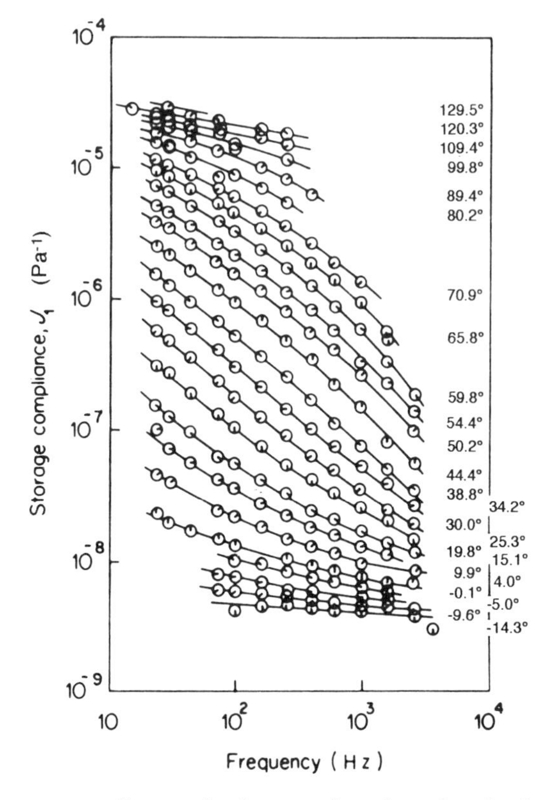

Figure 6.12 Storage compliance of poly-n-octyl methacrylate in the glass transition region plotted against frequency at 24 temperatures as indicated. (Reproduced from Ferry, *Viscoelastic Properties of Polymers* 1st edn, Wiley, New York, 1961, Ch. 11)

where C_1 and C_2 are constants. This relation, known as the WLF equation, applies over a temperature range $(T_S \pm 50)$ °C, where T_S is a reference temperature specific to a particular polymer. The constants C_1 and C_2 were originally determined by arbitrarily choosing $T_S = 243$ K for polyisobutylene.

Following this empirical discovery there was naturally some speculation as to whether the WLF equation has a fundamental interpretation. To proceed further we must consider the dilatometric glass transition, and its interpretation in terms of free volume.

The glass transition can be defined on the basis of dilatometric measurements. As shown in Figure 6.15, if the specific volume of the polymer is measured against temperature, a change of slope is observed at a characteristic temperature, which we may call T_g. In practice the change in slope is somewhat less sharp than the diagram suggests and T_g is determined

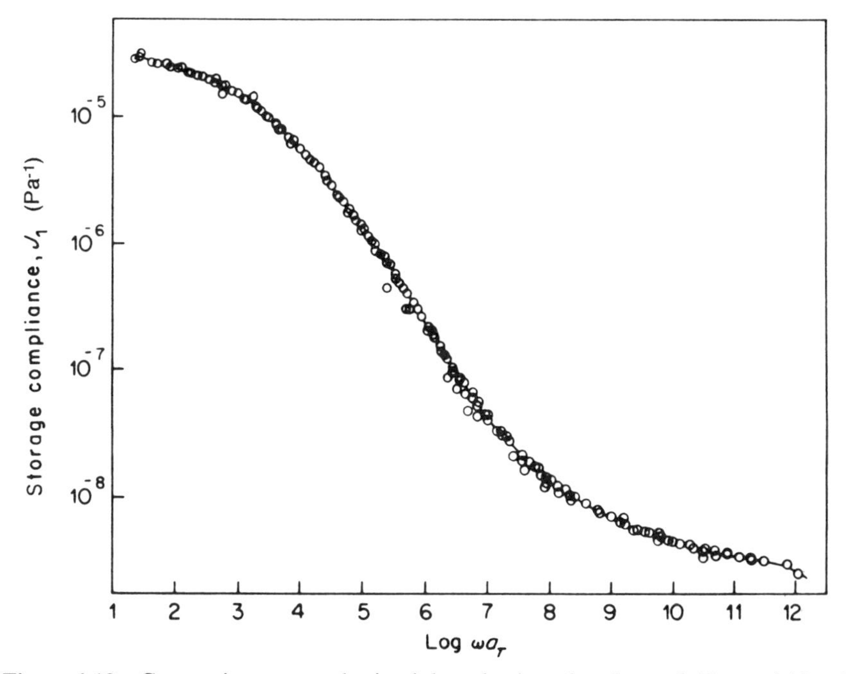

Figure 6.13 Composite curve obtained by plotting the data of Figure 6.12 with suitable shift factors, giving the behaviour over an extended frequency scale at temperature T_0. (Reproduced from Ferry, *Viscoelastic Properties of Polymers*, 1st edn, Wiley, New York, 1961, Ch. 11)

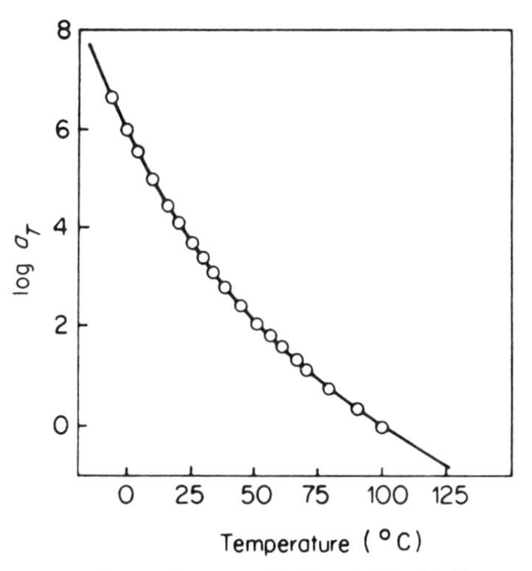

Figure 6.14 Temperature dependence of the shift factor α_T used in plotting Figure 6.13. Points, chosen empirically; curve is WLF equation with a suitable choice of T_g (or T_s). (Reproduced from Ferry, *Viscoelastic Properties of Polymers* 1st edn, Wiley, New York, 1961, Ch. 11)

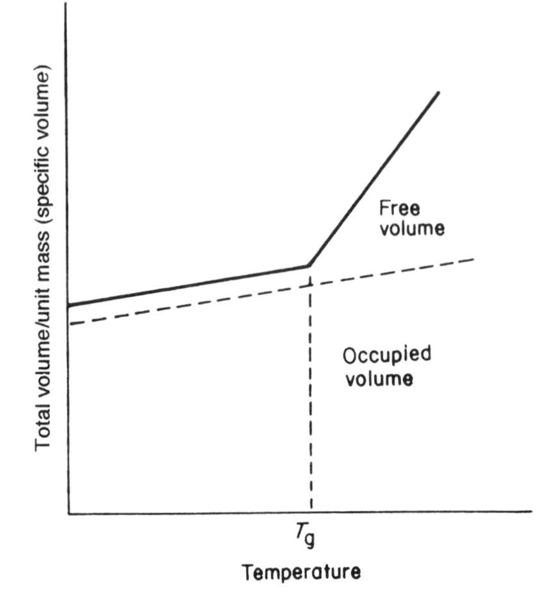

Figure 6.15 The volume–temperature relationship for a typical amorphous polymer

by extrapolation. When dilatometric measurements are carried out at very slow rates of temperature change, T_g approaches a constant value, and will vary by only 2–3 K when the heating rate is decreased from 1 K min^{-1} to 1 K per day. Thus it appears possible to define a rate-independent value of T_g to at least a very good approximation.

It has subsequently been shown that the original WLF equation can be rewritten in terms of this dilatometric transition temperature such that

$$\log a_T = \frac{C_1^g(T - T_g)}{C_2^g + (T - T_g)}$$

where C_1^g and C_2^g are new constants and $T_g = T_g - 50\ °\mathrm{C}$.

Moreover, it is now possible to give a plausible theoretical basis to the WLF equation in terms of free volume [6]. In liquids, this concept has proved useful in discussing transport properties such as viscosity and diffusion, which are considered to relate to the difference $v_f = v - v_0$, where v is the total macroscopic volume, v_0 is the actual molecular volume of the liquid molecules, the 'occupied volume', and v_f is the proportion of holes or voids, the 'free volume'.

Figure 6.15 shows the schematic division of the total volume of the polymer into both occupied and free volumes. It is argued that the occupied volume increases uniformly with temperature. The discontinuity in the expansion coefficient at T_g then corresponds to a sudden onset of expansion in the free volume, which suggests that certain molecular processes which control the viscoelastic behaviour commence at T_g, and not merely that T_g is the temperature when their time-scale becomes comparable with that of the measuring time-scale. This behaviour would seem to imply that T_g is a genuine thermodynamic second-order transition temperature. The point is not, however, completely resolved, and it has been shown by Kovacs [16] that the T_g measured dilatometrically is still sensibly dependent on the time-scale, i.e. the rate of heating. However, as already mentioned, this time dependence is small. Thus to a good approximation it can be assumed that the free volume is constant up to T_g and then increases linearly with increasing temperature.

The fractional free volume $f = v_f/v$ can therefore be written as

$$f = f_g + \alpha_f(T - T_g), \tag{6.19}$$

where f_g is the fractional free volume at the glass transition T_g and α_f is the coefficient of expansion of the free volume.

The WLF equation can now be obtained in a simple manner. The model representations of linear viscoelastic behaviour all show that the relaxation times are given by expressions of the form $\tau = \eta/E$ (see the Maxwell model in section 4.2.3.2 above), where η is the viscosity of a dashpot and E the modulus of a spring.

If we ignore the changes in the modulus E with temperature compared with changes in the viscosity η, this suggests that the shift factor a_T for changing temperature from T_g to T will be given by

$$a_T = \frac{\eta_T}{\eta_{T_g}} \tag{6.20}$$

At this juncture, we introduce Doolittle's viscosity equation [17], based on experimental data for monomeric liquids, which relates the viscosity to the free volume through

$$\eta = a \exp(bv/v_f), \tag{6.21}$$

where a and b are constants. Using (6.20) and (6.21) it can be shown that the Doolittle equation becomes

$$\ln a_T = b\left\{\frac{1}{f} - \frac{1}{f_g}\right\} \tag{6.22}$$

Substituting $f = f_g + \alpha_f(T - T_g)$ we have

$$\log a_T = -\frac{(b/2.303 f_g)(T - T_g)}{f_g/\alpha_f + T - T_g} \tag{6.23}$$

which is the WLF equation.

The fractional free volume at the glass transition temperature f_g is 0.025 ± 0.003 for most amorphous polymers. The thermal coefficient of expansion of free volume α_f is a more variable quantity, but has the physically reasonable 'universal' average value of 4.8×10^{-4} K^{-1}.

Substituting for a_T from (6.20) the WLF equation may be rewritten as

$$\log \eta_T = \log \eta_{T_g} + \frac{C_1^g(T - T_g)}{C_2^g + (T - T_g)} \tag{6.24}$$

which implies that at a temperature $T = T_g - C_2^g$, i.e. $T = T_g - 51.6$, the viscosity of the polymer becomes infinite. This feature has suggested that at the molecular level the WLF equation should be related to the temperature $T_2 = T_2 - 51.6$, rather than to the dilatometric glass transition T_g.

There have been two basic approaches along these lines:

1. The free volume theory is modified so that the changes in free volume with temperature relate to a discontinuity which occurs at T_2 rather than T_g.
2. It is considered that T_2 represents a true thermodynamic transition temperature. Adam and Gibbs [18] have developed a modified transition state theory in which the frequency of molecular jumps relates to the cooperative movement of a group of segments of the chain. The number of segments acting cooperatively is then calculated from statistical thermodynamic considerations.

Despite the successes mentioned above there have been objections to free volume theories [12, 19] based on observations that a few polymers behave similarly under both constant pressure and constant volume conditions, although in the latter case free volume must decrease with increasing temperature.

6.4 FLEXIBLE MOLECULAR CHAIN MODELS

Condensed matter physicists calculate many properties of crystalline solids in terms of a model, due initially to Debye, in which massive point atoms are connected by linear elastic springs. The dynamics of molecular chains can be considered from this starting-point. The theories discussed below, although initially derived for polymer solutions, can be used to predict relaxation spectra and time–temperature equivalence for amorphous solid polymers. As a full treatment involves quite advanced mathematics, we shall discuss the theories only in outline.

6.4.1 Normal Mode Theories

If a sinusoidal vibration is applied to a longitudinal rod, and both ends of the rod are then fixed, the wave will be reflected at these fixed points or nodes. A series of discrete standing waves, known as the *normal modes*, can be established in the rod, with wavelengths given by

$$\lambda_n = \frac{2l}{n}$$

where l is the length of the rod and n represents a series of integers. It is more usual to express normal modes in terms of the wave vector,

$$k = \frac{2\pi}{\lambda}, \quad \text{so that} \quad k_n = \frac{n\pi}{l}$$

To calculate the normal modes of vibration in a crystal it is necessary to take into account interatomic forces and crystal structure. The starting-point for such calculations is a model of an infinite one-dimensional linear chain in which point atoms of identical mass are joined by equal springs with a common force constant. It can be seen from Figure 6.16 that the equation of motion of the nth atom must taken into account the displaced positions of both the $(n-1)$th and $(n+1)$th atoms. A full analysis of this type of model can be found in a university-level text on solid state physics, for example that by Guinier and Jullien [20].

The similarities between a one-dimensional lattice analogue and a polymer chain are evident, and related models have been used to set up a series of linear differential equations which can represent the motion of polymer chains in a viscous solvent. In the simplest model, due to Rouse [21] each polymer chain is considered as built from submolecules, represented as beads, which are linked by springs whose behaviour is that of a freely jointed chain on the Gaussian theory of rubber elasticity. The molecular chains between beads are of equal length and are assumed to be long enough for the end-to-end separation to approximate to a Gaussian probability distribution. It is assumed that only the beads and not the chains

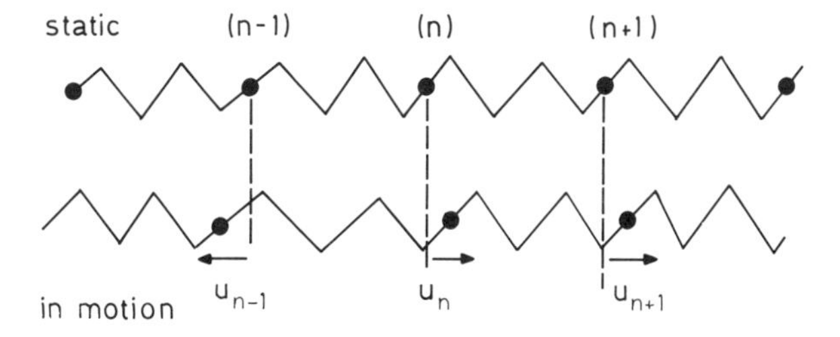

Figure 6.16 Vibrations in a linear chain. The displacement of the nth atom is U_n. The equation of motion for atom n must take into account U_{n-1} and U_{n+1}

interact directly with the solvent molecules. When a bead is displaced from its equilibrium position there are two types of forces acting on it: those that result from the viscous interaction with the solvent molecules, and those that represent the tendency of the molecular chains to return to a state of maximum entropy by Brownian diffusional movements.

Consideration of the restoring force when a bead is displaced from its equilibrium position leads to the expression

$$\eta \dot{x}_i + \frac{3kT}{zl^2}(2x_i - x_{i-1} - x_{i+1}) = 0,$$

where η is the coefficient defining the viscous interaction between the beads and the solvent, l the length of each link in a chain, z the number of links in a submolecule and $\dot{x}_i$ the time differential of the displacement of a bead situated between the $(i-1)$th and $(i+1)$th submolecules. For m submolecules there are $3m$ of these equations, each of which is equivalent to the equation of a kelvin unit (cf. $\eta\dot{e} + Ee = 0$).

The $3m$ equations have to be uncoupled using a normal coordinate transformation, to obtain eigenfunctions which are linear combinations of the positions of the submolecules. Each eigenfunction then describes a single viscoelastic element with characteristic time-dependent properties.

It can be shown, for example, that the stress relaxation modulus is given by

$$G(t) = NkT \sum_{p=1}^{m} \exp\left(-\frac{t}{\tau_p}\right) \tag{6.25}$$

and the real part of the complex modulus in dynamic experiments is

$$G_1(\omega) = NkT \sum_{p=1}^{m} \left(\frac{\omega^2\tau_p^2}{1 + \omega^2\tau_p^2}\right) \tag{6.26}$$

Where there are N molecules per unit volume and τ_p is the relaxation time of the pth mode, given by

$$\tau_p = zl^2\eta[24kT\sin^2\{p\pi/2(m+1)\}]^{-1}, \qquad p = 1, 2 \ldots, m. \tag{6.27}$$

The moduli are thus determined by a discrete spectrum of relaxation times, each of which characterizes a given normal mode of motion. These normal modes are shown schematically in Figure 6.17. In the first mode, corresponding to $p = 1$, the ends of the molecule move, with the centre of the molecule remaining stationary. In the second mode there are two nodes in the molecule, and in the general case the pth mode has p nodes, with the motion of the molecule occurring in $(p + 1)$ segments.

For the Rouse model the submolecule is the shortest length of chain that can undergo relaxation, and the motion of segments within a submolecule is

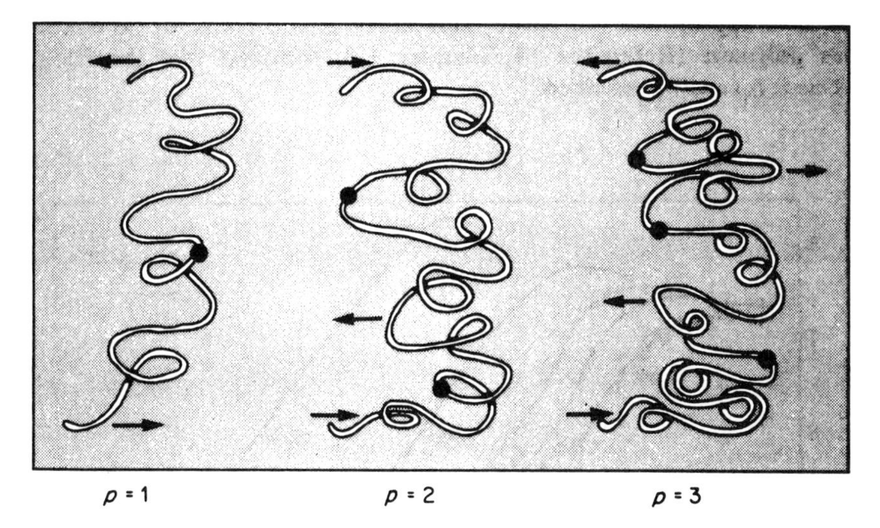

Figure 6.17 Illustration of the first three normal modes of a chain molecule

ignored. This limitation implies that the theory is applicable only when $m \gg 1$, and means that the equation for τ_p reduces to

$$\tau_p = \frac{n^2 l^2 \eta_0}{6\pi^2 p^2 kT} \tag{6.28}$$

where $\eta_0 = \eta/z$ is the friction coefficient per random link. The relaxation times depend on temperature through $1/T$, through nl^2, which defines the equilibrium mean-square separation of the chain ends, and through changes in η_0. This last parameter varies rapidly with temperature and is primarily responsible for the changes in τ_p. As each τ_p has the same temperature dependence the theory satisfies the requirements of thermorheological simplicity and gives justification for time–temperature equivalence.

A later theory due to Zimm [22] gives a modified relaxation spectrum through taking into account the way in which the solvent is affected by the movement of the polymer molecules, and by considering the hydrodynamic interaction between the moving submolecules.

6.4.2 The Dynamics of Highly Entangled Polymers

In a concentrated polymer solution, a melt, or a solid polymer, the molecular chains cannot pass through one another, a constraint that effectively confines each chain within a tube [23]. The centre line of this tube defines the overall path of the chain in space, and has been called by Edwards the primitive chain (Figure 6.18). Each chain 'sees' its environment

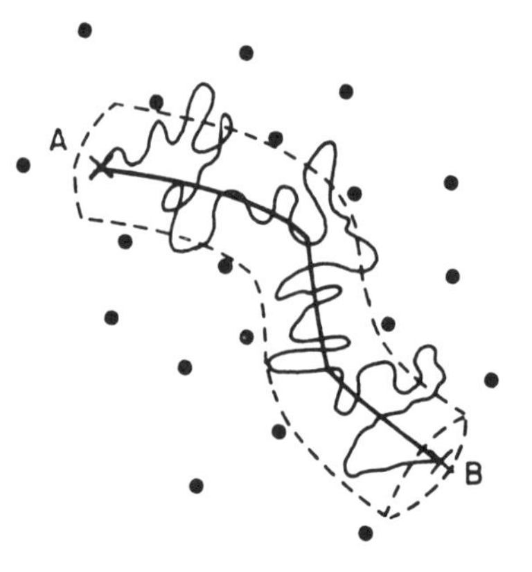

Figure 6.18 Chain segment AB in dense rubber. The points A and B denote the cross-linked points and the dots represent other chains which, in this drawing, are assumed to be perpendicular to the paper. Due to entanglements the chain is confined to the tube-like region denoted by the broken line. The bold line shows the primitive path. (Reproduced with permission from Doi and Edwards, *J. Chem. Soc. Faraday Trans.*, **74** 1802 (1978))

as a tube, because although all the other chains are moving, there are so many entanglements that at any one time the tube is well defined. De Gennes [24] has described the possible motions of a polymer chain confined to a tube as snake-like, and has called the phenomenon 'reptation'. He considered two distinct forms of motion. First, the comparatively short-term wriggling motions that correspond to the migration of a molecular kink along the chain, for which the longest relaxation time is proportional to the square of the molecular mass. Second, there is the much longer time associated with the movement of the chain as a whole through the polymer. This motion corresponds to an overall movement of the centre of gravity of the chain, and has a characteristic time proportional to the cube of the molecular mass.

Doi and Edwards [23] have extended the work of de Gennes, and have derived mathematical expressions for features such as the stress relaxation that occurs after a large strain. Their explanation for the physical situation is illustrated in Figure 6.19, in which the hatched area indicates the deformed part of the tube. Here (a) represents the tube before deformation, when the conformation of the primitive chain is in equilibrium. The deformation is considered to be affine, so that each molecule deforms to the same extent as the macroscopic body. In (b) the situation immediately after the step deformation is given, with the primitive chain in the affinely deformed conformation. In (c) the situation after a characteristic time τ_R (called by de Gennes the disengagement time) is given, with the primitive chain recovering to its equilibrium contour length by contracting along the tube.

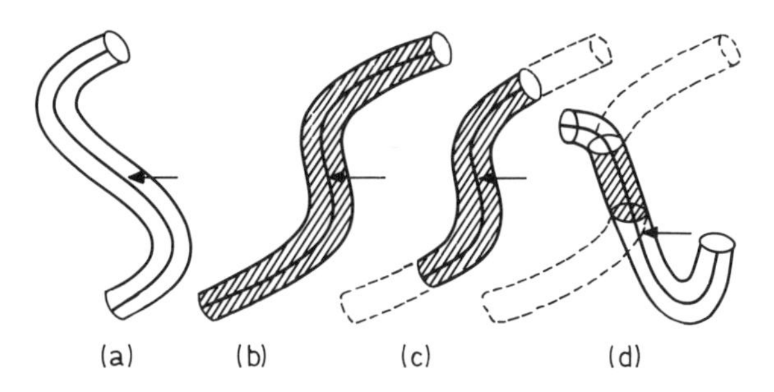

Figure 6.19 Explanation of the stress relaxation after large step strain. (a) Before deformation the conformation of the primitive chain is in equilibrium ($t = -0$). (b) Immediately after deformation, the primitive chain is in the affinely deformed conformation ($t = +0$). (c) After time τ_R, the primitive chain contracts along the tube and recovers the equilibrium contour length ($t \approx \tau_R$). (d) After the time τ_d, the primitive chain leaves the deformed tube by reptation ($t \approx \tau_d$). The oblique lines indicate the deformed part of the tube. (Reproduced from Doi and Edwards, *The Theory of Polymer Dynamics*, Oxford University Press, 1986)

After a longer characteristic time τ_d (d) the primitive chain leaves the deformed tube by reptation. For a fuller discussion refer to the advanced text by Doi and Edwards [23].

Graessley [25] has reviewed the Doi–Edwards theory, and concludes that the agreement with experiment is good, apart from the effects of the distribution of the relative molecular mass. The shortcoming probably arises because chains of all lengths are compelled to diffuse by reptation, whereas the cage lifetimes may be very small compared with the disengagement time of the longest chains. The motions of the longest chains mays thus be better modelled by Rouse theory, where tube constraints are absent.

REFERENCES

1. R. S. Marvin, in *Proceedings of the Second International Congress of Rheology*, Butterworths, London, 1954.
2. R. S. Marvin and H. Oser, *J. Res. Natl Bur. Stand. B*, **66**, 171 (1962).
3. R. S. Marvin and J. T. Berger, *Viscoelasticity: Phenomenological Aspects*, Academic Press, New York, 1960, p. 27.
4. K. Schmieder and K. Wolf, *Kolloidzeitschrift*, **134**, 149 (1953).
5. A. B. Thompson and D. W. Woods, *Trans. Faraday Soc.*, **52**, 1383 (1956).
6. M. L. Williams, R. F. Landel and J. D. Ferry, *J. Amer. Chem. Soc.*, **77**, 3701 (1955).
7. N. G. McCrum and E. L. Morris, *Proc. Roy. Soc.*, **A281**, 258 (1964).
8. S. Glasstone, K. J. Laidler and H. Eyring, *The Theory of Rate Processes*, McGraw-Hill, New York, 1941.

9. S. Glasstone, *Textbook of Physical Chemistry*, 2nd edn, Macmillan, London, 1953.
10. P. Debye, *Polar Molecules*, Dover, New York, 1945.
11. H. Fröhlich, *Theory of Dielectrics*, Oxford University Press, Oxford, 1949.
12. J. D. Hoffman, G. Williams and E. Passaglia, *J. Polymer. Sci. C*, **14**, 173 (1966).
13. Z. Arrhenius, *J. Phys. Chem.*, **4**, 226 (1889).
14. J. B. Wachtman, *Phys. Rev.*, **131**, 517 (1963).
15. J. D. Ferry, *Viscoelastic Properties of Polymers*, Wiley, New York, 1961, Chapter 11.
16. A. Kovacs, *J. Poymer Sci.*, **30**, 131 (1958).
17. A. K. Doolittle, *J. Applied Phys.*, **22**, 1471 (1951).
18. G. Adam and J. H. Gibbs, *J. Chem. Phys.*, **43**, 139 (1965).
19. G. Williams, *Trans. Faraday Soc.*, **60**, 1556 (1964).
20. A. Guinier and R. Jullien, *The Solid State*, Oxford University Press, Oxford, p. 13, 1989.
21. P. E. Rouse, *J. Chem. Phys.*, **21**, 1272 (1953).
22. B. H. Zimm, *J. Chem. Phys.*, **24**, 269 (1956).
23. M. Doi and S. F. Edwards, *The Theory of Polymer Dynamics*, Oxford University Press, 1986.
24. P. G. de Gennes, *J. Chem. Phys.*, **55**, 572 (1971).
25. W. W. Graessley, *J. Polym. Sci., Polymer Phys.*, **18**, 27, (1980).

Chapter 7

Anisotropic Mechanical Behaviour

Although oriented solid polymers are in general anisotropic non-linear viscoelastic materials, the present discussion is restricted to small strains and short times, where it can be assumed that stress is directly proportional to strain, i.e. linear viscoelastic behaviour is observed. This simplification enables us to identify the character of the anisotropy and its relation to the geometry of the orienting process. Following a brief mention of experimental methods, we report measurements of the form of the anisotropy, and then describe a model that seeks to explain the development of anisotropy through the orientation of an aggregate of pre-existing subunits having an initially random disposition.

7.1 ELASTIC CONSTANTS AND POLYMER SYMMETRY

The mechanical properties of an isotropic linear elastic solid are defined by the generalized Hooke's law, which is conveniently written in the abbreviated notation

$$\sigma_p = c_{pq} e_q$$

or equivalently

$$e_p = s_{pq} \sigma_q$$

where $p, q = 1, 2 \ldots 6$ and

$$\sigma_1 = \sigma_{xx}, \quad \sigma_2 = \sigma_{yy}, \quad \sigma_3 = \sigma_{zz}, \quad \sigma_4 = \sigma_{yz}, \quad \sigma_5 = \sigma_{xz}, \quad \sigma_6 = \sigma_{xy}$$

so that $\sigma_1 \ldots \sigma_6$ are the six independent components of the stress tensor, and

$$e_1 = e_{xx}, \quad e_2 = e_{yy}, \quad e_3 = e_{zz}, \quad e_4 = e_{yz}, \quad e_5 = e_{xz}, \quad e_6 = e_{xy}$$

are the six engineering components of strain.

The constants c_{pq} are called the stiffness constants and the s_{pq} are called the compliance constants. Because the most straightforward experimental procedure is to apply a single tensile stress or a single shear stress and measure the corresponding strains, it is often more convenient to work in terms of compliance rather than stiffness constants.

For example, application of a stress σ_1 along the 1 direction gives

$$e_3 = s_{33}\sigma_3 \qquad e_1 = s_{13}\sigma_3.$$

The engineering elastic constants, Young's modulus E_1 and Poisson's ratio ν_{31} are then given by

$$E_3 = \frac{1}{S_{33}} = \frac{\sigma_3}{e_3}$$

and

$$\nu_{13} = -\frac{e_1}{e_3} = -\frac{S_{13}}{S_{33}}$$

where the negative sign is introduced because e_3 is a contraction, i.e. a negative strain, whereas Poisson's ratios are by convention positive.

The elastic constants are then defined by the symmetric matrix

$$\begin{bmatrix} s_{11} & s_{12} & s_{13} & s_{14} & s_{15} & s_{16} \\ s_{12} & s_{22} & s_{23} & s_{24} & s_{25} & s_{26} \\ s_{13} & s_{23} & s_{33} & s_{34} & s_{35} & s_{36} \\ s_{14} & s_{24} & s_{34} & s_{44} & s_{45} & s_{46} \\ s_{15} & s_{25} & s_{35} & s_{45} & s_{55} & s_{56} \\ s_{16} & s_{26} & s_{36} & s_{46} & s_{56} & s_{66} \end{bmatrix}$$

It is important to note that in this simplified notation, which is the most usual notation, neither the strain components nor the elastic constants form the elements of tensor. For an account of elastic anisotropy in terms of tensor notation and for the transformation rules which permit the elastic constants to be transformed from one set of Cartesian coordinates to another, the reader is referred to Appendix 1.

The use of the generalized Hooke's law is not restricted to time-independent behaviour. The compliance and stiffness constants can be time dependent, defining creep compliances and stress relaxation stiffnesses or complex compliances and complex stiffnesses in dynamic mechanical measurements. For simplicity, the methods of measurement are usually carefully standardized, e.g. by measuring each creep compliance after the same loading programme and the same time interval. It will be assumed that for such measurements there is an exact equivalence between elastic and linear viscoelastic behaviour, as proposed by Biot [1].

7.1.1 Specimens Possessing Fibre Symmetry

The simplest form of anisotropy in polymers results when fibres or films are prepared by deforming them plastically using a uniaxial stretching process known as drawing (see Chapter 11), and so possess isotropy in planes perpendicular to the direction of drawing. In this case the number of independent elastic constants is reduced to five [2]. With the z direction as the axis of symmetry the compliance matrix reduces to

$$s_{pq} = \begin{bmatrix} s_{11} & s_{12} & s_{13} & 0 & 0 & 0 \\ s_{12} & s_{11} & s_{13} & 0 & 0 & 0 \\ s_{13} & s_{13} & s_{33} & 0 & 0 & 0 \\ 0 & 0 & 0 & s_{44} & 0 & 0 \\ 0 & 0 & 0 & 0 & s_{44} & 0 \\ 0 & 0 & 0 & 0 & 0 & 2(s_{11} - s_{12}) \end{bmatrix}$$

The various compliance constants are illustrated diagramatically for a fibre specimen in Figure 7.1; a uniaxially oriented sheet possesses Young's and shear moduli and Poisson's ratio as follows:

1. Consider a tensile stress σ_{zz} (or σ_3) applied along the axial or draw direction. Then $e_{zz} = s_{33}\sigma_{zz}$, i.e. $e_3 = s_{33}\sigma_3$ and the Young's modulus,

$$E_3 = \frac{\sigma_{zz}}{e_{zz}} \frac{(\sigma_3)}{(e_3)} = \frac{1}{s_{33}}$$

2. Similarly a tensile stress σ_{zz} causes strain in the plane transverse to the z axis: $e_{xx} = e_{yy} = s_{13}\sigma_{zz}$; giving the Poisson's ratio,

$$\nu_{13} = -\frac{e_{xx}}{e_{zz}} = -\frac{s_{13}}{s_{33}}$$

3. In a similar manner the transverse modulus E_1 and the corresponding Poisson's ratios $\nu_{21} = \nu_{12}$ and $\nu_{31} = \nu_{13}$ for a stress applied perpendicular to the fibre axis are given in terms of s_{11}, s_{12} and s_{13} by

$$E_1 = \frac{1}{s_{11}}, \qquad \nu_{21} = -\frac{s_{21}}{s_{11}} = -\frac{s_{12}}{s_{11}}, \qquad \nu_{31} = -\frac{s_{31}}{s_{11}} = -\frac{s_{13}}{s_{11}}$$

4. The shear torsional modulus is given in terms of the equivalent shear compliances $s_{44} = s_{55} = 1/G$, which relate to torsion about the symmetry axis z, i.e. shear in the yz or xz planes.

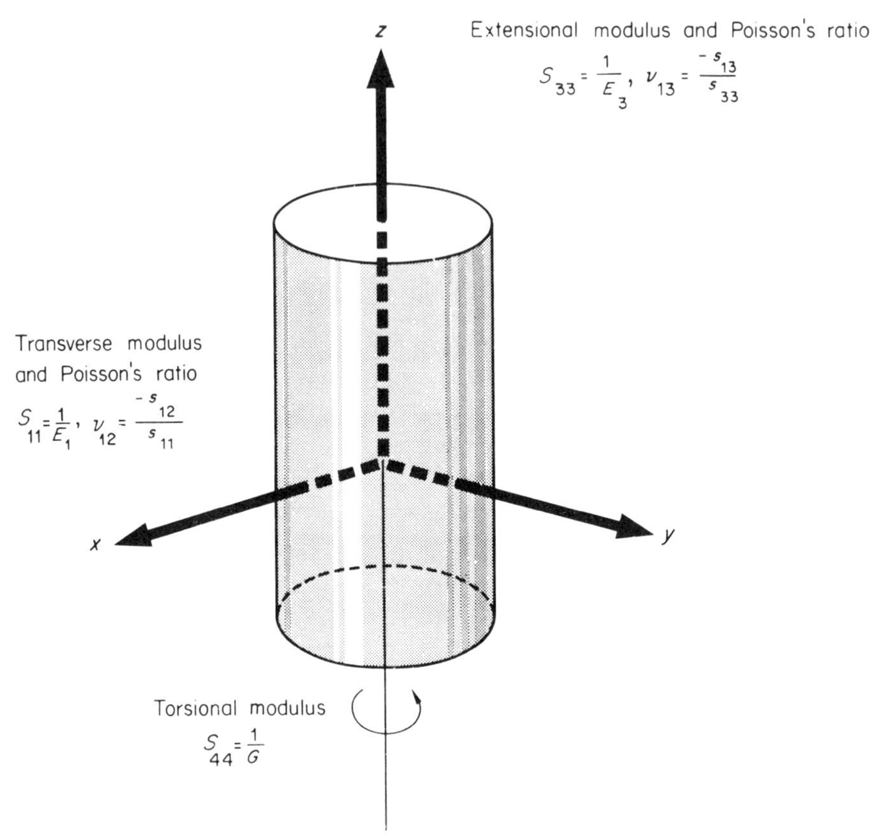

Figure 7.1 The fibre compliance constants

7.1.2 Specimens Possessing Orthorhombic Symmetry

Films prepared by rolling, with or without annealing, or using certain commercial draw processes, may possess orthorhombic symmetry[†]. Let the z axis for a system of rectangular Cartesian coordinates coincide with the initial drawing or rolling direction, and let the x axis lie in the plane of the film, with the y axis normal to the film plane (Figure 7.2). The compliance matrix then contains nine independent elastic constants.

[†]Orthorhombic symmetry means that there are three orthogonal planes of symmetry, but the properties are anisotropic in each plane.

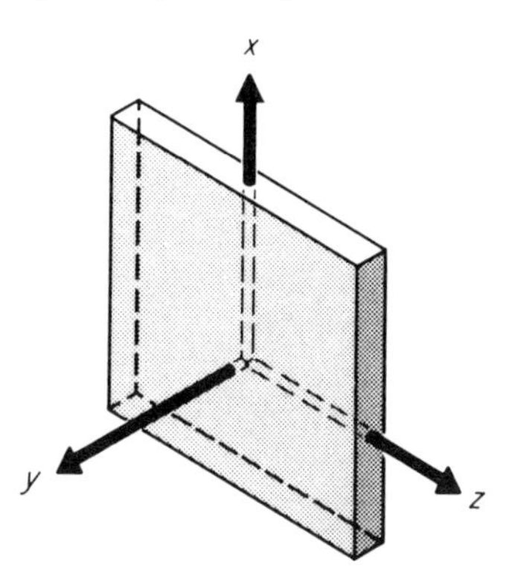

Figure 7.2 Choice of axes for a polymer sheet possessing orthorhombic symmetry

$$s_{pq} = \begin{bmatrix} s_{11} & s_{12} & s_{13} & 0 & 0 & 0 \\ s_{12} & s_{22} & s_{23} & 0 & 0 & 0 \\ s_{13} & s_{23} & s_{33} & 0 & 0 & 0 \\ 0 & 0 & 0 & s_{44} & 0 & 0 \\ 0 & 0 & 0 & 0 & s_{55} & 0 \\ 0 & 0 & 0 & 0 & 0 & s_{66} \end{bmatrix}$$

There are three Young's moduli:

$$E_1 = \frac{1}{s_{11}}, \qquad E_2 = \frac{1}{s_{22}}, \qquad E_3 = \frac{1}{s_{33}}$$

and six Poisson's ratios:

$$\nu_{21} = -\frac{s_{21}}{s_{11}}, \qquad \nu_{31} = -\frac{s_{31}}{s_{11}}, \qquad \nu_{32} = -\frac{s_{32}}{s_{22}}$$

$$\nu_{12} = -\frac{s_{12}}{s_{22}}, \qquad \nu_{13} = -\frac{s_{13}}{s_{33}}, \qquad \nu_{23} = -\frac{s_{23}}{s_{33}}$$

Note that in each of these expressions the denominator indicates the direction x, y, z (or 1, 2, 3) in which the tensile stress is applied.

There are also three independent shear moduli:

$$G_1 = \frac{1}{s_{44}}, \qquad G_2 = \frac{1}{s_{55}}, \qquad G_3 = \frac{1}{s_{66}}$$

corresponding to shear in the yz, zx and xy planes respectively. Torsion experiments where the sheet is twisted about the x, y or z axis will in general involve a combination of shear compliances (see section 7.2).

7.2 MEASURING ELASTIC CONSTANTS

Here we describe only the principal features of methods of measuring the elastic constants, and refer the reader either to the references cited or to Ward's more advanced text [3] for details of the experimental arrangements.

The experimental methods employed for filaments are different from those used for flat sheets, and the two cases will therefore be discussed separately.

7.2.1 Measurements on Films or Sheets

In all cases the draw or roll direction is taken as the z axis, with the x axis in the plane of the sheet and the y axis perpendicular to that plane.

7.2.1.1 Extensional moduli

The principle here is to measure the Young's modulus for long narrow strips cut at various angles to the z axis. End effects that arise from the non-uniform stresses near the clamps are very severe [4, 5], so that it is desirable to measure specimens with the highest possible aspect ratio, i.e. length to width and thickness ratios. A good general rule is that the aspect ratio, i.e. ratio of length of specimen to the longest lateral dimension, should be much greater than

$$10 \times \sqrt{\frac{E}{G}}$$

where E is the relevant Young's modulus and G is the appropriate shear modulus (i.e. either G_1 or G_2).

It is convenient to measure strips cut at 0°, 45° and 90° to the z axis. As shown in Appendix 1 the required relations are as follows:

$$E_0 = \frac{1}{s_{33}}, \qquad E_{90} = \frac{1}{s_{11}}, \quad \text{and} \quad \frac{1}{E_{45}} = \frac{1}{4}[s_{11} + s_{33} + (2s_{13} + s_{55})].$$

For a transversely isotropic sheet $s_{55} = s_{44}$.

7.2.1.2 Transverse stiffness

The stiffness c_{22} normal to the plane of the sheet has been determined by compressing narrow strips in a compressional creep apparatus [6]. A lever device is used to determine the deformation, and precision can be improved by compressing a sandwich of sheets separated by effectively rigid spacers.

Wilson *et al.* [6] found that for sheets of polyethylene terephthalate frictional constraints prevented strain in the x or z directions. In this case $e_x = e_z = 0$ and $\sigma_y = c_{22}e_y$, or

$$e_y = \left[s_{22} + s_{12}(s_{13}s_{23} - s_{12}s_{33}) - \frac{s_{23}(s_{11}s_{23} - s_{12}s_{23})}{(s_{11}s_{33} - s_{13}^2)} \right] \sigma_y$$

7.2.1.3 *Lateral compliances and Poisson's ratio*

For a polymer film possessing orthorhombic symmetry there are three lateral compliances: s_{12}, s_{13} and s_{23}, which relate to the six Poisson's ratios previously defined.

The compliance s_{13}, which defines the contraction in the x direction for a stress applied along the z direction, has been determined from the change in shape of a grid of straight lines parallel with the x and z axes vacuum-deposited on the surface of the long thin specimen [7]. The grid was photographed under the smallest load that held the specimen taut, and again at a fixed time after the application of further loads.

Compliances s_{12} and s_{23} relate to contraction in the thickness (y) direction for stresses along the x and z axes respectively. For sufficiently thick specimens the changes in thickness, claimed accurate to within 0.25 μm, can be measured by extensometers, fitted with lever arms, which bear against the faces of the polymer sheet. Thin cover slides are inserted between the extensometer elements and the specimen faces to remove any possibility of indentation [8]. For thinner specimens, of materials of high optical clarity, the sample can be inserted in one arm of a Michelson interferometer and immersed in turn in two fluids of different refractive index. The fringe shift Δm can be related to the change in thickness Δt by

$$\Delta m = \frac{2}{\lambda}[(n_{\mathrm{i}} - 1)\Delta t + t\,\Delta n_{\mathrm{i}}],$$

where λ is the wavelength, n_{i} the refractive index and Δn_{i} the change in refractive index [9].

A Hall effect extensometer (Figure 7.3), due to Richardson and Ward [10] is capable of recording strains down to 10^{-3} in sheets of 0.5 mm thickness. Two permanent magnets A_1 and A_2 are mounted in tubes T with their like poles adjacent, so that the intermediate field has a null point and the field gradient is twice that of a single magnet. Springs P_1 and P_2 enable the polymer sheet S to be held against the face of A_1 by the stainless steel plate B and its cover slip C, which contain the Hall effect device H. The Hall plate is positioned so that its sensing element E lies initially along the null field axis.

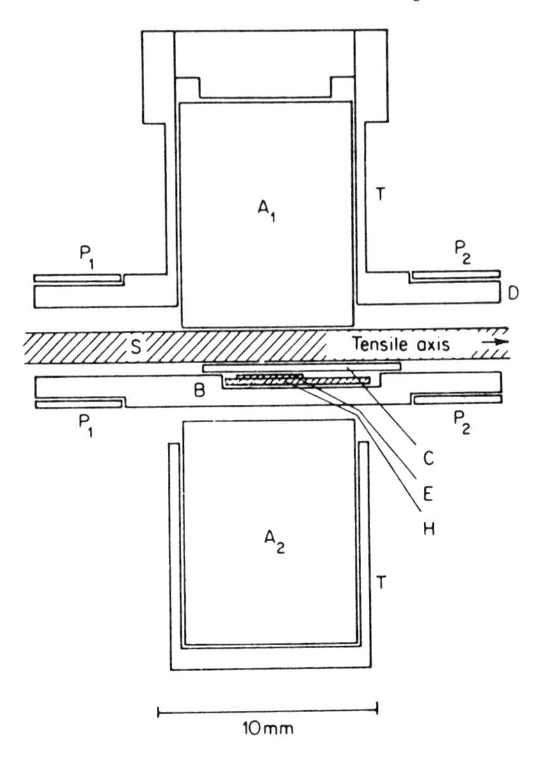

Figure 7.3 Scale diagram of the Hall effect lateral extensometer. (Reproduced with permission from Richardson and Ward, *J. Polymer Sci., Polymer Phys. Ed.*, **16**, 667 (1978))

7.2.1.4 Measurement in torsion

For sheets of orthorhombic symmetry a solution to the elastic torsion problem is possible only when the sheets are cut as rectangular prisms with their surfaces normal to the three axes of symmetry, and where the torsion axis coincides with one of these three axes. For example, torsion about the z axis involves compliances s_{44} in the yz plane and s_{55} in the xz plane.

In this case, for a specimen of length l, thickness a and width b (Figure 7.4), the torsional rigidity, which is the ratio of the torque Q_z to the angle of twist T, is given by theory due to St Venant (see [11] p. 201) as

$$\frac{Q_z}{T} = \frac{ab^3}{s_{55}l}\,\beta(c_z) = \frac{a^3b}{s_{44}l}\,\beta(\overset{+}{c}_z),$$

where

$$c_z = \frac{1}{\overset{+}{c}_z} = \frac{a}{b}\left(\frac{s_{55}}{s_{44}}\right)^{1/2}$$

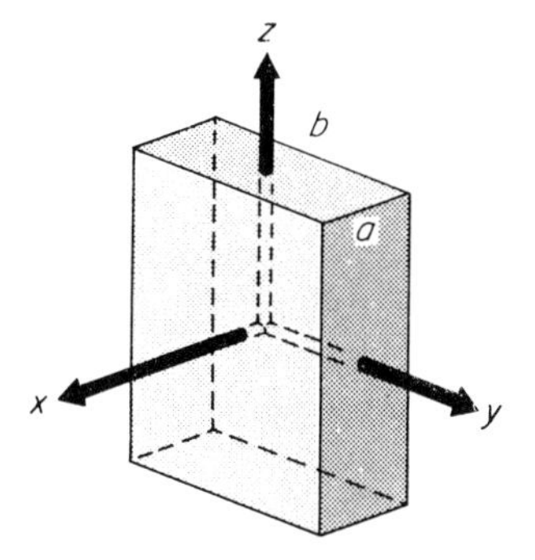

Figure 7.4 The orthorhombic sheet

and $\beta(c_z)$ is a rapidly converging function of c_z, which for $c_z > 3$ can be approximated to

$$\beta(c_z) = \frac{1}{3}\left\{1 - \frac{0.630}{c_z}\right\}$$

For a sheet with isotropy perpendicular to the z axis, $s_{44} = s_{55}$, so that the torque Q_z about an axis perpendicular to the symmetry axis is given by

$$Q_z = \frac{bc^3 T}{s_{66} l}\,\beta(\overset{*}{c})\,\frac{cb^3 T}{s_{44} l}\,\beta(\overset{+}{c}),$$

where

$$\overset{+}{c} = \frac{1}{\overset{*}{c}} = \frac{c}{b}\left(\frac{s_{44}}{s_{66}}\right)^{1/2}$$

with b the thickness, c the width and l the length of the specimen. Both compliances have been obtained from measurements on sheets of a range of aspect ratios [12].

For torsion about the symmetry axis z, the torque for transversely isotropic sheets is given by

$$Q_z = \left(\frac{ab^3}{s_{44} l}\right)\beta(c),$$

where $c = a/b$; $\beta(c)$ is now the same function of $c = a/b$ only.

Practical difficulties arise because the specimen is normally deformed additionally by an axial stress, and warping of the twist axes is restrained in the vicinity of the end clamps. The former effect can be compensated for by carrying out experiments over a range of axial stresses and extrapolating to zero stress. Folkes and Arridge [5] have considered end effects as confined to a block with compliance s' and length p at each end of the specimen of total length l, the central region of which has the true compliance s_0. The overall sample compliance s is then

$$s = s_0 + \left(\frac{2p}{l}\right)(s' - s_0),$$

which means that s_0 can be found by extrapolating measurements on samples of different length to zero reciprocal length.

Because of these complicating effects it can be advantageous to determine s_{44} and s_{66} from experiments in simple shear. Lewis, Richardson and Ward [13] have used a Hall effect transducer to measure this type of deformation, and obtain shear compliances that agree with those measured in torsion.

7.2.2 Measurements on Filaments

Extensional modulus

$$E = \frac{1}{s_{33}}$$

has been measured statically, and also dynamically over a wide range of frequencies. The shear modulus can be obtained either from free torsional vibrations, as described in Chapter 5, or from a forced vibration pendulum. The Poisson's ratio

$$\nu_{13} = -\frac{s_{13}}{s_{33}}$$

has been measured directly, using a microscope to record the radial contraction corresponding to a known change in length of a marked section of a much longer filament [14]. For strains of 0.01–0.02 the uncertainty was in the region of 10%. The transverse modulus

$$\frac{1}{s_{11}}$$

and the transverse Poisson's ratio

$$\nu_{12} = -\frac{s_{12}}{s_{11}}$$

have both been measured for filaments compressed under known loads between parallel glass flats (Figure 7.5(a)) using a simple device attached to a microscope stage [14, 15].

The filament's transverse isotropy implies that under compressive loading normal to the fibre axis the stresses in the transverse plane are identical in form to those for the compression of an isotropic cylinder and, provided that the length of filament under compression is long compared with the width of the contact strip $2b$, friction ensures that compression occurs under plane strain conditions. As $e_{zz} = 0$ only a normal stress acts along the filament axis, which can be found in terms of the normal stresses σ_{xx} and σ_{yy} in the perpendicular plane, i.e.

$$\sigma_{zz} = -\frac{s_{13}}{s_{33}}(\sigma_{xx} + \sigma_{yy}).$$

The stresses can therefore be obtained from the solution to the problem of compression of an isotropic cylinder, and the corresponding strains derived from $e_p = s_{pq}\sigma_q$.

Through an extension of Hertz's solution for the compression of an *isotropic* cylinder Ward was able to derive an analytic solution for the contact width $2b$, of the form

$$b^2 = \frac{4FR}{\pi}\left(s_{11} - \frac{s_{13}^2}{s_{33}}\right) \tag{7.1}$$

where F is the compressive force per unit length of filament of radius R (Figure 7.5(a)). Rewriting (7.1) as

$$b^2 = \frac{4FR}{\pi}(s_{11} - \nu_{13}^2 s_{33}),$$

we can consider the relative importance of the two terms inside the brackets. Poisson's ratio is typically close to 0.5, and s_{33} is normally small compared with s_{11}, as highly oriented polymers are usually much stiffer along the z

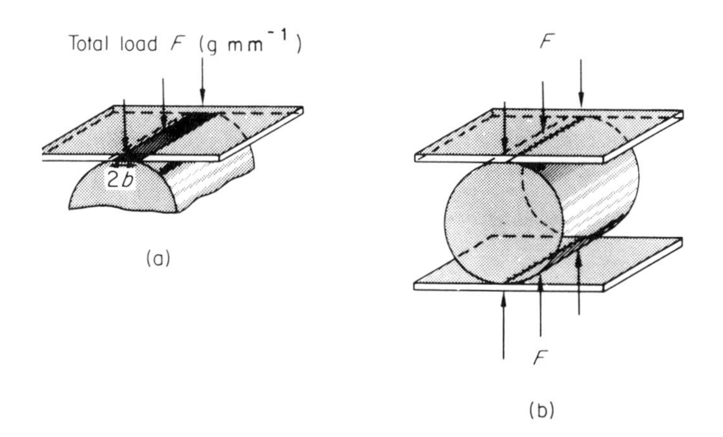

Figure 7.5 The contact zone in the compression of a fibre monofilament (a); for consideration of deformation in the central zone of the compressed monofilament it is sufficient to assume line contacts (b); F is the force acting on unit length of the filament

axis than transverse to it. Hence the contact problem provides a good method for determining s_{11}.

Further, it can be shown that the diametral compression u_1 parallel to the direction of the applied load is

$$u_1 = -\frac{4F}{\pi}\left(s_{11} - \frac{s_{13}^2}{s_{33}}\right)\left[0.19 + \sinh^{-1}\left(\frac{R}{b}\right)\right] \tag{7.2}$$

Provided that the width of the contact zone is small compared with the filament radius, the change in diameter parallel with the plane of contact u_2 can be calculated by considering the deformation of a cylinder under concentrated loads (Figure 7.5b). The stresses here correspond exactly with those for the isotropic case, whose solution can be found in texts on elasticity (e.g. [16] p. 107). When the strains have been calculated it is possible to derive the diametral expansion as

$$u_2 = F\left[\left(\frac{4}{\pi} - 1\right)\left(s_{11} - \frac{s_{13}^2}{s_{33}}\right) - \left(s_{12} - \frac{s_{13}^2}{s_{33}}\right)\right] \tag{7.3}$$

Thus measurement of the diametral expansion provides a method of determining s_{12}, once s_{11} has been derived from a measurement of the contact width.

7.3 EXPERIMENTAL STUDIES OF MECHANICAL ANISOTROPY IN TRANSVERSELY ISOTROPIC POLYMERS

We shall concentrate here on selected early studies because this work highlighted several unexpected features of the anisotropy, indicated the differences between individual polymers and provided the basis for a theoretical model which has been tested against subsequent measurements on a range of materials.

7.3.1 Sheets of Low-density Polyethylene

Raumann and Saunders [17] uniaxially stretched initially isotropic sheets of low-density (i.e. branched) polyethylene to varying final extensions, and measured the tensile modulus in directions over a range of angles to the initial draw direction. For a highly oriented sample the plot of Young's modulus ($1/s_\theta$) against angle with the draw direction (Figure 7.6) shows the lowest stiffness at an angle close to 45° to that direction.

The general compliance equation for transverse isotropy (see A1, section A1.7) is

$$s_\theta = s_{11}\sin^4\theta + s_{33}\cos^4\theta + (2s_{13} + s_{44})\sin^2\theta\cos^2\theta.$$

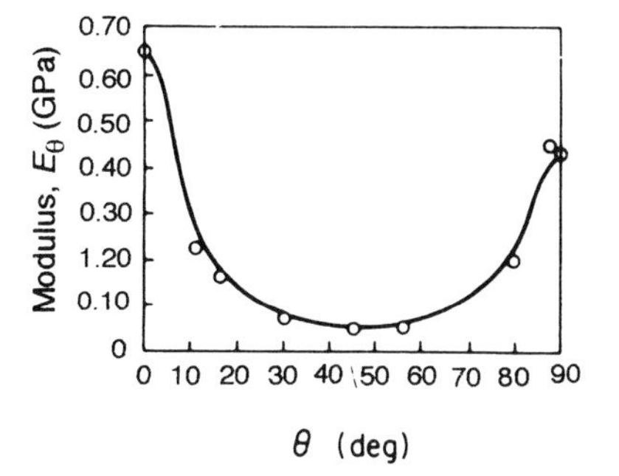

Figure 7.6 Comparison of the observed variation of modulus E_0 with angle θ to draw direction and the theoretical relation (i.e. full curve) calculated from E_0. E_{45} and E_{90} for low-density polyethylene sheet drawn to a draw ratio of 4.65. (Reproduced with permission from Raumann and Saunders, *Proc. Phys. Soc.*, **77**, 1028 (1961))

The experimental result implies that $(2s_{13} + s_{44})$ is much greater than either s_{11} or s_{33}, for when $\theta = 45°$ the terms will be equally weighted.

Replotting, to obtain the modulus at a given angle as a function of draw ratio (Figure 7.7) the results are again somewhat unexpected: E_0 initially falls with increasing draw ratio, so that at low draw ratios $E_{90} > E_0$. Subsequently, Gupta and Ward [18] showed that this unusual behaviour was specific to room-temperature measurements, and at a sufficiently low temperature the behaviour resembled that of most other polymers (Figure 7.8).

7.3.2 Filaments Tested at Room Temperature

In a comprehensive study at room temperature, Hadley, Pinnock and Ward [19] determined the five independent elastic constants for oriented filaments of polyethylene terephthalate, nylon 6:6, low and high-density polyethylene and polypropylene. The orientation was determined in terms of draw ratio and optical birefringence. Subsequent studies indicated that it would have been appropriate to record not only the overall orientation, as derived from birefringence, but also the crystal orientation, obtainable from X-ray measurements. The results are summarized in Table 7.1 and Figures 7.9–7.13 (see section 7.5 for discussion of the aggregate theory predictions).

Although the detailed development of mechanical anisotropy in these particular filaments must depend on their exact chemical composition and subsequent processing, several general features can be distinguished. The principal effect of drawing (i.e. increasing molecular orientation) is to increase the Young's modulus E_3 measured along the filament axis. In nylon 6:6 and polyethylene terephthalate there is a corresponding but small

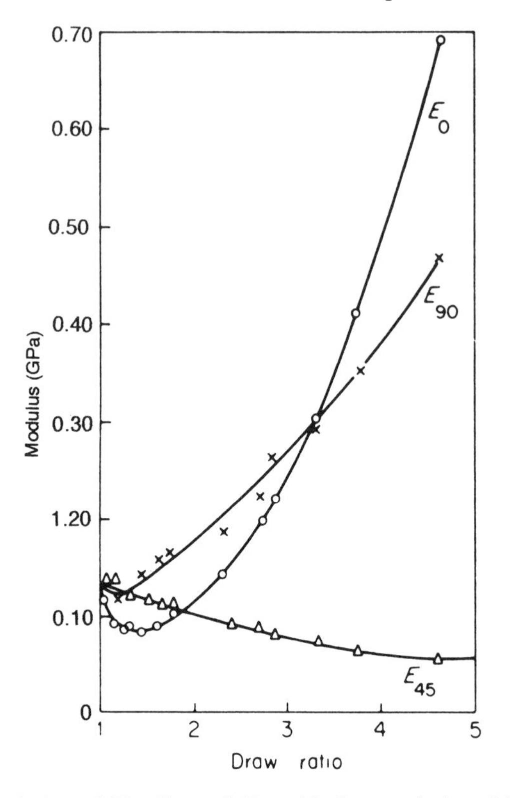

Figure 7.7 The variation of E_0, E_{45} and E_{90} with draw ratio in cold-drawn sheets of low-density polyethylene. Modulus measurements taken at room temperature (Reproduced with permission from Raumann and Saunders, *Proc. Phys. Soc.*, **77**, 1028 (1961))

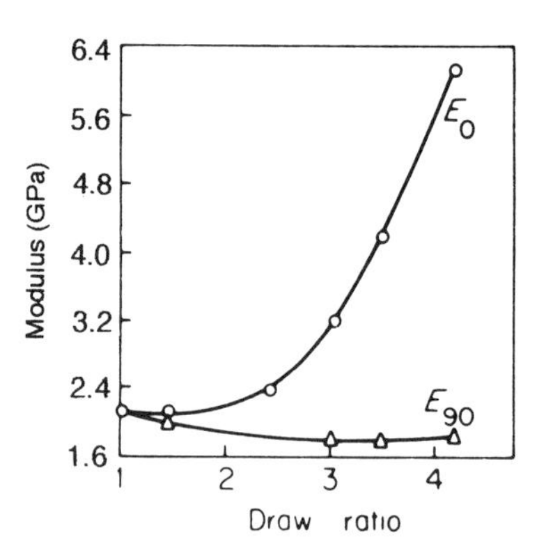

Figure 7.8 The variation of E_0 and E_{90} with draw ratio in cold-drawn sheets of low-density polyethylene. Modulus measurements taken at −125 °C.

Table 7.1 Elastic compliances of oriented fibres (units of compliance are 10^{-10} Pa^{-1}; errors quoted are 95% confidence limits) [1]

Material	Birefringence (Δn)	s_{11}	s_{12}	s_{33}	s_{13}	s_{44}		
Low-density polyethylene film 14		22	−15	14	−7	680	0.50	0.68
Low-density polyethylene 1	0.0361	40 ± 4	−25 ± 4	20 ± 2	−11 ± 2	878 ± 56	0.55 ± 0.08	0.61 ± 0.20
Low-density polyethylene 2	0.0438	30 ± 3	−22 ± 3	12 ± 1	−7 ± 1	917 ± 150	0.58 ± 0.08	0.73 ± 0.20
High-density polyethylene 1	0.0464	24 ± 2	−12 ± 1	11 ± 1	−5.1 ± 0.7	34 ± 1	0.46 ± 0.15	0.52 ± 0.08
High-density polyethylene 2	0.0594	15 ± 1	−16 ± 2	2.3 ± 0.3	−0.77 ± 0.3	17 ± 2	0.33 ± 0.12	1.1 ± 0.14
Polypropylene 1	0.0220	19 ± 1	−13 ± 2	6.7 ± 0.3	−2.8 ± 1.0	18 ± 1.5	0.42 ± 0.16	0.68 ± 0.18
Polypropylene 2	0.0352	12 ± 2	−17 ± 2	1.6 ± 0.04	−0.73 ± 0.3	10 ± 2	0.47 ± 0.17	1.5 ± 0.3
Polyethylene terephthalate 1	0.153	8.9 ± 0.8	−3.9 ± 0.7	1.1 ± 0.1	−0.47 ± 0.05	14 ± 0.5	0.43 ± 0.06	0.44 ± 0.09
Polyethylene terephthalate 2	0.187	16 ± 2	−5.8 ± 0.7	0.71 ± 0.04	−0.31 ± 0.03	14 ± 0.2	0.44 ± 0.07	0.37 ± 0.06
Nylon 6:6	0.057	7.3 ± 0.7	−1.9 ± 0.4	2.4 ± 0.3	−1.1 ± 0.15	15 ± 1	0.48 ± 0.05	0.26 ± 0.08

$\nu_{13} = -\frac{s_{13}}{s_{33}}$ $\nu_{12} = -\frac{s_{12}}{s_{11}}$

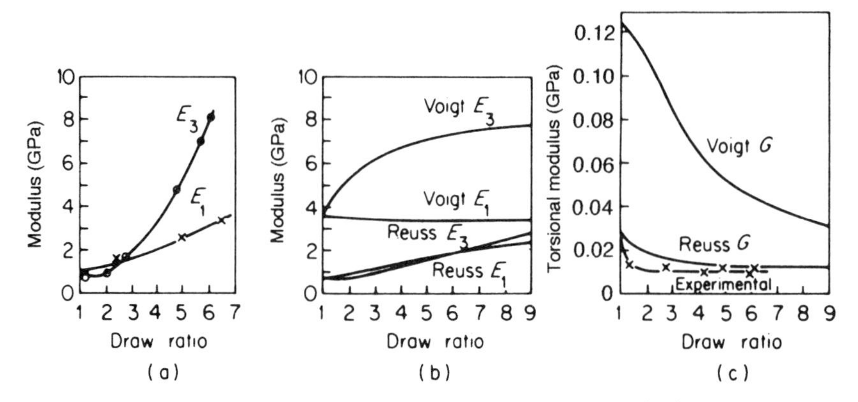

Figure 7.9 Low-density polyethylene filaments: extensional (E_3), transverse (E_1) and torsional moduli (G); comparison between experimental results and simple aggregate theory for E_3 and E_1 ((a) and (b)) and for G (c)

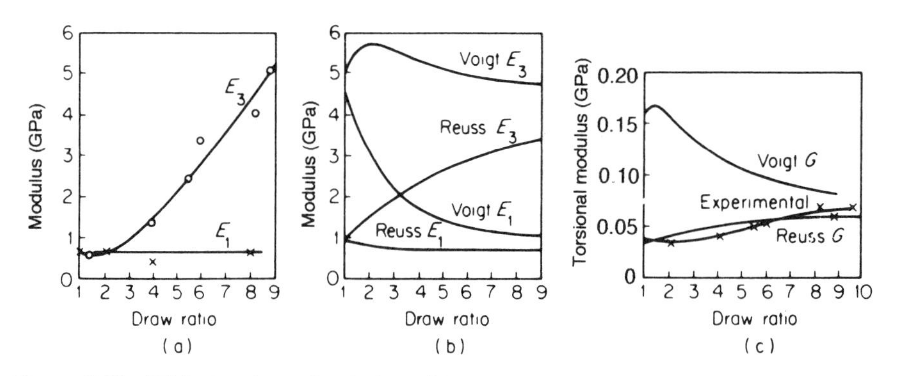

Figure 7.10 High-density polyethylene filaments: extensional (E_3), transverse (E_1) and torsional moduli (G); comparison between experimental results and simple aggregate theory for E_3 and E_1 ((a) and (b)) and for G (c)

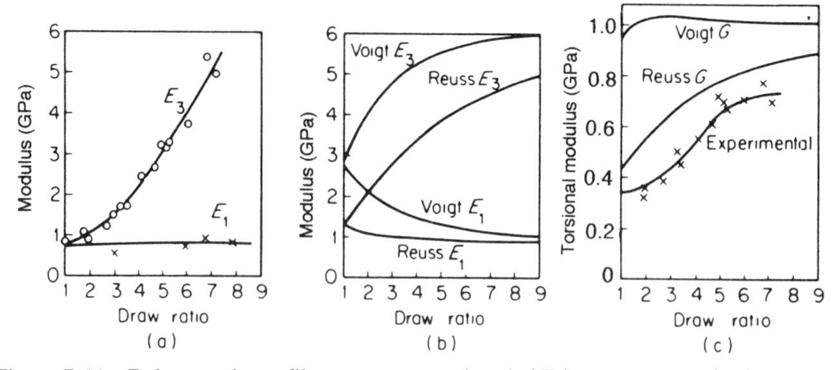

Figure 7.11 Polypropylene filaments; extensional (E_3), transverse (E_1) and torsional moduli (G); comparison between experimental results and simple aggregate theory for E_3 and E_1 ((a) and (b)) and for G (c)

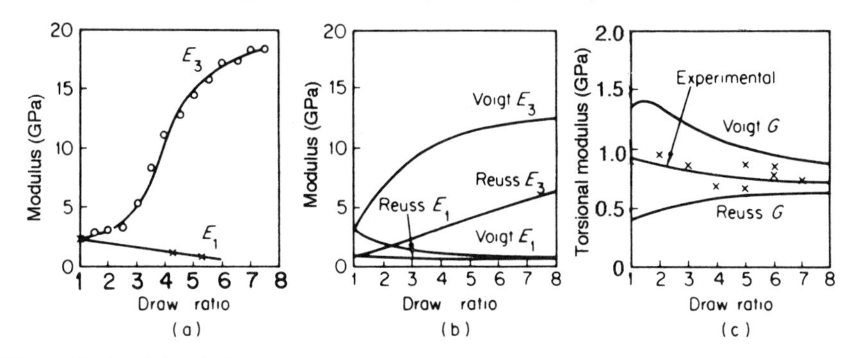

Figure 7.12 Polyethylene terephthalate filaments: extensional (E_3), transverse (E_1) and torsional moduli (G); comparison between experimental results and simple aggregate theory for E_3 and E_1 ((a) and (b)) and for G (c)

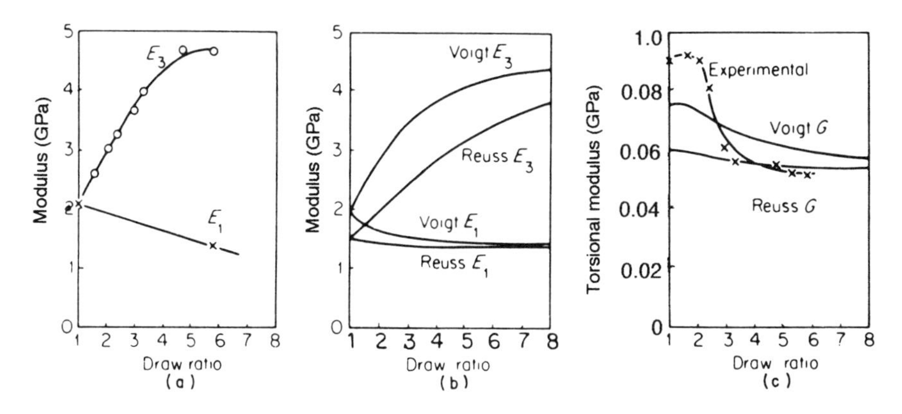

Figure 7.13 Nylon filaments: extensional (E_3), transverse (E_1) and torsional moduli (G); comparison between experimental results and simple aggregate theory for E_3 and E_1 ((a) and (b)) and for G (c)

decrease in the transverse modulus E_1, for polypropylene and high-density polyethylene E_1 is almost independent of draw ratio, and for low-density polyethylene E_1 increases significantly, in agreement with the results of Raumann and Saunders. Note here, too, the anomalous behaviour of this polymer at low draw ratios. Overall, E_3 for highly oriented filaments is greater than E_1, with the anisotropy being greatest for polyethylene terephthalate:

$$\frac{E_3}{E_1} = \frac{s_{11}}{s_{33}} \sim 27.$$

The shear modulus G of low-density polyethylene, and its change with orientation, provides another striking contrast with the other materials examined. A decrease of G by more than a factor of three over the range of orientation used compares with only small changes for the other filaments. In polyethylene terephthalate, high-density polyethylene and polypropylene $s_{44} \sim s_{11}$; in nylon $s_{44} \sim 2s_{11}$. By contrast, low-density polyethylene, at least as regards room-temperature behaviour, is exceptional, with the extensional compliance s_{33} having the same order of magnitude as the transverse compliance s_{11}, and with the shear compliance s_{44} being more than an order of magnitude greater than either s_{11} or s_{33}. Such measurements provided the basis for the discussion of relaxation transitions in Chapter 9.

In all cases the compliance s_{13} is low, and appears to decrease fairly rapidly with increasing draw ratio, in a manner comparable with s_{33}. Hence the extensional Poisson's ratio $\nu_{13} = s_{13}/s_{33}$ is rather insensitive to draw ratio and, with the exception of high-density polyethylene does not differ significantly from 0.5. It is thus generally a valid approximation to assume that the filaments are incompressible. [For anisotropic bodies ν_{13} is not confined to a maximum value of 0.5, but is limited solely by the inequalities necessary for a positive strain energy [2]

$$s_{12}^2 < s_{11}^2;\ s_{13}^2 < \tfrac{1}{2}s_{33}(s_{11} + s_{12})].$$

7.4 INTERPRETATION OF MECHANICAL ANISOTROPY: GENERAL CONSIDERATIONS

The mechanical anisotropy of oriented polymers is determined by the following factors, which will be discussed in turn: (1) the structure of the molecular chain and, where the polymer crystallizes, the crystal structure; (2) The molecular orientation and, in a crystalline polymer, the morphology; (3) thermally activated relaxation processes in both the crystalline and non-crystalline regions.

7.4.1 Chain Structure and Crystal Structure

The molecular chain in the crystalline regions of polyethylene is able to take the form of a planar zigzag, owing to the small size of an individual hydrogen atom. For many other polymers, where the side groups are larger, the chain takes a helical form. In isotactic polypropylene, for instance, three monomer units constitute a single turn of the helix. For both cases there is a large difference between the forces involved when the structure is deformed parallel with and perpendicular to the axis of the chain or helix.

Theoretical estimates for the stiffness constants of the polyethylene crystal [20] lead to a compliance matrix:

$$s_{ij} = \begin{bmatrix} 14.5 & -4.78 & -0.019 & 0 & 0 & 0 \\ -4.78 & 11.7 & -0.062 & 0 & 0 & 0 \\ -0.019 & -0.062 & 0.317 & 0 & 0 & 0 \\ 0 & 0 & 0 & 31.4 & 0 & 0 \\ 0 & 0 & 0 & 0 & 61.7 & 0 \\ 0 & 0 & 0 & 0 & 0 & 27.6 \end{bmatrix} \times 100\ \text{GPa}^{-1}.$$

The high value of the crystal modulus along the chain direction, $E_3^c = 1/s_{33} \sim 300$ GPa, arises because the deformation involves primarily the bending and stretching of covalent bonds. By contrast the tensile modulus perpendicular to the chain direction, and the shear modulus, are much lower ($\sim$1–10 GPa) because they relate to the van der Waals' or dispersion forces between the chains. For polypropylene the tensile modulus in the axial direction is, at $\sim$50 GPa, much lower than that for polyethylene, because it involves rotation around bonds as well as the bending of bonds. It is still much larger than the perpendicular tensile modulus and the shear modulus, which are of similar magnitude and origin to those in polyethylene.

Only in a very few cases, notably ultra-high modulus polyethylene (see section 8.6 below) have extensional moduli in line with theoretical expectations been achieved (Table 7.2).

The determination of the elastic constants for the crystalline regions of polymers is of importance for providing a baseline against which practical achievements can be judged. Comprehensive reviews have been published that deal not only with theoretical calculations but with moduli determined from the changes in the X-ray diffraction pattern on stressing, by Raman spectroscopy and from the inelastic scattering of neutrons [21].

7.4.2 Orientation and Morphology

The degree of mechanical anisotropy is usually much less than considerations of the molecular chain would imply; and, in particular, the very high

Table 7.2 Elastic constants of ultra-high modulus polyethylene

	20 °C	−196 °C	Theoretical (Tadokoro)
Axial modulus (GPa)	70	160	316
Transverse modulus (GPa)	1.3	—	8–10
Shear modulus (GPa)	1.3	1.95	1.6–3.6
Poisson's ratio	0.4	—	0.5

intrinsic modulus along the axial direction is not achieved. Crystalline polymers are essentially composite materials with alternating crystalline and amorphous regions. The former regions can be highly aligned during processing, but the latter are less well oriented. Even when the overall orientation of the chain segments, measured for instance by birefringence, appears to be quite high, there are still very few chains where long lengths of a molecule are axially aligned. It is such molecules that are critical in increasing the stiffness, because there is such a large difference between the stresses involved in bond bending and stretching, and in other modes of deformation. Peterlin [22] has proposed that the Young's modulus of an oriented filament is essentially determined by the proportion of extended chain tie molecules that produce links in the axial direction between crystalline blocks. An alternative proposal of crystalline bridges linking the crystalline blocks is discussed in Chapter 8.

Polymers that do not crystallize, such as polymethyl methacrylate, show a good correlation between the (low) degree of mechanical anisotropy and molecular orientation determined from birefringence. There is so much disorder that it seems unlikely that a significant proportion of the chains can achieve the high alignment of a crystalline polymer such as polyethylene. Other polymers, such as polyethylene terephthalate, which have a comparatively low overall crystallinity ($\sim$30% is typical), may occupy an intermediate position. The mechanical anisotropy produced by drawing correlates well with overall molecular orientation, but this result may arise because tie molecules play a vital role, and their number increases with overall molecular orientation.

In conclusion, it must be emphasized that although it is convenient for the purpose of constructing models to assume a composite that comprises distinct crystalline and non-crystalline components, on the molecular level a gradual transition must occur, extending over a number of monomer units, between the well-orientated and ordered crystallites and the bulk of the remaining material.

7.4.3 Relaxation Processes

Thermally activated relaxation processes can, as discussed in Chapter 9, be associated with either main-chain segmental motions or side group motions, and in a crystalline material be associated with either the crystalline or amorphous phase. In terms of mechanical anisotropy the relaxation processes may reflect either molecular orientation or morphological structure; for example, the important feature could be the orientation of the chain axis, or else the lamellar orientation, where the non-crystalline interlamellar material softens preferentially.

7.5 A MODEL FOR THE DEVELOPMENT OF ANISOTROPY

In general it is to be expected that mechanical anisotropy will depend both on the crystalline morphology and the molecular orientation. However, Ward observed that for polyethylene terephthalate, as shown in Table 7.3, the influence of crystallinity on the extensional and torsional moduli is small compared with the effect of molecular orientation on the extensional modulus. It was proposed that as an approximation the polymer could be considered as an aggregate of identical units, which in the unstretched material are oriented randomly. As orientation develops the units rotate towards the draw direction and become completely aligned at the maximum achievable orientation. Hence the elastic properties of the units are those of the most highly drawn material [23].

There are two stages in testing the appropriateness of the model:

1. Can the elastic constants of the isotropic polymer be deduced from measurements on the most highly oriented sample?
2. Does the mechanical anisotropy develop with orientation in the predicted manner?

We shall consider each stage in turn.

7.5.1 Correlation Between the Elastic Constants of a Highly Oriented and an Isotropic Polymer

The average elastic constants for the isotropic aggregate can be obtained in two ways, either by assuming that the constituent units are in series or in parallel (Figure 7.14). The former case assumes a summation of strains for each unit subjected to the same stress (which implies a summation of compliance constants), and the latter assumes that each unit undergoes the same strain, with the stresses being summed (which implies a summation of stiffness constants). In general the principal axes of stress and of strain do

Table 7.3 Physical properties of polyethylene terephthalate fibres at room temperature

Birefringence	X-ray crystallinity	Extensional modulus (GPa)	Torsional modulus (GPa)
0	0	2.0	0.77
0	33	2.2	0.89
0.142	31	9.8	0.81
0.159	30	11.4	0.62
0.190	29	15.7	0.79

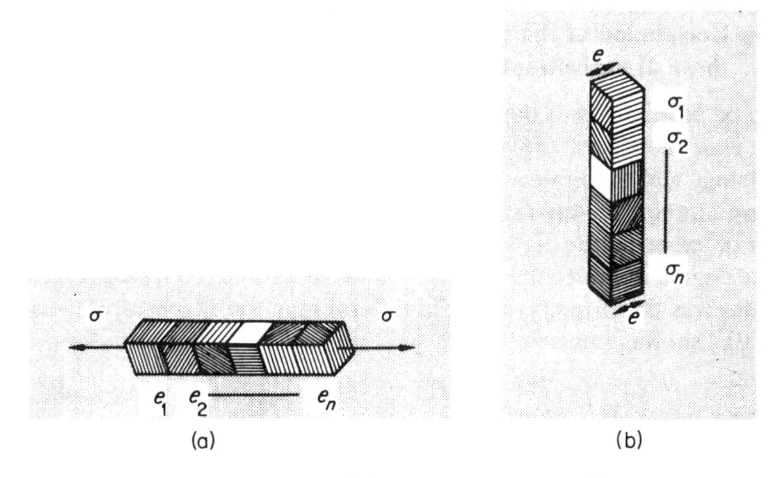

Figure 7.14 The aggregate model (a) for uniform stress; (b) for uniform strain

not coincide in an anisotropic solid, with the result that both the above approaches involve an approximation: for the assumption of uniform stress the strains throughout the aggregate are not uniform; for the alternative assumption of uniform strain, the stresses are non-uniform. Bishop and Hill [24] have shown that for a random aggreagate the correct response lies between the extremes predicted by the alternative schemes.

Consider the polymer to be comprised of N identical, transversely isotropic but randomly oriented cubes. For the case of uniform stress (Figure 7.14(a)) the cubes are arranged end-to-end forming a series model, in which the direction of elastic symmetry is defined by the angle θ which the cube axis makes with the direction of external stress σ. The strain in a single cube e_1 is given by the compliance formula

$$e_1 = [s_{11} \sin^4\theta + s_{33} \cos^4\theta + (2s_{13} + s_{44}) \sin^2\theta \cos^2\theta]\sigma,$$

where s_{11}, etc. are the compliance constants of the cube. We ignore the fact that the cubes will distort under the applied stress and do not satisfy compatibility of strain throughout the aggregate. Then the average strain e is

$$e = \frac{\sum e_1}{N} = [s_{11} \overline{\sin^4\theta} + s_{33} \overline{\cos^4\theta} + (2s_{13} + s_{44}) \overline{\sin^2\theta \cos^2\theta}]\sigma,$$

where $\overline{\sin^4\theta}$ etc. now define the average values of $\sin^4\theta$, etc. for the aggregate of units. Evaluating the average values of the trigonometrical functions we obtain for the average extensional compliance

$$\overline{s'_{33}} = \frac{e}{\sigma} = \frac{8}{15} s_{11} + \frac{1}{5} s_{33} + \frac{2}{15} (2s_{13} + s_{44}). \tag{7.4}$$

For the alternative uniform strain model, where the N elementary cubes are stacked in parallel (Figure 7.14(b)), the stress in a single cube is given by

$$\sigma_1 = [c_{11}\sin^4\theta + c_{33}\cos^4\theta + 2(c_{13} + 2c_{44})\sin^2\theta\cos^2\theta]e$$

where c_{11}, etc. are the stiffness constants of the cube. Taking average values as before, we obtain for a random aggregate

$$\overline{c'_{33}} = \frac{\sigma}{e} = \frac{8}{15}c_{11} + \frac{1}{5}c_{33} + \frac{4}{15}(c_{13} + 2c_{44}). \tag{7.5}$$

For an isotropic polymer there are two independent elastic constants, and the two alternative schemes predict a value for the isotropic shear compliance $\overline{s'_{44}}$ and the isotropic shear stiffness $\overline{c'_{44}}$ respectively.

$$\overline{s'_{44}} = \tfrac{14}{15}s_{11} - \tfrac{2}{3}s_{12} - \tfrac{8}{15}s_{13} + \tfrac{4}{15}s_{33} + \tfrac{2}{5}s_{44}, \tag{7.6}$$

$$\overline{c'_{44}} = \tfrac{7}{30}c_{11} - \tfrac{1}{6}c_{12} - \tfrac{2}{15}c_{13} + \tfrac{1}{15}c_{33} + \tfrac{2}{5}c_{44}. \tag{7.7}$$

Averaging the compliance constants defines the elastic properties of the isotropic aggregate in terms of $\overline{s'_{33}}$ and $\overline{s'_{44}}$, giving the 'Reuss average' [25]. Averaging the stiffness constants defines the properties in terms of $\overline{c'_{33}}$ and $\overline{c'_{44}}$, giving the 'Voigt average' [26]. In the latter case the matrix can be inverted to obtain the corresponding $\overline{s'_{33}}$ and $\overline{s'_{44}}$, in order to make a direct comparison between the two averaging procedures.

Table 7.4 compares the measured compliances for isotropic samples of five polymers with the Reuss and Voigt average compliances calculated from measurements on highly oriented specimens. For polyethylene terephthalate

Table 7.4 Comparison of calculated and measured extensional and torsional compliances (units are 10^{-10} Pa^{-1}) for unoriented fibres

	Extensional compliance ($\overline{s'_{11}} = \overline{s'_{33}}$)			Torsional compliance ($\overline{s'_{44}}$)		
	Calculated			Calculated		
	Reuss average	Voigt average	Measured	Reuss average	Voigt average	Measured
Low-density polyethylene	139	26	81	416	80	238
High-density polyethylene	10	2.1	17	30	6	26
Polypropylene	7.7	3.8	14	23	11	2.7
Polyethylene terephthalate	10.4	3.0	4.4	25	7.6	11
Nylon	6.6	5.2	4.8	17	13	12

and low-density polyethylene the measured isotropic compliances fall between the calculated bounds, suggesting that here molecular orientation could well be the principal factor that determines mechanical anisotropy. For nylon 6:6 the measured compliances fall just outside the bounds, which suggests that both molecular orientation and structural factors are important. In high-density polyethylene and in polypropylene the extensional compliances lie well outside the calculated bounds, suggesting that factors other than orientation play a major role in determining the mechanical anisotropy.

7.5.2 The Development of Mechanical Anisotropy With Molecular Orientation

The starting-point for the extension of aggregate theory to transversely isotropic polymers of intermediate orientation lay in earlier investigations of the manner in which birefringence increased as a function of the draw ratio imposed on initially isotropic specimens. Some results for low-density polyethylene are shown in Figure 7.15(a). Crawford and Kolsky [27] proposed a model of transversely isotropic rod-like units that rotated towards the draw axis. It was assumed that overall deformation occured uniaxially at constant volume (a reasonable assumption for most polymers, for which the extensional Poisson's ratio $\nu_{13} \sim 0.5$ (see section 7.3)) and that the symmetry axes of the anisotropic units rotated in the same manner as lines joining pairs of points in the macroscopic body. This assumption differs from the 'affine' deformation scheme for the optical anisotropy of rubbers (see section 3.1) in ignoring any change in length of the units on deformation, and has been called 'pseudo-affine' deformation. As the result of drawing the angle between the rod axis and the direction of draw changes

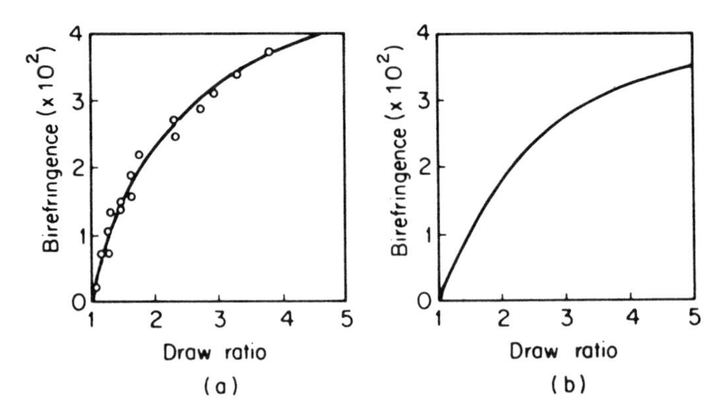

Figure 7.15 (a) Experimental, and (b) theoretical curves for the birefringence of low-density polyethylene as a function of draw ratio

from θ to θ', and it can be shown that for a draw ratio λ,

$$\tan\theta' = \frac{\tan\theta}{\lambda^{3/2}}$$

which leads to a birefringence Δn, given by

$$\Delta n = \Delta n_{\max}\,(1 - \tfrac{3}{2}\overline{\sin^2\theta})$$

where $\overline{\sin^2\theta}$ is the average value of $\sin^2\theta$ for the aggregate of rod-like units and $\Delta n_{\max}$ is the birefringence at full orientation.

The pseudo-affine deformation scheme gives a reasonable approximation to experimental data for low-density polyethylene (Figure 7.15(b)), nylon, polyethylene terephthalate and polypropylene despite ignoring the distinction between crystalline and disordered regions. It thus provides a basis for extending Ward's aggregate model to predict the compliance constants s'_{ij} and stiffness constants c'_{ij} of the partially oriented polymer in terms of the constants s_{ij} and c_{ij} for the anisotropic elastic unit (in practice those of the most highly oriented sample obtainable).

$$s'_{11} = \tfrac{1}{8}(3I_2 + 2I_5 + 3)s_{11} + \tfrac{1}{4}(3I_3 + I_4)s_{13} + \tfrac{3}{8}I_1 s_{33} + \tfrac{1}{8}(3I_3 + I_4)s_{44},$$

$$c'_{11} = \tfrac{1}{8}(3I_2 + 2I_5 + 3)c_{11} + \tfrac{1}{4}(3I_3 + I_4)c_{13} + \tfrac{3}{8}I_1 c_{33} + \tfrac{1}{2}(3I_3 + I_4)c_{44},$$

$$s'_{12} = \tfrac{1}{8}(I_2 - 2I_5 + 1)s_{11} + I_5 s_{12} + \tfrac{1}{4}(I_3 + 3I_4)s_{13} + \tfrac{1}{8}I_1 s_{33} + \tfrac{1}{8}(I_3 - I_4)s_{44},$$

$$c'_{12} = \tfrac{1}{8}(I_2 - 2I_5 + 1)c_{11} + I_5 c_{12} + \tfrac{1}{4}(I_3 + 3I_4)c_{13} + \tfrac{1}{8}I_1 c_{33} + \tfrac{1}{2}(I_3 - I_4)c_{44},$$

$$s'_{13} = \tfrac{1}{2}I_3 s_{11} + \tfrac{1}{2}I_4 s_{12} + \tfrac{1}{2}(I_1 + I_2 + I_5)s_{13} + \tfrac{1}{2}I_3 s_{33} - \tfrac{1}{2}I_3 s_{44},$$

$$c'_{13} = \tfrac{1}{2}I_3 c_{11} + \tfrac{1}{2}I_4 c_{12} + \tfrac{1}{2}(I_1 + I_2 + I_5)c_{13} + \tfrac{1}{2}I_3 c_{33} - 2I_3 c_{44},$$

$$s'_{33} = I_1 s_{11} + I_2 s_{33} + I_3(2s_{13} + s_{44}),$$

$$c'_{33} = I_1 c_{11} + I_2 c_{33} + 2I_3(c_{13} + 2c_{44}),$$

$$s'_{44} = (2I_3 + I_4)s_{11} - I_4 s_{12} - 4I_3 s_{13} + 2I_3 s_{33} + \tfrac{1}{2}(I_1 + I_2 - 2I_3 + I_5)s_{44},$$

$$c'_{44} = \tfrac{1}{4}(2I_3 + I_4)c_{11} - \tfrac{1}{4}I_4 c_{12} - I_3 c_{13} + \tfrac{1}{2}I_3 c_{33} + \tfrac{1}{2}(I_1 + I_2 - 2I_3 + I_5)c_{44}.$$

The terms $I_1 \ldots I_5$ are orientation functions that define average values of $\sin^4\theta(I_1)$, $\cos^4\theta(I_2)$, $\cos^2\theta\sin^2\theta(I_3)$, $\sin^2\theta(I_4)$ and $\cos^2\theta(I_5)$ for the aggregate. Only two of these orientation functions are independent parameters (e.g. $I_4 = I_1 + I_3$, $I_5 = I_2 + I_3$, $I_4 + I_5 = 1$).

In Figures 7.9–7.13 the predictions of the aggregate model assuming the pseudo-affine deformation scheme are compared with experimental measurements. For low-density polyethylene the Reuss average predictions show the correct general form, and provide an explanation for the minimum in the extensional modulus. With increasing orientation θ decreases, so that $\overline{\sin^4\theta}$ decreases monotonically and $\overline{\cos^4\theta}$ increases with increasing draw ratio, whereas $\overline{\sin^2\theta\cos^2\theta}$ shows a maximum at a draw ratio of about 1.2. Thus s'_{33} can pass through a maximum (corresponding to a minimum in the Young's modulus E_0) provided that $(2s_{13} + s_{44})$ is sufficiently large compared with s_{11} and s_{33} (which differ only slightly at low draw ratios). For low-density polyethylene at room temperature s_{44} is much larger than either s_{11} or s_{33}, so fulfilling the condition. At much lower temperatures s_{44} is no longer much greater than the other compliances, so that a more conventional pattern of mechanical anisotropy is observed.

The theoretical curves of Figures 7.9–7.13 differ from those obtained experimentally in two ways. Some features of detail, such as a small minimum in the transverse modulus of low-density polyethylene and a small minimum in the extensional modulus of high-density polyethylene, are not predicted at all. However, it has been shown that such effects may be associated with mechanical twinning [28]. A second deficiency is that the predicted development of mechanical anisotropy with draw ratio is considerably less rapid than is observed. The quantities $\overline{\sin^4\theta}$, $\overline{\cos^4\theta}$ and $\overline{\sin^2\theta\cos^2\theta}$ can also be determined by X-ray diffraction and nuclear magnetic resonance methods, and for low-density polyethylene a much improved fit was obtained by these means. For this material it appears that the anisotropy relates to the orientation of the crystalline regions, and is predicted to a good approximation by the Reuss averaging scheme. A similar explanation is unlikely for polyethylene terephthalate where other evidence suggests that the anisotropy resembles that of a deformed network. In this material, moreover, the experimental compliances lie approximately midway between the two bounds, with the median condition applying almost exactly for cold-drawn filaments [29].

In nylon the Voigt average lies closer to the experimental data and, although the quantitative fit is poor, both averaging schemes predict a maximum in the torsional modulus at an intermediate draw ratio. For polypropylene the aggregate model appears appropriate only at low draw ratios, which is consistent with evidence for simultaneous changes in morphology and molecular mobility for highly drawn specimens. The model does not appear appropriate for high-density polyethylene, a highly crystalline material in which a two-phase lamellar texture is often evident (see section 8.4.1).

In particular polymers either the Reuss or the Voigt averages or a mean of the two lie closest to measured values. It is likely that these differences relate to details of the stress and strain distributions at a molecular level, which should in turn be related to the structure.

Kausch, who has applied the aggregate model to a range of both crystalline and amorphous polymers [30], also noted that for some materials the experimental increase in extensional modulus with stiffness was significantly greater than predicted, and has suggested that the effect may be due to an additional orientation of segments within each unit of the aggregate. Another possibility is that the increase in stiffness of crystalline polymers may be enhanced at the higher draw ratios through the competing process of pulling out more intercrystalline tie molecules. Subsequently [31] Kausch emphasized the value of reformulating the model in terms of a molecular network rather than orienting rods, which would allow the representation of high strain properties.

In summary it is evident that despite the highly simplistic nature of the assumptions made the aggregate model provides an appropriate model for a number of important polymers. For such materials details of the crystal structure can play no more than a subsidiary role in the development of mechanical anisotropy, and the deformation is essentially that expected for a single-phase texture or a distorted network.

7.5.3 The Anisotropy of Amorphous Polymers

Relatively few measurements have been performed on amorphous polymers, but some typical data are summarized in Table 7.5 [32, 33] and Figure 7.16 [34].

Table 7.5 Elastic compliances of oriented amorphous polymers (units of compliance are 10^{-9} Pa^{-1})

Material	Draw ratio	s_{33}	s_{11}	s_{44}
Polyvinyl chloride	1	0.313	0.313	0.820
	1.5	0.276	0.319	0.794
	2.0	0.255	0.328	0.781
	2.5	0.243	0.337	0.769
	2.8	0.238	0.341	0.763
	∞	0.204	0.379	0.730
Polymethyl methacrylate	1	0.214	0.214	0.532
	1.5	0.208	0.215	0.524
	2.0	0.204	0.215	0.518
	2.5	0.200	0.216	0.510
	3.0	0.196	0.217	0.505
Polystyrene	1	0.303	0.303	0.769
	2.0	0.296	0.304	0.769
	3.0	0.289	0.305	0.769
Polycarbonate	1	0.376	0.376	1.05
	1.3	0.314	0.408	0.980
	1.6	0.268	0.431	0.926

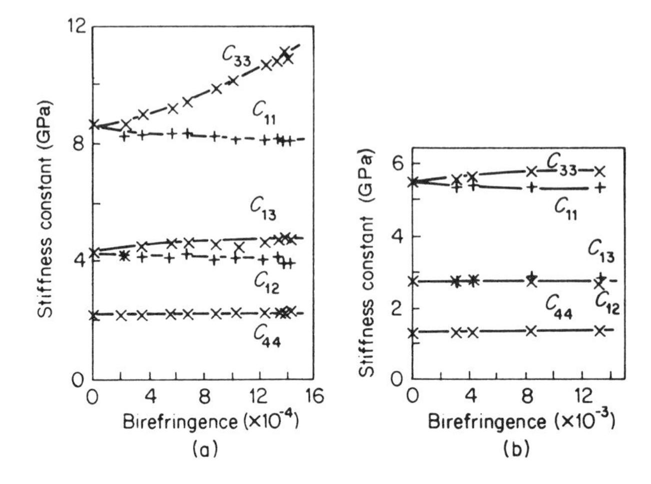

Figure 7.16 Stiffness constants of uniaxially drawn amorphous polymers, measured at room temperature, as a function of birefringence: (a) polymethyl methacrylate, (b) polystyrene. (Reproduced with permission from Wright *et al.*, *J Phys. D*, **4**, 2002 (1971))

The measurements, which were made at widely different frequencies, all indicate, as expected, that the mechanical anisotropy is much smaller than for partially crystalline polymers. Ward and co-workers [35], Kausch [30] and Rawson and Rider [36] have found the aggregate model appropriate for discussing the mechanical anisotropy of non-crystalline polymers, and the development of anisotropy with draw ratio can often be described in terms of pseudo-affine deformation.

7.5.4 Later Applications of the Aggregate Model

All nine independent elastic constants have been measured for polyethylene terephthalate sheet drawn uniaxially at constant width to provide a material with single crystal texture [37]. The high degree of mechanical anisotropy (Table 7.6) is related to the high chain axis orientation and the preferential orientation of the terephthalate residues. The extensional compliance s_{33} is low, presumably because the deformation involves bond bending and stretching in extended chain molecules. By contrast the much higher transverse compliances s_{11} and s_{22} relate principally to dispersion forces. It follows that if the material is stretched perpendicular to the draw direction, the principal contraction occurs in the second perpendicular direction rather than in the draw direction. The shear compliances s_{44} and s_{66} are large compared with s_{55}, reflecting easy shear in the 23 and 12 planes respectively, presumably as the result of planar terephthalate chains sliding over one

Table 7.6 Full set of compliances for an oriented polyethylene terephthalate sheet with orthorhombic symmetry

Compliance	Value ($\times 10^{-10}$ Pa^{-1})
s_{11}	3.61 ± 0.12
s_{22}	9.0 ± 1.6
s_{33}	0.66 ± 0.01
s_{12}	-3.8 ± 0.4
s_{13}	-0.18 ± 0.01
s_{23}	-0.37 ± 0.05
s_{44}	97 ± 3
s_{55}	5.64 ± 0.25
s_{66}	141 ± 8

another constrained only by weak dispersion forces. The shear compliance s_{55} involves distortion of the plane of the poyester molecules, and is of the same order as s_{11} and s_{22}.

It is interesting to apply the aggregate model to these data, calculating bounds for the elastic constants of an 'equivalent fibre' by averaging the sheet constants in the plane normal to the sheet draw direction. This requires an extension of the mathematical treatment of section 7.5.2 to deal with the case of a transversely isotropic aggregate of orthorhombic units. The basic equations are given in detail elsewhere so only the key results will be summarized here. If the orthorhombic unit constants are $s_{11}, s_{13} \ldots, s_{66}$, the Reuss average fibre constants $s'_{33}, s'_{13}, \ldots, s'_{44}$ obtained by averaging in the 12 plane are given by

$$s'_{33} = s_{33},$$

$$s'_{11} = \tfrac{3}{8}s_{11} + \tfrac{1}{4}s_{12} + \tfrac{3}{8}s_{22} + \tfrac{1}{6}s_{66},$$

$$s'_{12} = \tfrac{1}{8}s_{11} + \tfrac{3}{4}s_{12} + \tfrac{1}{8}s_{22} - \tfrac{1}{8}s_{66},$$

$$s'_{13} = \tfrac{1}{2}(s_{13} + s_{23}),$$

$$s'_{44} = \tfrac{1}{2}(s_{44} + s_{55})$$

and the Voigt average fibre constants in similar terms are $c'_{33}, c'_{13}, \ldots, c'_{44}$ where

$$c'_{33} = c_{33},$$

$$c'_{11} = \tfrac{3}{8}c_{11} + \tfrac{1}{4}c_{12} + \tfrac{3}{8}c_{22} + \tfrac{1}{2}c_{66},$$

$$c'_{12} = \tfrac{1}{8}c_{11} + \tfrac{3}{4}c_{12} + \tfrac{1}{8}c_{22} - \tfrac{1}{2}c_{66},$$

$$c'_{13} = \tfrac{1}{2}(c_{13} + c_{23}),$$

$$c'_{44} = \tfrac{1}{2}(c_{44} + c_{55}).$$

Table 7.7 Comparison of calculated and measured compliance constants ($\times 10^{-10}$ Pa^{-1}) for polyethylene terephthalate fibres based on the sheet compliances

Compliance constant	Calculated bounds		Experimental value
	Reuss	Voigt	
Highly oriented fibres			
s_{11}	21	7.3	16.1
s_{12}	−19	−5.5	−5.8
s_{13}	−0.28	−0.25	−0.31
s_{33}	0.66	0.66	0.71
s_{44}	51	10.7	13.6
Isotropic fibres			
s_{33}	18	2.4	4.4
s_{44}	53	6.4	11

The results of this calculation are shown in Table 7.7 together with the experimental value obtained for a highly oriented fibre monofilament. Although the experimental values do not always lie exactly within the predicted bounds, they are always in the correct range. In Table 7.7 a comparison is given between the calculated and measured compliance constants for isotropic polyethylene terephthalate based on the sheet data. Again the measured values lie between the Reuss and Voigt bounds. Taking into account the very large degree of anisotropy and the very simplistic nature of these calculations, these results afford support for the contention that to a first approximation the mechanical anisotropy can be considered in terms of the single-phase aggregate model.

7.5.4.1 Liquid crystalline polymers

The aggregate model has been used with success to describe the mechanical anisotropy of several liquid crystalline polymers. Ward and co-workers [38] examined the dynamic mechanical behaviour of several thermotropic copolyesters in tension and shear over a wide temperature range, and used the single-phase aggregate model to relate quantitatively the fall in tensile modulus with temperature to the corresponding fall in shear modulus. On the compliance averaging scheme (see section 7.5.2) the tensile modulus of the oriented polymer E_3 is given by

$$\frac{1}{E_3} = s'_{33} = s_{11}\overline{\sin^4\theta} + s_{33}\overline{\cos^4\theta} + (2s_{13} + s_{44})\overline{\sin^2\theta\cos^2\theta}. \tag{7.8}$$

As before, θ represents the angle between the unit of the aggregate and the axis representing total alignment, and $\overline{\sin^4\theta}$, etc. represent average values.

For the very high degree of molecular orientation found in these polymers $\overline{\sin^4\theta} \ll 1$ and $\overline{\cos^4\theta} \equiv \overline{\cos^2\theta} \equiv 1$. We can then rewrite (7.8) as

$$\frac{1}{E_3} = s'_{33} = \frac{1}{E_c} + s_{44}\overline{\sin^2\theta}, \tag{7.9}$$

where E_c is the tensile modulus of the aggregate unit. To a similar degree of approximation it can be shown that

$$s'_{44} = s_{44} = \frac{1}{G} \tag{7.10}$$

where G is the shear modulus of the polymer.

Combining (7.9) and (7.10)

$$\frac{1}{E_3} = \frac{1}{E_c} + \frac{\overline{\sin^2\theta}}{G} \tag{7.11}$$

It can be seen from Figure 7.17, obtained from data on a highly oriented thermotropic copolyester, that a plot of $1/E$ versus $1/G$ gives a reasonable straight line extrapolating to a value of 173 GPa for the tensile modulus of a

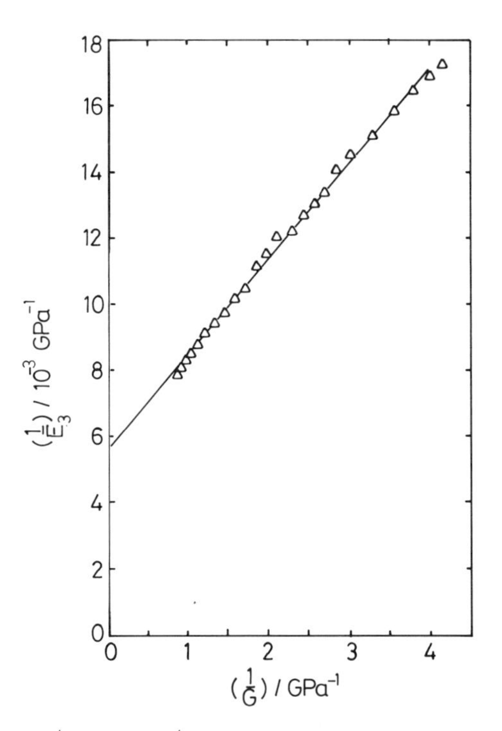

Figure 7.17 Plot of $1/E_3$ versus $1/G$ for a highly oriented thermotropic copolyester

unit of the aggregate. This value compares favourably with theoretical estimates based on bond stretching and bond bending modes of deformation.

In a second version of the aggregate model Ward and his co-workers assumed, on the basis of observation of the X-ray diffraction pattern, that the aggregate unit averages the deformation over a length of 8–10 monomer units. The chain modulus can in this case be determined experimentally by measuring the change in the X-ray diffraction pattern under stress, and a temperature-dependent E_c was observed. By rearrangement of (7.11) a further plot of

$$\left(\frac{1}{E_3} - \frac{1}{E_c}\right) \text{versus } \frac{1}{G}$$

was obtained (Figure 7.18), and gave a good straight line through the origin. The value of $\sin^2\theta$ obtained from the slope of this line is less than that found from the fitting procedure of Figure 7.17, which reflects the misalignment along the length of the liquid crystalline polymer chain. Because of the sinuous nature of the polymer chains the authors have called the effect the 'sinuosity'.

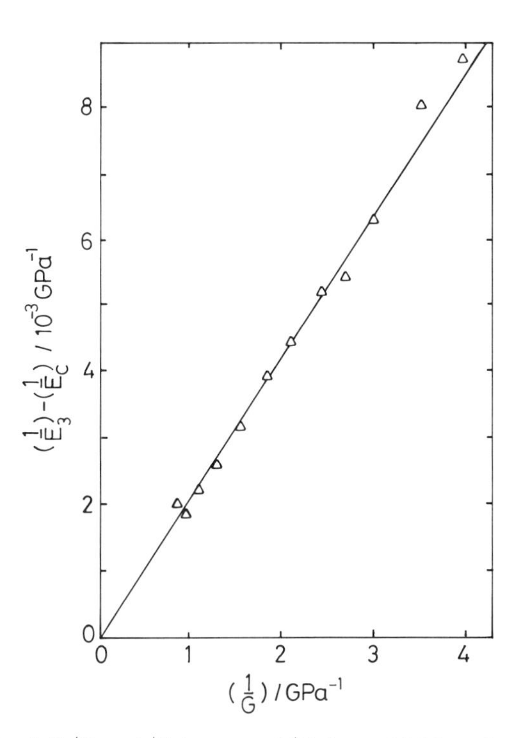

Figure 7.18 Plot of $(1/E_3 - 1/E_c)$ versus $1/G$ for a highly oriented thermotropic copolyester

The aggregate model has also been used by Northolt and Von Aartsen [39] to explain the rapid increase in Young's modulus with increasing crystalline orientation in the lyotropic crystal polymer polyparaphenylene terephthalamide as a consequence of the reduced contribution of shear deformation to the overall deformation. In addition it is interesting to note that the aggregate model has been applied with success to graphite [40].

The above experiments, which suggest that from the viewpoint of mechanical properties a range of liquid crystal polymers can be considered as single phase, considerably extend the scope of a model initially considered most appropriate for polymers with a relatively low degree of crystallinity. It has been suggested by Read and Dean [41] that the aggregate model might be extended further by considering separate contributions from the crystalline and amorphous regions, which are coupled together in a manner dependent on the morphology. As the symmetry of the crystal must be taken into account in all cases, nine rather than five elastic constants would be required, even for transverse isotropy on a macroscopic scale. These proposals are moving away from essence of the aggregate model, which is a single-phase model embodying the assumptions of perfectly elastic behaviour together with no changes of morphology with orientation, and it is appropriate to consider them as a link with models based on the theory of macroscopic composites that are discussed in Chapter 8.

REFERENCES

1. M. A. Biot, in *International Union of Theor. and Appl. Mechs. Colloquium (Madrid)*, Springer-Verlag, Berlin, 1955, p. 251.
2. J. F. Nye, *Physical Properties of Crystals*, Clarendon Press, Oxford, 1957, p. 138.
3. I. M. Ward, *Mechanical Properties of Solid Polymers*, Wiley, Chichester, 1983.
4. C. O. Horgan, *J. Elasticity*, **2**, 169, 335 (1972); *Int. J. Solids Structure*, **10**, 837 (1974).
5. M. J. Folkes and R. G. C. Arridge, *J. Phys. D*, **8**, 1053 (1975).
6. I. Wilson, A. Cunningham, R. A. Duckett and I. M. Ward, *J. Mater. Sci.*, **11**, 2189 (1976).
7. N. H. Ladizesky and I. M. Ward., *J. Macromol. Sci. B*, **5**, 661 (1971).
8. D. Clayton, M. W. Darlington and M. M. Hall., *J. Phys. E.*, **6**, 218 (1973); M. W. Darlington and D. W. Saunders, in *Structure and Properties of Oriented Polymers* (ed. I. M. Ward), Applied Science Publishers, London, 1975, ch. 10.
9. I. Wilson, A. Cunningham and I. M. Ward., *J. Mater. Sci.*, **11**, 2181 (1976).
10. I. D. Richardson and I. M. Ward, *J. Polym. Sci., Polymer Phys.*, **16**, 667 (1978).
11. S. G. Lekhnitskii, *Theory of Elasticity of an Anisotropic Elastic Body*, Holden Day, San Francisco, 1964.
12. N. H. Ladizesky and I. M. Ward., *J. Macromol. Sci. B*, **5**, 759, (1971); **9**, 565 (1974); E. L. V. Lewis and I. M. Ward, *J. Mater. Sci.*, **15**, 2354 (1980).
13. E. L. V. Lewis, I. D. Richardson and I. M. Ward, *J. Phys. E.*, **12**, 189 (1979).

14. D. W. Hadley, I. M. Ward and J. Ward., *Proc. Roy. Soc.*, **A285**, 275 (1965).
15. P. R. Pinnock, I. M. Ward and J. M. Wolfe, *Proc. Roy. Soc.*, **A291**, 267 (1966).
16. S Timoshenko and J. N. Goodier, *Theory of Elasticity*, McGraw-Hill, New York, 1951.
17. G. Raumann and D. W. Saunders, *Proc. Phys. Soc.*, **77**, 1028 (1961).
18. V. B. Gupta and I. M. Ward., *J. Macromol. Sci. B*, **1**, 373 (1967).
19. D. W. Hadley, P. R. Pinnock and I. M. Ward., *J. Mater. Sci.*, **4**, 152 (1969).
20. K. Tashiro, M. Kobayashi and H. Tadokoro, *Macromolecules*, **11**, 914 (1978).
21. I. Sakurada, T. Ito and K. Nakamae, *J. Polymer Sci. C*, **15**, 75 (1966); R. F. Shaufele and T. Shimanouchi, *J. Chem. Phys.*, **47**, 3605 (1967); L. A. Feldkamp, G. Venkateraman and J. S. King, in *Neutron Inelastic Scattering*, Vol. 2, IAEA, Vienna, 1968, p. 159.
22. A. Peterlin, in *Ultra-High Modulus Polymers*, (A. Ciferri and I. M. Ward eds), Applied Science Publishers, London, 1979, ch. 10.
23. I. M. Ward, *Proc. Phys. Soc.*, **80**, 1176 (1962).
24. J. Bishop and R. Hill, *Phil. Mag.*, **42**, 414, 1248 (1951).
25. A. Reuss, *Zeit. Angew. Math. Mech.*, **9**, 49 (1929).
26. W. Voigt, *Lehrbuch des Krystallphysik*, Teubner, Leipzig, 1928, p. 410.
27. S. M. Crawford and H. Kolsky, *Proc. Phys. Soc. B.*, **64**, 119 (1951).
28. F. C. Frank, V. B. Gupta and I. M. Ward., *Phil. Mag.*, **21**, 1127 (1970).
29. S. W. Allison and I. M. Ward, Brit. *J. Appl. Phys.*, **18**, 1151 (1967).
30. H. H. Kausch, *J. Appl. Phys.*, **38**, 4213 (1967); *Kolloidzeitschrift.*, **234**, 1148 (1969); 237, 251 (1970); *Polymer Fracture*, Springer, Berlin, 1978, p. 33.
31. H. H. Kausch, *J. Macromol. Sci. B.*, **5**, 269 (1971).
32. J. Hennig, *Kolloidzeitschrift*, **200**, 46 (1964).
33. R. E. Robertson and R. J. Buenker, *J. Polymer Sci. A2*, **2**, 4889 (1964).
34. H. Wright, C. S. N. Faraday, E. F. T. White and L. R. G. Treloar., *J. Phys. D.*, **4**, 2002 (1971).
35. M. Kashiwagi, M. J. Folkes and I. M. Ward., *Polymer*, **12**, 697 (1971).
36. F. F. Rawson and J. G. Rider, *J. Phys. D.*, **7**, 41 (1974).
37. E. L. V. Lewis and I. M. Ward., *J. Mater. Sci.*, **15**, 2354 (1980).
38. G. R. Davies and I. M. Ward, in *High Modulus Polymers* (eds. A. E. Zachariades and R. S. Porter), Marcel Dekker, New York, 1988, ch. 2; M. J. Troughton, G. R. Davies and I. M. Ward., *Polymer*, **30**, 58 (1989). D. I. Green, A. P. Unwin, G. R. Davies and I. M. Ward., *Polymer*, **31**, 579 (1990).
39. M. G. Northolt and J. J. Van Aartsen, *J. Polymer Sci. C,* **58**, 283 (1978).
40. P. R. Goggin and W. N. Reynolds, *Phil. Mag.*, **16**, 317 (1967).
41. B. E. Read and G. Dean, *Polymer*, **11**, 597 (1970).

Chapter 8

Polymer Composites: Macroscale and Microscale

In this chapter we first give a brief survey of the advantages to be gained by using a composite material, whose components often have contrasting but complementary properties, e.g. ductile fibres reinforcing a brittle matrix. We then discuss two distinct applications of these general principles, macroscopic composites, composed of a polymeric matrix in which a second component is embedded, and microscale composites used to model the morphology of partially crystalline polymers.

8.1 COMPOSITES: A GENERAL INTRODUCTION

Many useful engineering materials have a heterogeneous composition. Metals for instance, are often used in the form of alloys. The addition of a small percentage of another metal, such as copper, magnesium or manganese, is necessary to prevent plastic deformation occurring in aluminium at very low stresses. An increase in carbon content from 0.1% to 3% is a primary determinant in whether a ferrous alloy becomes a mild steel or a cast iron. Concrete, which like cast iron, has good compressive but poor tensile properties, consists of a hard aggregate embedded in a metal silicate network.

Both animal and vegetable life are dependent on natural composites. Bones must be stiff and yet able to absorb significant amounts of energy without fracturing; they also provide anchor points for muscles, which are composite. The skeletal material of plants, and in particular wood, provides a splendid example of the desirable properties of a composite. As a gross simplification, its structure can be considered in terms of an array of relatively stiff fibres embedded in a more compliant matrix. The matrix permits stresses to be redistributed among the fibres, so retarding the onset of fracture at stress concentrations. Wood fails in compression when its fibres buckle. The fracture stress is higher in tension, as a large amount of work must be done in pulling the fibres out of the matrix.

Reinforced concrete has practically a century of use as a building material. Continuous steel rods, prestressed under tension, pass completely through each structural element, and enable the material to withstand tensile as well as compressive stresses, to give a combination that combines the desirable features of each component.

A further form of composite is one where the second component acts as a filler. Carbon black in vehicle tyres is an example of a filler needed to provide the required properties. Each carbon particle provides an anchorage for many rubber molecules, and so assists in the redistribution of stress; and the carbon is also essential to obtain the desired hysteresis behaviour and abrasion resistance. A much simpler application of a filler is the use of sawdust or other cheap powder in mouldings made from a thermosetting plastic. Although the mechanical properties of the base material are degraded (except possibly for impact resistance), they are still adequate for the proposed application, and the product cost is reduced. We shall not consider fillers in the discussion that follows.

Another desirable property of a composite that will not be considered further is the protection that a compliant matrix affords to a brittle reinforcing fibre. Glass and other brittle materials fracture in tension due to the deepening of pre-existing cracks. Because of the absence of plastic flow (unlike the situation for polymers discussed in Chapter 11), blunting of the crack tip cannot occur, and so the stress rapidly approaches that required for fracture. If glass fibres are encapsulated in a soft plastic matrix, the possibility of surface scratches is reduced and the fracture stress is thereby increased. Good adhesion between fibre and matrix will assist in reducing stress concentrations, and transverse cracks will grow only with difficulty across a fibrous composite.

8.2 THE YOUNG'S MODULUS FOR LAMELLAR COMPOSITES

Here we model an idealized composite that consists of alternating layers of a high modulus reinforcement and a more compliant and ductile matrix. Provided that the bonding between layers remains intact, the volume fraction of each component and not the thickness of individual layers is the important feature. As with the model aggregate discussed in section 7.5, different values of overall stiffness are obtained when the components are in parallel and in series, yielding Voigt and Reuss average moduli respectively.

The maximum stiffness is obtained when a uniaxial stress is applied parallel with the layers, as indicated in Figure 8.1. It is assumed that the strain is the same in all the composite layers, a form of loading known as the isostrain (or homogeneous strain) condition.

The force acting on the composite (F_c) is equal to the sum of the forces acting on the fibre and matrix layers.

$$F_c = F_f + F_m. \tag{8.1}$$

Force is equal to stress times area. Hence

$$\sigma_c A_c = \sigma_f A_f + \sigma_m A_m,$$

where A_f and A_m represents the areas of the end faces occupied by each component. As both components are of length l, areas can be represented by volumes, or rather by volume fractions V_f and V_m. The volume fraction of the composite (V_c) is unity. Hence

$$\sigma_c V_c = \sigma_f V_f + \sigma_m V_m. \tag{8.2}$$

Under isostrain conditions this expression can be rewritten in terms of Young's modulus (E) as

$$E_c = E_f V_f + E_m V_m \tag{8.3}$$

which is a Voigt average modulus (see section 7.5.1).

The modulus is, however, much lower in the direction transverse to the layered structure (Figure 8.2). In this case each layer is subjected to the same force, and hence to the same stress, as the area remains constant through the stack. Loading of this form is known as the isostress (or homogeneous stress) condition.

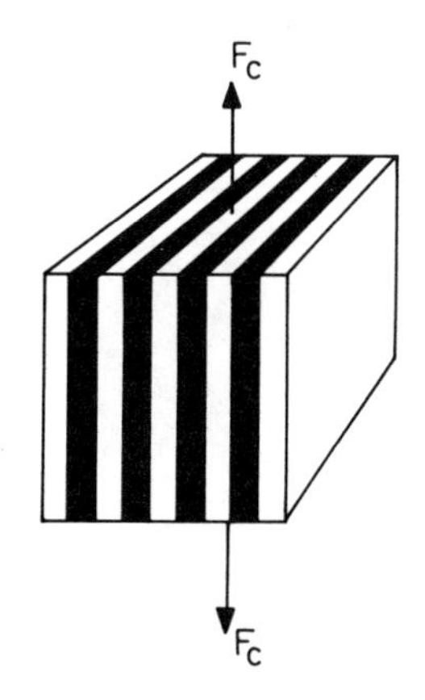

Figure 8.1 Isostrain condition for layered composite

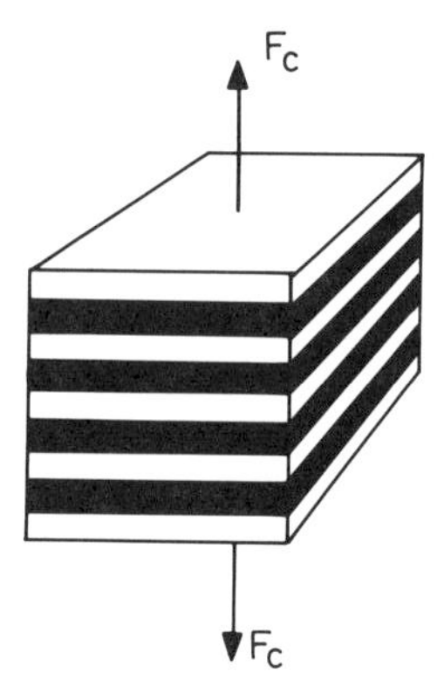

Figure 8.2 Isostress condition for layered composite

The total deformation δl_c is equal to the sum of the deformations in each component:

$$\delta l_c = \delta l_f + \delta l_m.$$

Length changes can be converted to strains, using $\varepsilon = \delta l/l$:

$$\varepsilon_c l_c = \varepsilon_f l_f + \varepsilon_m l_m \tag{8.4}$$

substituting modulus E as the ratio of (uniform) stress to strain, we obtain

$$\frac{\sigma l_c}{E_c} = \frac{\sigma l_f}{E_f} + \frac{\sigma l_m}{E_m}$$

As the cross-section of the composite is assumed to be uniform, the length of a component is proportional to its volume fraction. Again take V_c as unity to give

$$\frac{1}{E_c} = \frac{V_f}{E_f} + \frac{V_f}{E_m} \tag{8.5}$$

This expression can be rewritten as

$$E_c = \frac{E_f E_m}{E_m V_f + E_f V_m} \tag{8.6}$$

and is a Reuss average modulus (see section 7.5.1).

A further very important example of a composite material is a fibre reinforced composite. A simple theory proposed by Brody and Ward [1] extended the assumptions of equations (8.3) and (8.5) to this system, and calculated the elastic constants for a unidirectional fibre composite (i.e. a composite which is assumed to consist of perfectly aligned fibres of infinite length).

Choosing the fibre direction as the 3 axis, the assumptions of homogeneous strain in the 3 direction and homogeneous stress in the 1 direction imply, in terms of the nomenclature proposed in section 7.1.1, that the five independent elastic constants for the composite, E_1^c, E_3^c, ν_{13}^c, ν_{12}^c and G_4^c are given by

$$\frac{1}{E_1^c} = \frac{V_f}{E_1^f} + \frac{V_m}{E^m} \tag{8.7a}$$

$$E_3^c = V_f E_3^f + V_m E^m \tag{8.7b}$$

$$\nu_{12}^c = \frac{V_f \nu_{12}^f E^m + V_m \nu^m E_1^f}{V_f E^m + V_m E_1^f} \tag{8.7c}$$

$$\nu_{13}^c = V_f \nu_{13}^f + V_m \nu^m \tag{8.7d}$$

and

$$\frac{1}{G_4^c} = \frac{V_f}{G_4^f} + \frac{V_m}{G^m} \tag{8.7e}$$

where V_f, V_m are the fibre and matrix volume fractions respectively, E_1^f, E_3^f, ν_{13}^f, ν_{12}^f and G_4^f the fibre elastic constants and E^m, G^m and ν^m the elastic constants of the isotropic matrix. Recently, more vigorous calculations have been undertaken for the Reuss and Voigt averages (which provide lower and upper bounds respectively for the elastic constants) based on the self-consistent field approach of Eshelby [2–4]. Exact analytical expressions for all the elastic constants have also been obtained by Wilczynski [5], and confirmed by finite element calculations [6].

Consider as an example a glass-fibre/resin composite containing 0.6 volume fraction of glass filaments with a tensile modulus of 70 GPa, and 0.4 volume fraction of a resin with a modulus of 3 GPa. On the simplest assumptions of equations (8.7) above, the predicted parallel (isostrain) modulus of the composite will be 43.2 GPa, but the predicted transverse (isostress) modulus will be 7.0 GPa. The composite thus has a highly anisotropic stiffness, which decreases very rapidly at quite small angles from the parallel direction. The variation of Young's modulus with fibre content alters in quite different ways for isostrain and isostress conditions, as is indicated in Figure 8.3, which is plotted for the filament and resin indicated above.

It is interesting to consider the mode of failure under isostrain conditions for a typical continuous filament composite, in which the fracture strain for the reinforcing filaments is considerably lower than that for the more

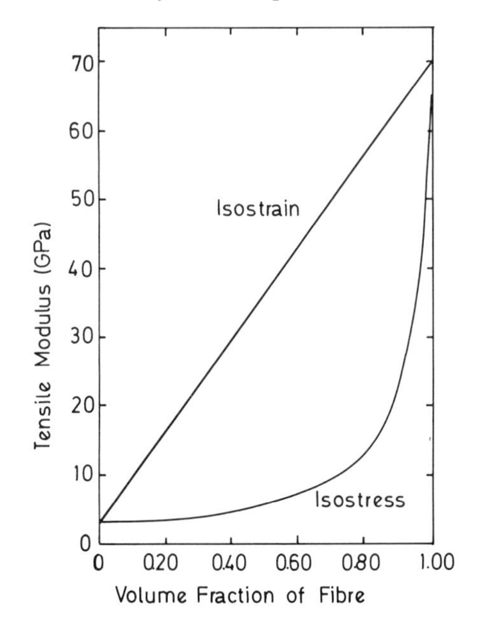

Figure 8.3 Variation of tensile modulus with fibre content

compliant matrix. For the case of a very small volume fraction of filaments the overall tensile modulus will be low, and so the extension will be large for comparatively low values of σ_c. The filaments will fracture when their breaking strain is reached, but the matrix will initially be able to carry the total load, and final failure is due to failure of the matrix. The filaments do not contribute to the ultimate properties, so that the failure stress can be represented by

$$\sigma_c{}^* = \sigma_m{}^* V_m, \tag{8.8}$$

where $\sigma_m{}^*$ is the fracture stress for the matrix. At higher V_f the stress which results in the filaments breaking is so high that the matrix is unable to sustain the total load; consequently the composite fails as a whole at a stress corresponding to

$$\sigma_c{}^* = \sigma_f{}^* V_f + \sigma_m V_m, \tag{8.9}$$

where σ_m is the stress in the matrix at the fracture strain of the composite. The variation of failure stress with composition is then given by the envelope of the straight lines defined by equations (8.8) and (8.9). It can be seen from Figure 8.4 that for a low fraction of filaments the tensile stress at failure is below that for the matrix alone.

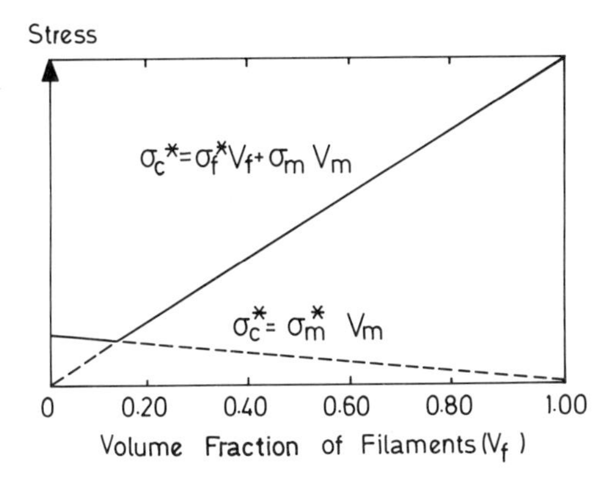

Figure 8.4 The variation of fracture stress with filament content for a continuous filament composite in which the fracture strain for the filaments is lower that that for the matrix

8.3 DISCONTINUOUS FILAMENT COMPOSITES

Although continuous filament composites are of considerable commercial importance, their fabrication is rather a complex process, and a cheaper though mechanically inferior product is obtained by mixing short lengths of fibre with a thermoplastic polymer.

A prime requirement is good adhesion between fibre and matrix, a condition that will be dependent on features such as chemical bonding and surface cleanliness as well as mechanical factors. The ratio of surface area to fibre volume should be as high as possible. Considering cylinders of length l and radius r:

$$\text{Area } A = 2\pi r^2 + 2\pi rl \quad \text{and} \quad \text{volume } V = \pi r^2 l$$

$$\therefore \quad \frac{A}{V} = \frac{2}{l} + \frac{2}{r} \tag{8.10}$$

Expressed in terms of the aspect ratio $(a = \frac{1}{2}r)$ the above expression becomes

$$\frac{A}{V} + \sqrt[3]{\frac{2\pi}{V}}(a^{-2/3} + a^{1/3}). \tag{8.11}$$

It can be seen, therefore, that for optimum adhesion the aspect ratio should be either very small, where $a^{-2/3}$ becomes very large, corresponding to flat platelets (minerals such as talc and mica have been used) or very high, where $a^{1/3}$ is very large, corresponding to fibres. The latter case will now be examined in more detail.

8.3.1 The Influence of Fibre Length

Consider a short length of fibre aligned with the tensile stress direction. The stiff fibre will tend to restrain the deformation of the matrix, and so a shear stress will be set up in the matrix at its interface with the fibre, which will be a maximum at the ends of the fibre and a minimum in the middle (Figure 8.5(a)). This shear stress then transmits a tensile stress to the fibre, but as the fibre–matrix bond ceases at the fibre ends there can be no load transmitted from the matrix at each fibre extremity. The tensile stress is thus zero at each end of the fibre and rises to an intermediate maximum or plateau over a critical length $l_0/2$ (Figure 8.5(b)). For effective reinforcement the fibre length must be greater than the critical value l_0, otherwise the stress will be less than the maximum possible.

The reduction in tensile stress towards the ends of each fibre inevitably leads to a decrease in the tensile modulus compared with the continuous filament case. Consider a plane drawn perpendicular to the stress direction in an aligned discontinuous fibre composite (Figure 8.6), which must intercept individual filaments at random positions along their length. Hence the stress carried by the composite must be lower than that for the continuous filament case, and is dependent on the length of each fibre. Cox [7] predicted a correction factor η_l, for the tensile modulus in the axial

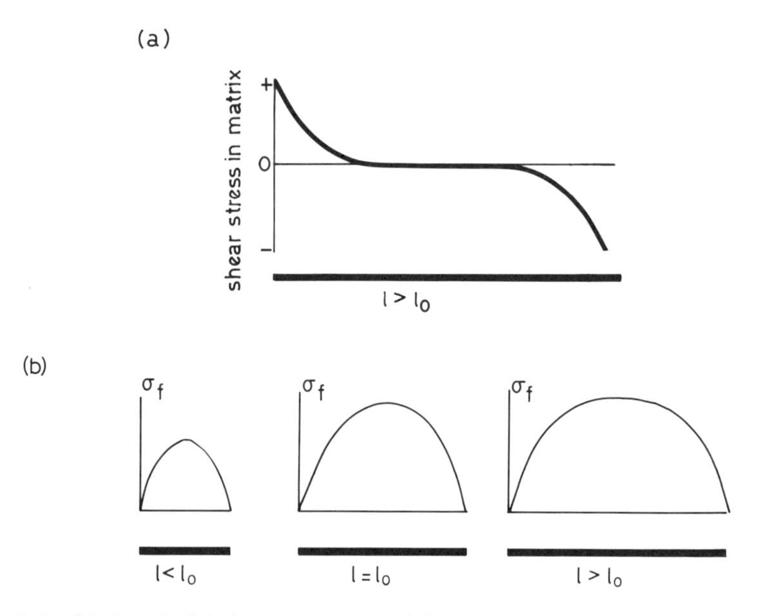

Figure 8.5 (a) Interfacial shear stress, and (b) fibre tensile stress as a function of fibre length (schematic)

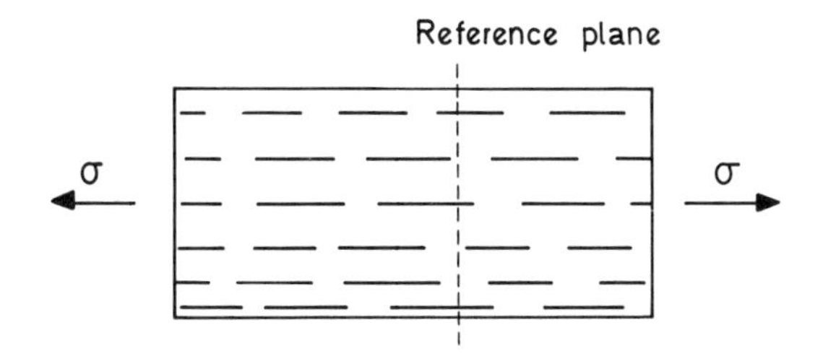

Figure 8.6 Schematic section through a discontinuous fibre composite (fibre fraction shown very low for clarity)

direction which takes into account the finite length of the fibres so that equation (8.3) is modified to

$$E = \eta_l E_f V_f + E_m V_m,$$

where

$$\eta_l = 1 - \left[\frac{\tanh ax}{ax}\right] \tag{8.12}$$

where a is the aspect ratio $l/2r$ and x is a dimensionless factor,

$$x = \left[\frac{2G_m}{E_f \ln(R/r)}\right]^{1/2} \tag{8.13}$$

where G_m is the shear modulus of the matrix and R is half the separation between the nearest fibres. The basis of this expression is the assumption that the deformation of the complete composite can be modelled by considering a fibre to be embedded in a cylinder of matrix of radius R. For a fuller discussion see for example the text by Kelly [8].

The factor x (8.13) depends on two key ratios: G_m/E_f, which is typically 0.01–0.02, and R/r which is not much greater than unity. Figure 8.7 indicates that the length correction factor becomes significant for values of ax less than 10. In practice the corresponding aspect ratio for effective reinforcement is usually greater than 100.

8.3.2 Debonding and Pull-out

A composite frequently fails as a result of debonding between fibre and matrix. New interfaces are created, which involve the expenditure of energy, as is discussed in Chapter 11. The basic process can be modelled by considering a single fibre embedded in a block of matrix to a depth x, and it can be shown that the debonding energy is a maximum when x is equal to half the critical length. If the embedded length is less than $l_0/2$ the fibre will

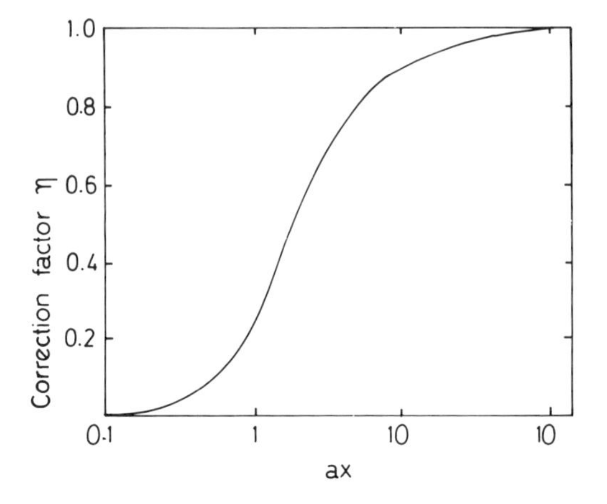

Figure 8.7 Correction factor for the tensile modulus of short fibres as a function of aspect ratio

be pulled out of the matrix rather than fracturing, so involving the expenditure of further energy.

A stress–strain plot can be derived from a tensile load–extension experiment, as depicted schematically in Figure 8.8. The energy of debonding is obtained from the area OAB, and the usually larger pull-out energy is associated with the area OBCD. A strong fibre, which has not fractured after some debonding has occurred, will bridge the newly formed surfaces in the wake of a propagating crack, and thereby hinder crack opening. This toughening process has a microscopic analogy in the role of extended-chain bridging filaments in semicrystalline polymers.

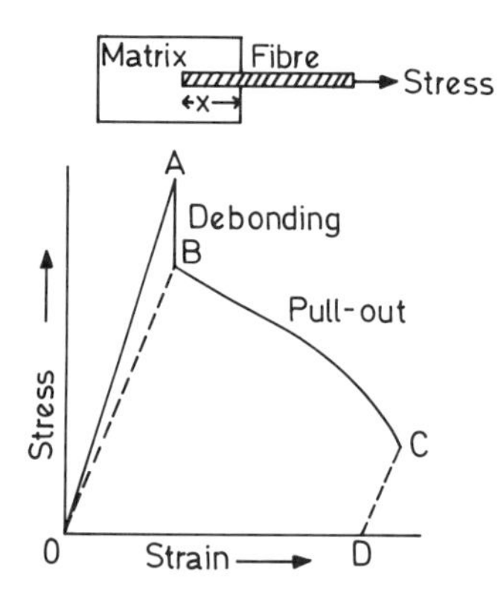

Figure 8.8 Pull-out test and the resulting stress–strain curve showing the difference in magnitude of the energies of debonding (area OAB) and pull-out (area OBCD). (From Anderson et al., *Materials Science*, 4th ed, Chapman and Hall, London, 1990, Ch. 11)

8.3.3 Practical Considerations

Discontinuous fibre composites are normally extruded or moulded, using conventional processing equipment, and so in practice the distribution of the fibres will be either random or no more than partially aligned along flow lines, with the result that the stiffness of the composite will be reduced still further compared with that for a continuous filament material in the isostrain situation. Partially aligned fibre composites have been modelled by Ward and co-workers [1, 6] using the aggregate model of section 7.5.2, together with the elastic constants for a fully aligned fibre composite using the methods of section 8.2. Additionally, in order that the molten material shall have suitable flow properties the volume fraction of fibres is usually less than 33%. Nevertheless, as seen in Figures 8.9 and 8.10, both tensile modulus and fracture stress can be several times greater than for the polymer component alone.

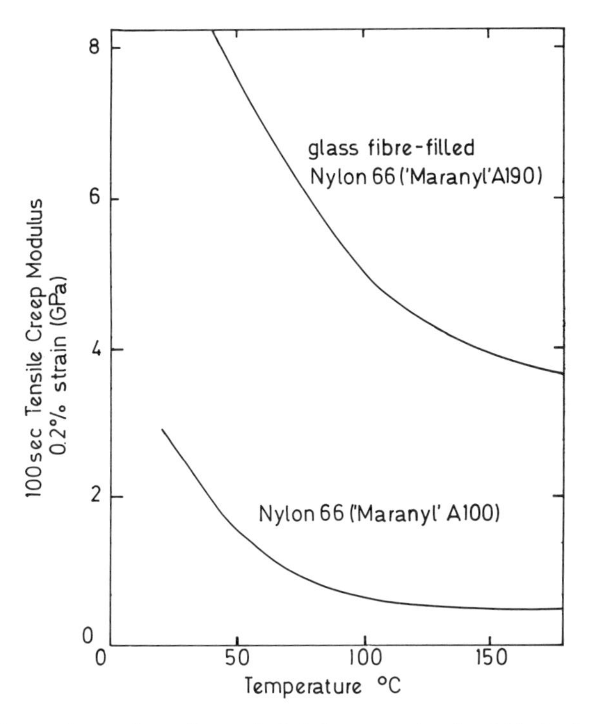

Figure 8.9 Comparison of tensile moduli (after creeping for 100 s) of nylon 6:6 and a similar material filled with 33% (mass) of glass fibre. (R. M. Ogorkiewicz, *Engineering Properties of Thermoplastics,* Wiley-Interscience, 1970)

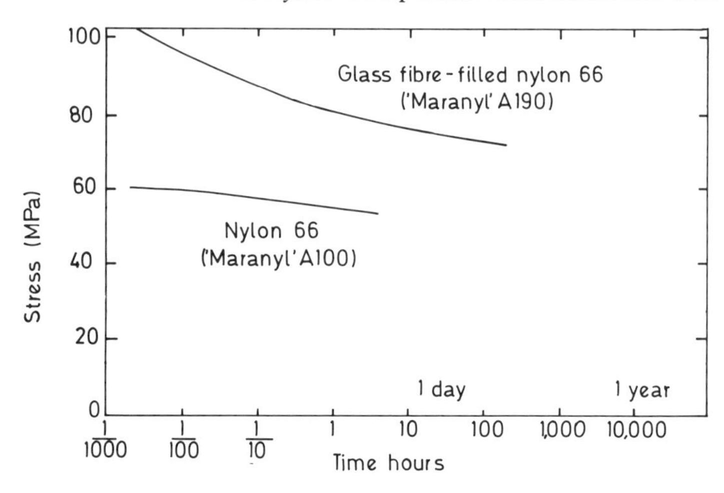

Figure 8.10 The rupture stress in tensile creep of nylon 6:6 and a similar material filled with 33% (mass) of glass fibre. (R. M. Ogorkiewicz, *Engineering Properties of Thermoplastics*, Wiley-Interscience, 1970)

8.4 TAKAYANAGI MODELS FOR SEMICRYSTALLINE POLYMERS

It was realized by Takayanagi [9] that oriented highly crystalline polymers with a clear lamellar texture might be modelled in terms of a two-component composite, in which the alternating layers corresponded to the crystalline and amorphous phases [3]. The model was later extended to include a parallel component in addition to that in series, and was applied first to describe the relaxation behaviour of amorphous polymers with two distinct phases, and later to crystalline polymers, in which the parallel component represented either interlamellar crystalline bridges or amorphous tie molecules threading through the amorphous phase.

8.4.1 The Simple Takayanagi Model

High-density polyethylene that has been uniaxially drawn (i.e. with fibre symmetry) and then annealed has a distinct lamellar texture, and we shall show that the orientation of the lamellae, as distinct from the molecular orientation, plays the dominant role in determining the mechanical anisotropy. The temperature variation of the in-phase (E_1) and out-of-phase (E_2) components of the dynamic modulus, both parallel with ($\|$) and perpendicular to ($\perp$) the draw direction is shown in Figure 8.11, with the parallel modulus ($E_0 = \|$) crossing the perpendicular modulus ($E_{90} = \perp$)

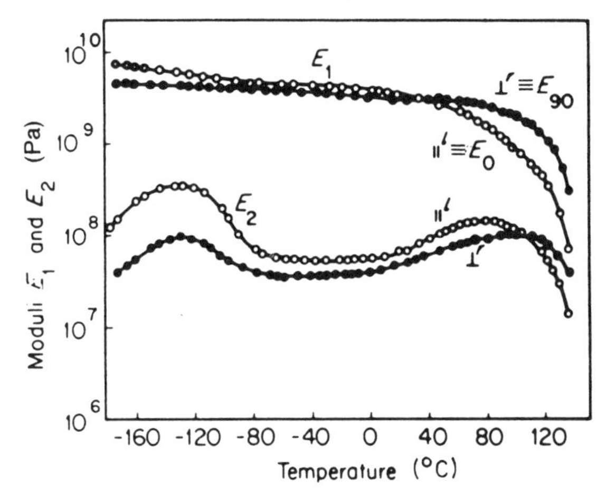

Figure 8.11 Temperature dependence of E_1 and E_2, the components of the dynamic modulus, in direction parallel ($\|'$) and perpendicular ($\perp'$) to the initial draw direction for annealed samples of high-density polyethylene. (Reproduced with permission from Takayanagi, Imada and Kajiyama, *J. Polymer Sci. C,* **15**, 263 (1966)

at high temperatures; at low temperatures $E_0 > E_{90}$, but at high temperatures $E_{90} > E_0$. The principal features can be explained in terms of a simple model of amorphous (A) and crystalline (C) components, which are in series in the draw direction, but in parallel in the transverse direction (Figure 8.12). In the orientation direction each component is subject to the same stress, so that compliances are added as in the Reuss averaging scheme. The stiffness above the relaxation transition is primarily determined by the compliant amorphous regions (equation 8.6), so giving a large fall in modulus as the temperature is increased. In the perpendicular direction the parallel components are each subject to the same strain, and stiffnesses are added as in the Voigt scheme. The crystalline regions support the applied stress at temperatures above the relaxation transition, and so a comparatively high stiffness is maintained. Takayanagi and his colleagues considered that appropriate values of modulus might be $E_c(\|) = 100$ GPa, $E_c(\perp) = 1$ GPa, E_A (low T) = 1 GPa, E_A (high T) = 0.01 GPa.

8.4.2 Takayanagi Models for Dispersed Phases

Takayanagi [9] devised series-parallel and parallel-series models as an aid to understanding the viscoelastic behaviour of a blend of two isotropic amorphous polymers in terms of the properties of the individual components. For an A phase dispersed in a B phase there are two extreme possibilities for the stress transfer. For efficient stress transfer perpendicular to the direction of tensile stress we have the series-parallel model (Figure

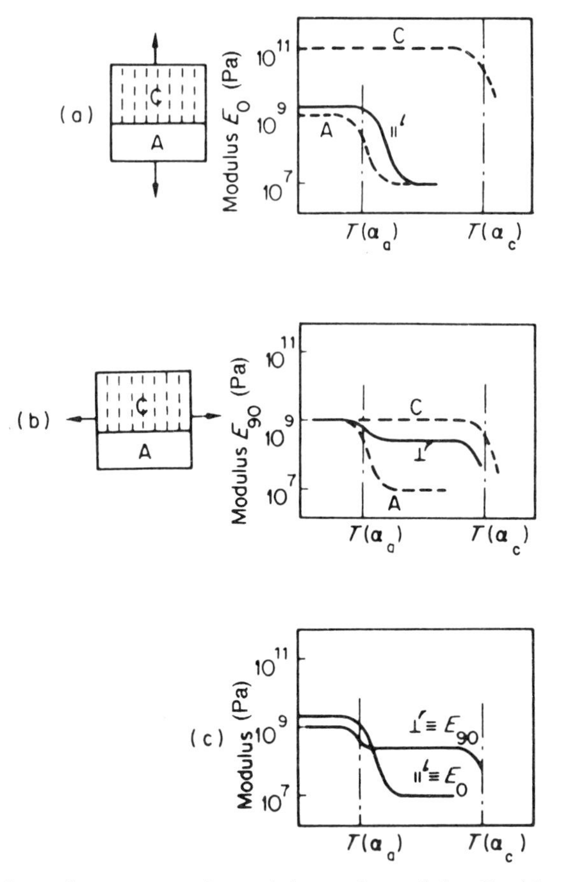

Figure 8.12 Schematic representations of change in modulus E with temperature on the Takayanagi model for (a) the $\|'$, and (b) the $\perp'$ situations corresponding to E_0 and E_{90} respectively. Calculations assume amorphous relaxation at temperature $T(\alpha_a)$ and crystalline relaxation at temperature $T(\alpha_c)$ and (c) shows combined results. C, crystalline phase; A, amorphous phase. (Reproduced with permission from Takayanagi, Imada and Kajiyama, *J. Polymer Sci. C*, **15**, 263 (1966)

8.13(a)) in which the overall modulus is given by the contribution for the two lower components in parallel (as in equation 8.3) in series with the contribution for the upper component (as in equation 8.5):

$$\frac{1}{E^*} = \frac{\phi}{\lambda E_A^* + (1 - \lambda)E_B^*} + \frac{1 - \phi}{E_B^*} \tag{8.14}$$

where E^*, etc. represent the complex moduli associated with dynamic experiments. If the stress transfer across planes containing the tensile stress is weak, a parallel-series model (Figure 8.13(b)) is appropriate, for which

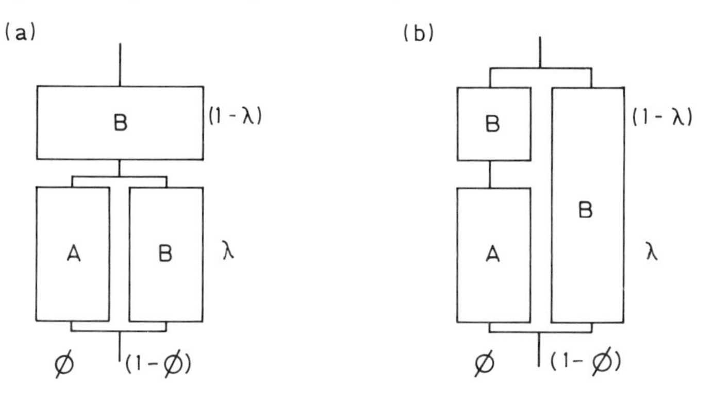

Figure 8.13 Takayanagi models for polymer blends: (a) the series-parallel model; (b) the parallel-series model

the two left-hand components combine in series (as in equation 8.5) before combining in parallel with the right-hand component (equation 8.3), giving a modulus

$$E^* = \lambda\left(\frac{\phi}{E^*_{\mathrm{A}}} + \frac{1-\phi}{E^*_{\mathrm{B}}}\right)^{-1} + (1-\lambda)E^*_B \qquad (8.15)$$

Note that $\lambda\phi$ corresponds to the volume fraction V_{f} in the earlier notation.

The predictions of both models were then compared with measurements of the temperature variation of storage and loss moduli for a film made from a blend of polyvinyl chloride and nitrile butadiene rubber (Figure 8.14). It is seen that the series-parallel model (a) gives the better fit. The performance of polymer blends was well represented by a series-parallel model in which the relative values of λ and ϕ were related to the shape of the dispersed phase: $\lambda = \phi$ for homogeneous dispersions, and $\lambda \geqslant \phi$ for dispersions in the form of elongated aggregates. For semicrystalline polymers in general, however, with A and B representing the crystalline and amorphous components, experimental dispersions were usually broader than the predictions, suggesting that at least some of the unordered material was not identical with that in a completely amorphous state.

Gray and McCrum [10] have criticized the Takayanagi model as applied to partially crystalline polymers, and refute the assertion that if mechanical relaxation occurs in the amorphous phase, the peak value of the out-of-phase modulus is proportional to the volume fraction of the amorphous phase. They assert that as the model represents a Voigt average solution it can give only upper bounds to moduli. Stress and strain fields must differ between the crystalline and amorphous components, so a Reuss-type average is equally inadmissible, and a correct solution must lie between the

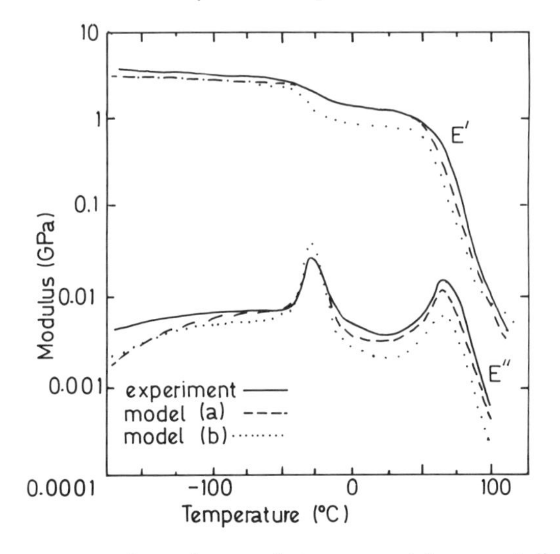

Figure 8.14 Temperature dependence of storage and loss moduli for PVC–NBR film bonded in parallel to a PVC film. Takayanagi model type (a) gives better fit to experiment. (Reproduced with permission from Takayanagi, *Mems. Fac. Eng. Kyushu Univ.*, **23**, 41 (1963))

two limits. An empirical logarithmic mixing hypothesis is advanced as an acceptable law of mixing

$$\log G^* = V_A \log G_A^* + V_c \log G_c^* \qquad (8.16)$$

where G^* represents the complex shear modulus and V represents a volume fraction. The logarithmic decrement of the polymer (Λ) is then given as a weighted combination of the logarithmic decrement of the two phases:

$$\Lambda = V_A \Lambda_A + V_c \Lambda_c \qquad (8.17)$$

which is a mathematical statement of the assumptions of some earlier workers.

8.4.3 Modelling Polymers with Single Crystal Texture

Rolling and annealing processes established by Hay and Keller [11] enable sheets of low-density polyethylene to be produced with well-defined crystallographic and lamellar orientations (Figure 8.15). Ward and his co-workers [12] studied the behaviour of three types of special structure sheet illustrated in Figure 8.16: '*bc* sheet', in which the *c* axes of the crystallites lie along the initial draw direction, the *b* axes lie in the plane of the sheet, and the *a* axes are normal to the plane of the sheet; '*ab* sheet', in which the *a* axes lie along the draw direction, the *b* axes again lie in the

Figure 8.15 Model of morphology of oriented and annealed sheets of low-density polyethylene. This photograph shows the structure of *b-c* sheet. a, *b* and *c* axes indicate the crystallographic directions in the crystalline regions. (Reproduced with permission from Stachurski and Ward, *J. Polymer Sci. A2*, **6**, 1817 (1968))

plane of the sheet, and the *c* axes are normal to the sheet; and 'parallel lamellae sheet', where the lamellar plane normals lie along the initial draw direction, and the *c* axes make an angle of about 45° with this direction. For specimens of *bc* and *ab* sheet a four-point small angle X-ray diffraction pattern is shown, which is interpreted as indicating the presence of lamellae inclined at about 45° to the direction of the *c* axes. This type of morphology is represented schematically for the *bc* sheet by the model shown in Figure 8.15, in which the solid blocks represent crystalline lamellae, and the intermediate spaces are occupied by disordered material and interlamellar tie molecules, which are relaxed as a result of the annealing treatment. In contrast, the parallel lamellae sheet has a twinned structure with regard to the crystallographic orientation, but a single texture structure (only a two-point low-angle diffraction pattern) as far as lamellar orientation is concerned.

In contrast with the Takayanagi model, which considers only extensional strains, a major deformation process involves *shear* in the amorphous

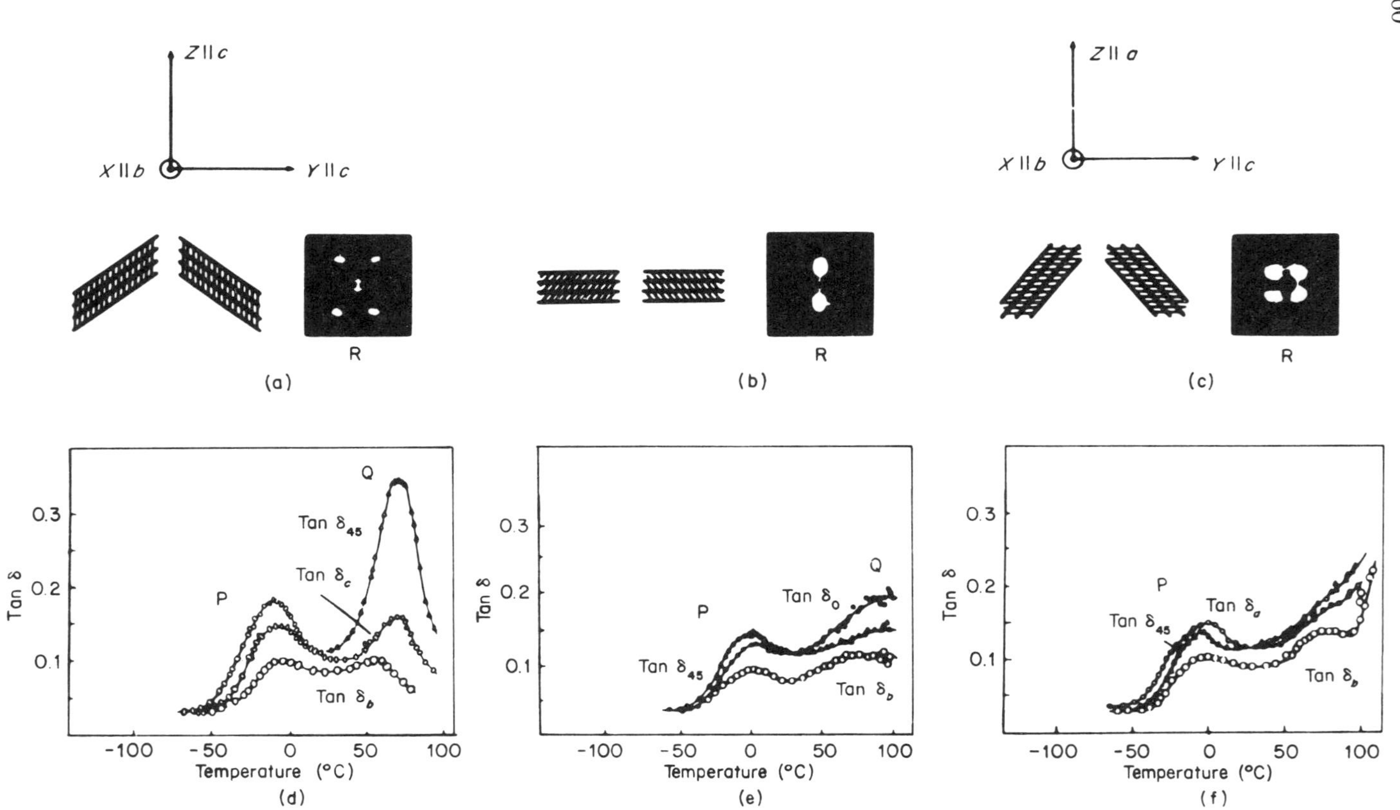

Z ll c
X ll b
Y ll c
R
(a)
R
(b)
Z ll a
X ll b
Y ll c
R
(c)
Q
Tan δ_{45}
Tan δ_c
P
Tan δ_b
Tan δ
0.3
0.2
0.1
-100
-50
0
50
100
Temperature (°C)
(d)
Q
Tan δ_0
P
Tan δ_{45}
Tan δ_b
(e)
P
Tan δ_a
Tan δ_{45}
Tan δ_b
(f)

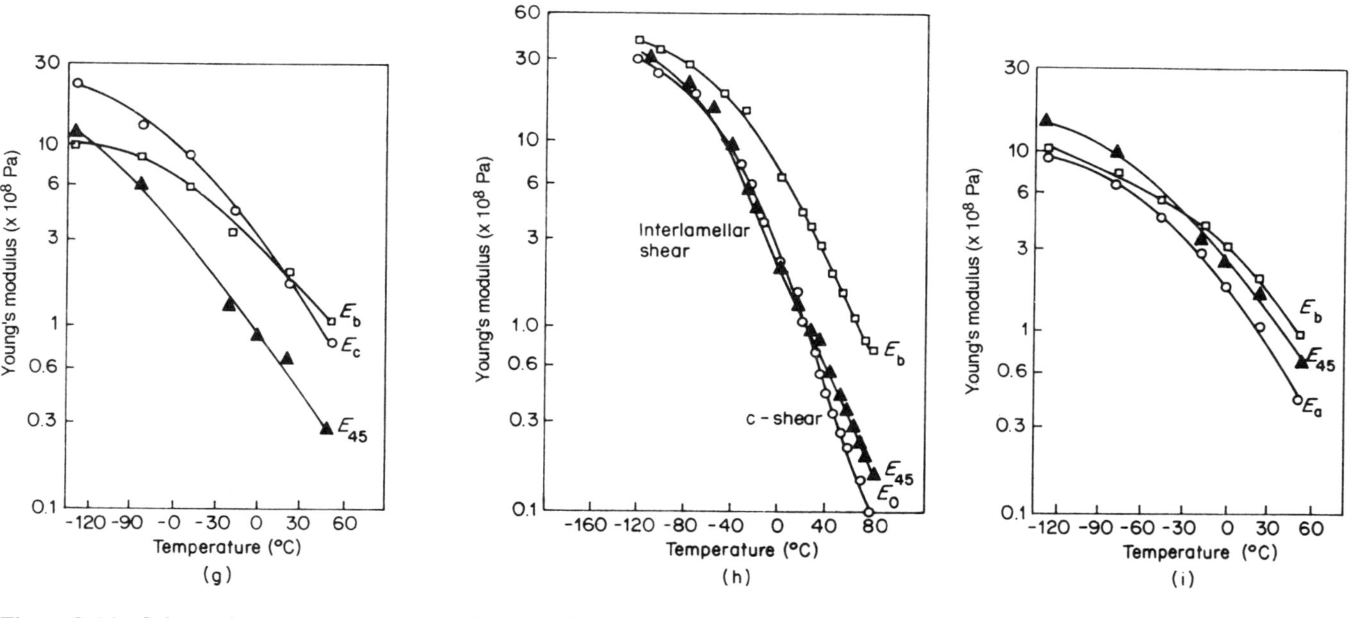

Figure 8.16 Schematic structure diagrams of mechanical loss spectra and 10 s isochronal creep moduli: (a), (d) and (g) for b–c sheet; (b), (e) and (h) for parallel lamellae sheet; (c), (f) and (i) for a–b sheet. P, interlamellar shear process; Q, c-shear process (note absence of c-shear process in (f)); R, small angle X-ray diagram, beam along X

regions. Rigid lamellae move relative to each other by a shear process in a deformable matrix. The process is activated by the resolved shear stress $\sigma \sin\gamma \cos\gamma$ on the lamellar surfaces where γ is the angle between the applied tensile stress σ and the lamellar plane normals, which reaches a maximum value for $\gamma = 45°$ (see Chapter 11 for discussion of resolved shear stress in plastic deformation processes).

Gupta and Ward found cross-over points in the extensional moduli for *bc* and *ac* sheets, similar to those found by Takayanagi in high-density polyethylene, but at lower temperatures corresponding to the β relaxation (see Chapter 9 for a discussion of relaxation processes). The fall in modulus in the *c* direction in *bc* sheet, and the *a* direction in *ab* sheet can be attributed to an interlamellar shear process. As the lamellar planes are approximately parallel to the *b* axis, a tensile stress in this direction will not favour interlamellar shear, so that at temperatures above the relaxation transition $E_b > E_a \sim E_c$ (Figure 8.16). Dynamic mechanical loss spectra show significant anisotropy, both for the lower temperature β relaxation, which corresponds to the cross-over in tensile moduli previously discussed, and for the higher temperature α relaxation, where the results are consistent with the proposal that the α process involves shear in the *c*-axis direction on planes containing the *c* axis of the crystallites (the *c*-shear process). In the *bc* sheet $\tan\delta_{45}$ is larger than $\tan\delta_0$ and $\tan\delta_{90}$, (angles being measured from the original draw direction) because it represents the situation in which there is maximum resolved shear stress parallel to the *c*-axis directions. Similarly, for the parallel lamellae sheets, the greatest losses for the α process occur when the stress is applied parallel to the initial draw direction. Finally, in *ab* sheet the α relaxation is very small, as there are no planes containing the *c*-axis which will shear in the *c* direction when a tensile stress is applied in the plane of the sheet.

Other applications of the Takayanagi model to oriented polymers have included linear polyethylene that was cross-linked and then crystallized by slow cooling from the melt under a high tensile strain [13], and sheets of nylon with orthorhombic elastic symmetry [14]. A fuller discussion is given in the more advanced text by Ward [15].

8.5 MODELLING ISOTROPIC CRYSTALLINE POLYMERS

Halpin and Kardos [16] developed a semi-empirical model to predict the moduli of isotropic crystalline polymers by a simple extension of an earlier analysis for fibrous composites. They emphasize that specifying the elastic moduli and volume concentrations of the two components is insufficient to define the properties of the composite, because geometrical factors are crucial in determining where the moduli lie between the Voigt and Reuss bounds. They present evidence to show, in a manner comparable with that for a discontinuous fibre composite, that for a given crystallinity the shear modulus is greatest when the crystalline phase has the largest aspect ratio ($a = l/2r$), with the extensional modulus given by

$$E = E_{\mathrm{m}}(1 + a\eta V_{\mathrm{f}})(1 - \eta V_{\mathrm{f}})^{-1} \tag{8.18}$$

where

$$\eta = \left(\frac{E_{\mathrm{f}}}{E_{\mathrm{m}}} - 1\right)\left(\frac{E_{\mathrm{f}}}{E_{\mathrm{m}}} + a\right)^{-1}$$

The subscripts m and f refer to matrix and fibre (i.e. amorphous and crystalline components), and V_{f} is the volume fraction of fibres. Similarly, the in-phase shear modulus is

$$G = G_{\mathrm{m}}(1 + a\eta V_{\mathrm{f}})(1 - \eta V_{\mathrm{f}})^{-1}. \tag{8.19}$$

The model has been applied to isotropic butadiene acrylonitrile co-polymers in which acetanilide was polymerized *in situ* to provide crystalline filler particles [17], and the results for low filler concentrations were consistent with aspect ratios measured for the crystalline component.

In contrast, as discussed in section 8.3.3 above, Brody and Ward [1] have applied the single-phase Ward aggregate model to composites reinforced with short fibres of carbon or glass. The aspect ratio of the fibres is not a parameter of this model, yet for all fibre lengths investigated the moduli lay close to the lower Reuss bounds, implying that where fibres lay at appreciable angles to the stressed direction they contributed only as a filler, with the high modulus of the fibres not utilized. In this experiment the fibres were much larger than the microscopic crystals in the Halpin and Kardos experiment, and the lengths were always adequate for efficient transfer of stress, so the two experiments may not be directly comparable.

8.6 ULTRAHIGH MODULUS POLYETHYLENE

Conventional drawing processes, which usually involve the polymer being uniaxially extended between two sets of rollers rotating at different speeds, rarely permit a draw ratio greater than 10 × (see Chapter 11). Such materials show an extensional modulus which is only a comparatively small fraction (~ 10%) of the chain modulus, due to the dominant effect of the relatively compliant unordered component. In the case of polyethylene, however, it is possible to produce an oriented polymer whose Young's modulus at low temperatures approaches the theoretical value of the crystal chain modulus of about 300 GPa (by comparison the Young's modulus of ordinary steel is about 210 GPa). Several production methods have been used, including solution spinning techniques [18] and a two-stage draw process [19], in which an initial stage of drawing (d.r. 8 ×) is followed by a continuing stage of extension so that the already drawn material thins down to achieve a final draw ratio of 30 × or more. These production processes are somewhat slower to operate than conventional methods, but nevertheless high stiffness polyethylene is produced commercially for specific end uses. Young's modulus as a function of draw ratio is shown in Figure 8.17 for a range of initially isotropic polyethylenes drawn at 75 °C. It is seen that the modulus, which even at room temperature can reach an appreciable fraction of the crystal modulus, depends only on the final draw ratio, and is independent of the relative molecular mass and the initial morphology, so that an appropriate model appears to be one that depends on the structure produced during deformation rather than on the starting material.

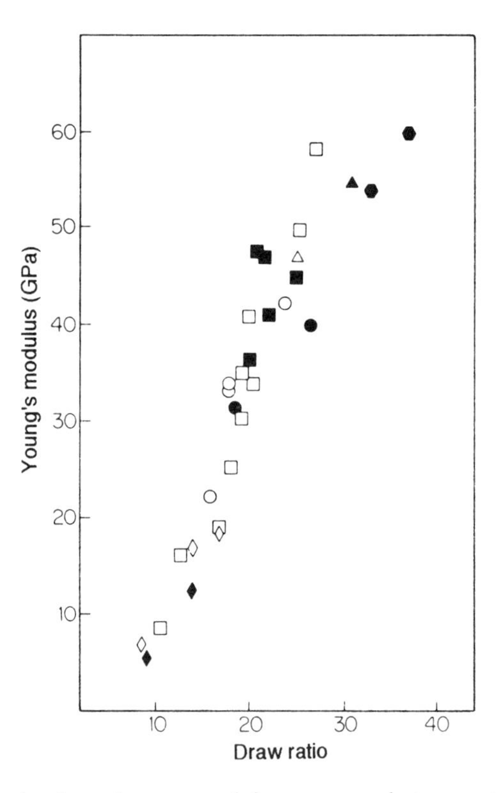

Figure 8.17 A 10 s isochronal creep modulus, measured at room temperature, as a function of draw ratio for a range of quenched (open symbols) and slowly cooled (closed symbols) samples of linear polyethlene drawn at 75 °C. (⬢), Rigidex 140–60; (△, ▲), Rigidex 25; (□. ■), Rigidex 50; (○, ●), P40; (◊, ⧫), H020–54P. (Reproduced with permission from Capaccio, Crompton and Ward, *J. Polymer Sci., Polymer Phys. Ed.*, **14**, 1641 (1976))

We shall outline two different models used to interpret the elastic behaviour of highly oriented linear polyethylene. Both models take macroscopic composite theory as their starting-point, but diverge in the way by which they account for the evidence from morphological studies for crystalline regions that can extend for more than 100 nm in the draw direction.

8.6.1 The Crystalline Fibril Model

This model, proposed by a group working at Bristol University [20], is a development of a larger-scale model that was used to account for the high mechanical anisotropy of certain copolymers [21]. Electron microscopy demonstrated that when a three-block polystyrene–polybutadiene–polystyrene copolymer was extruded into a mould, long and completely aligned filaments of glassy polystyrene, with a diameter about 15 nm were arranged with hexagonal symmetry in a rubber matrix. Despite the macroscopic anisotropy, for which the ratio of the longitudinal to the transverse Young's modulus was almost 100:1, both phases were comprised of randomly oriented molecular chains.

The model for highly oriented polyethylene similarly assumed that fibrils of high aspect ratio were arranged with hexagonal symmetry in a compliant matrix, but the discontinuous nature of the fibrils was now the determining factor for the extensional modulus. Fibrils observed in thin sections of the oriented polymer after staining with chlorosulphonic acid and uranyl acetate were considered to represent a needle-like crystal phase with the theoretical stiffness of the polyethylene chain ($E_c \sim 300$ GPa). These crystals, whose

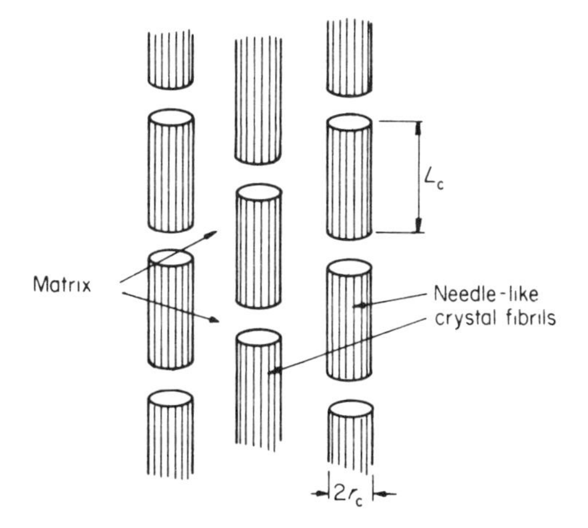

Figure 8.18 Schematic diagram of the Barham and Arridge model for ultrahigh modulus polyethylene

concentration (V_f) was estimated to be 0.75, were embedded in a partially oriented matrix, containing both amorphous and crystalline components, with a shear modulus $G_m \sim 1$ GPa (Figure 8.18).

Using the Cox model for a fibre composite, already discussed [7], and neglecting the very small contribution $E_m V_m$ arising from the tensile modulus of the compliant matrix, the extensional modulus E of the highly oriented polymer becomes

$$E = V_f E_c \left(1 - \frac{\tanh ax}{ax}\right) \tag{8.20}$$

where a is the fibre aspect ratio

$$\left(\frac{l_c}{2r_c}\right)$$

and

$$x = 2\left(\frac{G_m}{E_c \ln 2\pi/\sqrt{3V_f}}\right)^{1/2}$$

which is a restatement of equation (8.13).

The increased stiffness that results from post-neck drawing is postulated to be a direct consequence of the increased aspect ratio of the crystalline fibrils, (from slightly less than 2 to greater than 12 in extreme cases), which thus become more effective reinforcing elements. On the assumption that post-neck drawing is homogeneous on a structural level, so that the fibrils deform affinely, then the initial aspect ratio $(l_0/2r_0)$ transforms to $l_c/2r_c = (l_0/2r_0)t^{3/2}$, where t is the draw ratio in the post-neck region. Barham and Arridge [20] show that the observed change in modulus with draw ratio implies that x in equation (8.20) should depend on $t^{3/2}$. The good agreement (Figure 8.19) is advanced as a strong argument in favour of the model.

8.6.2 The Crystalline Bridge Model

An alternative approach, due to Gibson, Davies and Ward [22], is based on a Takayanagi model, which is then modified to include an efficiency ('shear-lag') factor that takes into account the discontinuous nature of the crystalline reinforcing component. The model was derived by comparing microstructural studies of conventionally drawn polyethylene and ultrahigh modulus material. At a draw ratio $\sim 10\times$, wide-angle X-ray diffraction indicates that the crystalline component is highly oriented, and together with a clear two-point small angle X-ray diffraction pattern suggests a regular stacking of crystal blocks whose length, to accommodate the non-crystalline regions, is less than the long period L, as shown in Figure 8.20. At

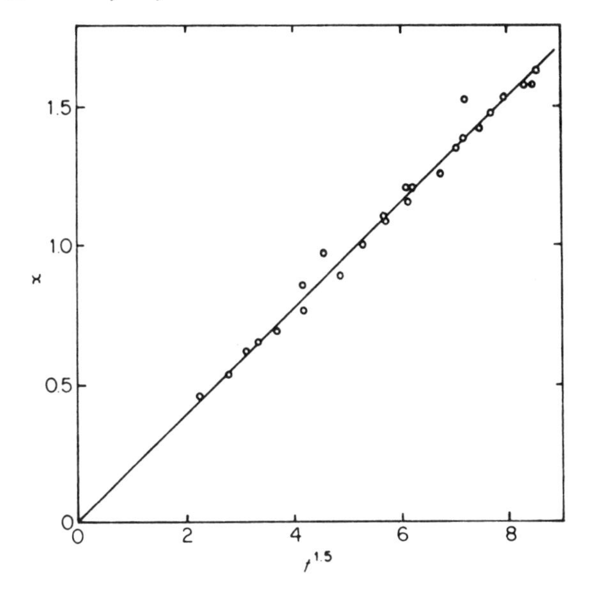

Figure 8.19 Parameter x in equation (8.20) as a function of the taper draw ratio t to the 3/2 power. (Reproduced with permission from Barham and Arridge, *J. Polymer Sci., Polymer Phys. Ed.*, **15**, 1177 (1977))

increasing draw ratio the small angle pattern retains the same periodicity, but diminishes in intensity, and a variety of techniques indicate an increase in the orientation of non-crystalline material. The average crystal length increases to about 50 nm, compared with the constant long period of about 20 nm. The concentration of crystals > 100 nm is low, in contrast with the implied lengths of 100–1000 nm for the crystalline fibrils in the model previously discussed (section 8.6.1). Reasons for this discrepancy have not been examined in detail, and the differences may be a consequence of the specific method used to produce the material, despite the inference from Figure 8.17 that the final drawing process dominates over differences in the initial morphology.

The large increase in stiffness is considered to be a consequence of the linking of adjacent crystal sequences by crystalline bridges (Figure 8.20). In this model the crystalline bridges play a similar role to the taut tie molecules earlier suggested by Peterlin [23], and are equivalent to the continuous phase of a Takayanagi model. The increase in modulus with increasing draw ratio is here considered to arise primarily from an increase in the *proportion* of fibre phase material, and not from the changing aspect ratio of a constant proportion of the fibre phase.

In the absence of information regarding the arrangement of the crystalline bridges, it is assumed that they are randomly placed, so that the probability

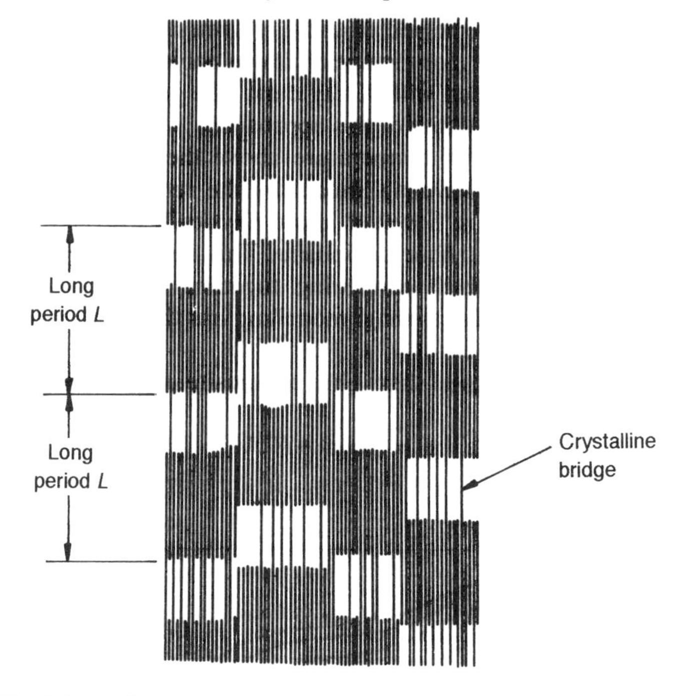

Figure 8.20 Schematic representation of the structure of the crystalline phase in ultrahigh modulus polyethylene (constructed for $p = 0.4$). (Reproduced with permission from Gibson, Davies and Ward, *Polymer*, **19**, 683 (1978) (C) IPC Business Press Ltd)

of a crystalline sequence traversing the disordered regions to link adjacent crystalline blocks is given in terms of a single parameter p, defined as

$$p = \frac{\bar{L} - L}{\bar{L} + L}$$

where $\bar{L}$ is the average crystal length determined from wide angle X-ray diffraction and L is the long period obtained from small angle X-ray scattering. It can be shown that the volume fraction of continuous phase V_{f} is given by

$$V_{\mathrm{f}} = Xp(2 - p),$$

where X is the crystallinity expressed as a fraction.

As a first stage, the contribution of the crystalline bridges can be considered as one element of a Takayanagi model (in Figure 8.13(b) this is the continuous phase) which is in parallel with the series combination of the remaining lamellar material and the amorphous component. Young's modulus would then be

$$E = E_c X p(2 - p) + E_a \frac{\{1 - X + X(1 - p)^2\}^2}{1 - X + X(1 - p)^2 E_a/E_c} \tag{8.21}$$

The first term in this expression, which corresponds to the crystalline bridge sequences, is next treated as an array of short fibres, so introducing the shear lag (efficiency) factor Φ, which is a function of the finite aspect ratio of the crystalline bridges. The analogous equation to (8.20) is

$$E = E_c X p(2 - p)\Phi' + E_a \frac{\{1 - X + X(1 - p)^2\}^2}{1 - X + X(1 - p)^2 E_a/E_c} \tag{8.22}$$

where Φ' is an average shear lag factor for all materials in the fibre phase. The advantage gained by converting from the one-dimensional Takayanagi model to the Cox short fibre model is that measurable tensile properties are able to yield information about the shear stress development in the matrix (see section 8.3.1). In particular it can be shown for the case of sinusoidally varying strain the out-of-phase component of the tensile modulus E'_c is related to the out-of-phase component of the shear modulus of the matrix G'_m by an expression that involves the volume fraction of fibres (i.e. crystalline bridges) and the fibre aspect ratio, together with the ratio G'_m/E'_f, where G'_m is the in-phase component of the shear modulus of the matrix. The geometric factors are constant for a given structure, but the modulus ratio varies with temperature because of the temperature dependence of G'_m.

The aspect ratio of the fibre component, which is a measure of the width of the crystalline bridges, is not directly accessible, but can be deduced from the value of the shear lag factor Φ' required to give the best match between the predicted and observed patterns of mechanical behaviour as a function of temperature. This exercise yields a radius of 1.5 nm for the crystalline bridge sequences, which suggests that each bridge is comprised of several extended polymer chains. Detailed considerations of the way in which the modulus increases at temperatures below −50 °C suggest that the modulus of the matrix increases with increasing draw ratio due to an increase in E_a in equation (8.21), which corresponds to an increase in the modulus of the non-crystalline material.

8.7 CONCLUSIONS

A major application of ultrahigh modulus polymers such as those discussed above is as an inert reinforcing fibre in composite materials [24]. A detailed analysis of the overall anisotropy will thus require appropriate theories on both microscopic and macroscopic scales. There are, however, important and highly significant differences between the two scales. In the macroscopic composite the fibre and the matrix are distinct entities, bonded only by weak secondary forces, whereas the crystalline and amorphous phases of an oriented semicrystalline homopolymer must blend gradually into one another: chain folds and chain ends are associated with crystallites, some of the less regular material is significantly oriented, and bridging molecules (either crystalline bridges or tie molecules) that link crystallites traverse amorphous material in the process.

Even for block copolymers, in which the phase separation can be distinguished in electron micrographs, there are problems in matching parameters such as Poisson's ratios of the two components: nevertheless the simple Takayanagi models, particularly when extended by a treatment to account for the finite length of the reinforcing component, can describe numerous features of static and dynamic elastic behaviour.

REFERENCES

1. H. Brody and I. M. Ward, *Polymer Enging. Sci.*, **11**, 139 (1971).
2. J. D. Eshelby, *Proc. Roy. Soc.*, **A241**, 376 (1957).
3. C-T. D. Wu and R. C. McCullough, in *Developments in Composite Materials*, (Holister ed.), Applied Science Publishers, London, 1977.
4. S. R. A. Dyer, D. Lord, I. J. Hutchinson, I. M. Ward and R. A. Duckett, *J. Phys. D. Appl. Phys.*, **25**, 66 (1991).
5. A. P. Wilczynski, *Comp. Sci. and Tech.*, **38**, 327 (1990).
6. P. J. Hine, R. A. Duckett and I. M. Ward, *Comp. Sci and Tech.*, **49**, 13 (1993).
7. H. L. Cox, *Brit. J. Appl. Phys.*, **3**, 72 (1952).
8. A. Kelly, *Strong Solids*, Clarendon Press, Oxford 1966.
9. M. Takayanagi, *Mems. Fac. Eng. Kyushu Univ.*, **23**, 41 (1963); M. Takayanagi, I. Imada and T. Kajiyama, *J. Polymer Sci.*, C., **15**, 263 (1966).
10. R. W. Gray and N. G. McCrum, *J. Polymer Sci., A2*, **7**, 1329 (1969).
11. I. L. Hay and A. Keller, *J. Mater Sci.*, **1**, 41 (1966).
12. V. B. Gupta and I. M. Ward, *J. Macromol. Sci. B.*, **2**, 89 (1968); Z. H. Stachurski and I. M. Ward, *J. Polymer Sci. A2*, **6**, 1083 (1969); *J. Macromol. Sci. B*, **3**, 427, 445 (1969); G. R. Davies, A. J. Owen, I. M. Ward and V. B. Gupta,
J. Macromol. Sci. B., **6**, 215 (1972); G. R. Davies and I. M. Ward, *J. Polymer Sci. B.*, **6**, 215 (1972).
13. M. Kapuscinski, I. M. Ward and J. Scanlan, *J. Macromol. Sci. B.*, **11**, 475 (1975).
14. E. L. V. Lewis and I. M. Ward, *J. Macromol. Sci. B.*, **18**, 1 (1980); **19**, 75 (1981).
15. I. M. Ward, *Mechanical Properties of Solid Polymers*, Wiley, 1983, Ch. 10.
16. J. C. Halpin and J. L. Kardos, *J. Appl. Phys.*, **43**, 2235 (1972).
17. J. L. Kardos, W. L. McDonnell and J. Raisoni, *J. Macromol. Sci. B.*, **6**, 397 (1972).
18. A. Zwijnenburg and A. J. Pennings, *J. Polymer Sci., Polymer Letters*, **14**, 339 (1976); P. Smith and P. J. Lemstra, *J. Mater. Sci.*, **15**, 505 (1980).
19. G. Capaccio and I. M. Ward, *Nature Phys. Sci.*, **243**, 143 (1973); *Polymer*, **15**, 223 (1974).

20. R. G. C. Arridge, P. J. Barham and A. Keller, *J. Polymer Sci., Polymer Phys.*, **15**, 389 (1977); P. J. Barham and R. G. C. Arridge, *J. Polymer Sci., Polymer Phys.*, **15**, 1177 (1977).
21. R. G. C. Arridge and M. J. Folkes, *J. Phys. D.*, **5**, 344 (1972).
22. A. G. Gibson, G. R. Davies and I. M. Ward, *Polymer*, **19**, 683 (1978).
23. A. Peterlin, in *Ultra-High Modulus Polymers* (A. Ciferri and I. M. Ward, eds.), Applied Science Publishers, London, 1979, Ch. 10.
24. N. H. Ladizesky and I. M. Ward, *Pure & Applied Chem.*, **57**, (1985).

PROBLEMS FOR CHAPTERS 7 AND 8

1. Explain how the generalized Hooke's law defines the compliance matrix S_{ij} for an anisotropic material.

   The non-zero components of this matrix for an oriented polymer are given in units of 10^{-10} m^2 Pa^{-1} by the following:

$$S_{11} = S_{22} = 10 \qquad S_{33} = 1,$$

$$S_{12} = S_{21} = -4,$$

$$S_{13} = S_{31} = S_{23} = S_{32} = -0.45,$$

$$S_{44} = S_{55} = 15,$$

$$S_{66} = \tfrac{1}{2}(S_{11} - S_{12}) = 7.$$

   Calculate the value of Young's modulus for the following directions in the polymer:

   (a) the Ox_3 and Ox_1 directions;
   (b) any direction making an angle of 45° with Ox_3.

2. The elastic properties of a [0, 90] cross-ply laminate constructed using pre-impregnated tapes are given by

$$\sigma_{11} = 45\varepsilon_{11} + 7\varepsilon_{22},$$

$$\sigma_{22} = 7\varepsilon_{11} + 12\varepsilon_{22}.$$

   Calculate the Young's modulus of the laminate in the direction of one set of fibres, explaining clearly your assumptions.

3. A composite of A and B, with $\lambda\phi$ being the volume fraction of A, is modelled as indicated below

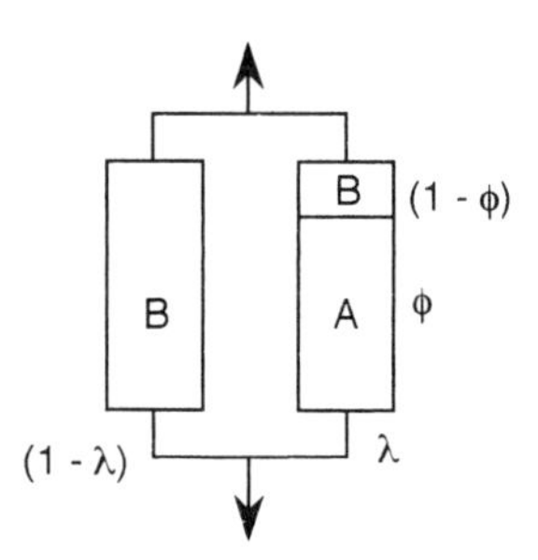

Show that the modulus of the composite is given by

$$E_c = \lambda\left(\frac{\phi}{E_a} + \frac{1-\phi}{E_b}\right)^{-1} + (1-\lambda)E_b,$$

where E_a, E_b represent the moduli of the components.

$$E_a = 10 \text{ GPa}; \qquad E_b = 1 \text{ GPa}.$$

What is the modulus of the composite if $\phi = 1$; $\phi = 0.8$? In each case $\lambda = 0.5$.

Chapter 9

Relaxation Transitions: Experimental Behaviour and Molecular Interpretation

We shall discuss the assignment of viscoelastic relaxations in a molecular sense to different chemical groups in the molecule, and in a physical sense to features such as the motion of molecules in crystalline or amorphous regions. Because amorphous polymers exhibit fewer structure-dependent features than those that are semicrystalline, we shall use these simpler materials to illustrate some general characteristics of relaxation behaviour.

9.1 AMORPHOUS POLYMERS: AN INTRODUCTION

It is customary to label relaxation transitions in polymers as α, β, γ, δ, etc. in alphabetical order with decreasing temperature. Three of the four transitions in polymethyl methacrylate (PMMA):

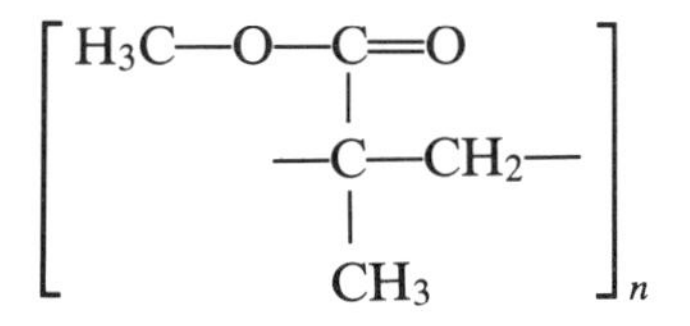

are shown in Figure 9.1, which summarizes data obtained using a low-frequency torsion pendulum. The highest temperature relaxation, the α relaxation, is the glass transition and is associated with a large change in modulus. Comparative studies on similar polymers, together with nuclear magnetic resonance (NMR) and dielectric measurements [1–5] have shown

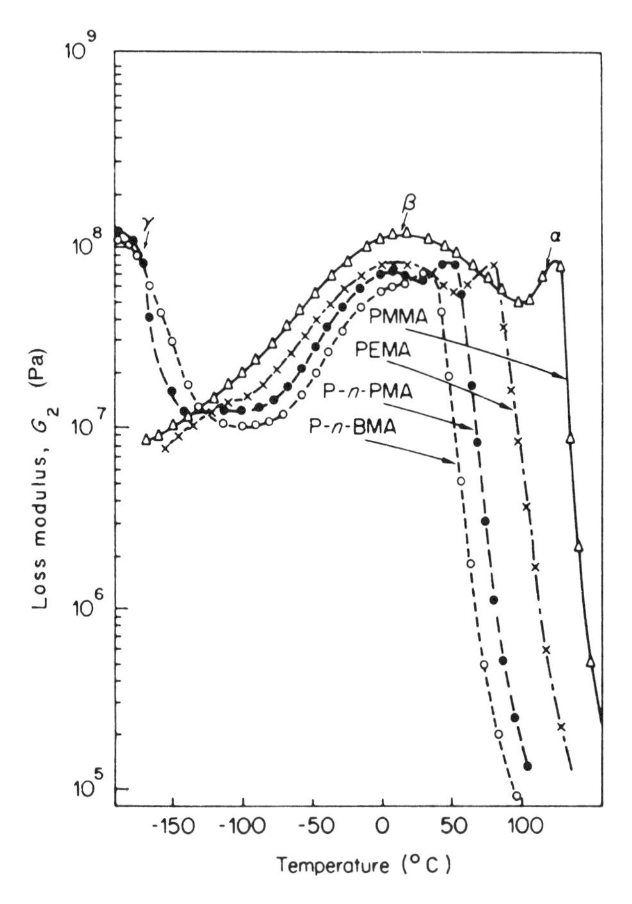

Figure 9.1 Temperature dependence of loss modulus G_2 for polymethyl methacrylate (PMMA), polyethyl methacrylate (PEMA), poly-n-propyl methacrylate (P-n-PMA) and poly-n-butyl methacrylate (P-n-BMA). (Reproduced with permission from Heijboer, in *The Physics of Non-crystalline Solids,* North-Holland, Amsterdam, 1965, p. 231)

the β relaxation to be associated with side-chain motions of the ester group. The γ and δ relaxations involve motion of the methyl groups attached to the main chain and the side chain respectively.

Many other amorphous polymers display a high-temperature transition, the glass transition, associated with the onset of main-chain segmental motion, and secondary transitions that have been assigned to either motion of side groups, restricted motion of the main-chain of end-group motions. It must be emphasized that in many amorphous polymers the assignment of a relaxation is by no means as straightforward as in PMMA and some related polymers. The case of amorphous polyethylene terephthalate (PET), where there are no side groups, is considered in section 9.3.2, in conjunction with semicrystalline forms of this polymer.

9.2 FACTORS AFFECTING THE GLASS TRANSITION IN AMORPHOUS POLYMERS

Two distinct models have been used for interpreting the influence of features such as chemical structure, molecular mass, cross-linking and plasticizers on the glass transition in amorphous polymers. The first approach considers changes in molecular flexibility, which modify the ease with which conformational changes can take place. The alternative approach relates all these effects to the amount of free volume, which is assumed to attain a critical value at the glass transition.

9.2.1 Effect of Chemical Structure

Although these factors have been intensively studied, because of their importance in selecting polymers for commercial exploitation, much of our knowledge is empirical in nature, due primarily to the difficulty in distinguishing between intra- and intermolecular effects. Some general features are, however, evident.

9.2.1.1 Main-chain structure

Flexible groups such as an ether link will enhance main-chain flexibility and reduce the glass transition temperature, with the opposite effect being shown by the introduction of an inflexible group, such as a terephthalate residue.

9.2.1.2 Side groups [6]

Bulky, inflexible side groups increase the temperature of the glass transition, as is illustrated in Table 9.1 for a series of substituted poly-α-olefins.

$$\left[\text{—CH}_2\text{—}\underset{\displaystyle \text{R}}{\underset{|}{\text{CH}}}\text{—} \right]_n$$

A difference between the effect of rigid and flexible side groups is shown in Table 9.2 for a series of polyvinyl butyl ethers

$$\left[\text{—CH}_2\text{—}\underset{\displaystyle \text{OR}_1}{\underset{|}{\text{CH}}}\text{—} \right]_n$$

Table 9.1. Glass transition of some vinyl polymers. [Reproduced with permission from Vincent, in *The Physics of Plastics* (P. D. Ritchie, ed.), 1965.]

Polymer	R	Transition temperature in °C at ~1 Hz
Polypropylene	CH_3	0
Polystyrene	C_6H_5	116
Poly-N-vinylcarbazole	N	211

All these polymers contain the same atoms in the side group OR_1 (where R_1 represents the butyl isomeric form), but more compact arrangements reduce the flexibility of the molecule and give a marked increase in the transition temperature.

Increasing the length of flexible side groups reduces the temperature of the glass transition, as is evident from Table 9.3 for a series of polyvinyl ethers,

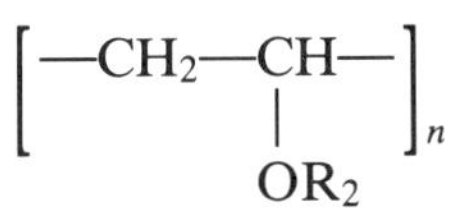

where R_2 represents the n-alkyl group. Here the increase in length is associated with an increase in free volume at a given temperature.

9.2.1.3 *Main-chain polarity*

In a series of polymers of similar main-chain composition, the temperature of the glass transition may be significantly depressed as the number of successive —CH_2 or —CH_3 groups in the side groups is increased (Figure 9.2). It is evident that the temperature of the glass transition

Table 9.2. Glass transition of some isomeric polyvinyl butyl ethers. [Reproduced with permission from Vincent, in *The Physics of Plastics* (P. D. Ritchie, ed.), 1965.]

Polymer	R_1	Transition temperature in °C at ~1 Hz
Polyvinyl n-butyl ether	$CH_2CH_2CH_2CH_3$	−32
Polyvinyl isobutyl ether	$CH_2CH(CH_3)_2$	−1
Polyvinyl t-butyl ether	$C(CH_3)_3$	+83

Table 9.3. Glass transition of some polyvinyl n-alkyl ethers. [Reproduced with permission from Vincent, in *The Physics of Plastics* (P. D. Ritchie, ed.), 1965.]

Polymer	R_2	Transition temperature in °C at ~1 Hz
Polyvinyl methyl ether	CH_3	−10
Polyvinyl ethyl ether	CH_2CH_3	−17
Polyvinyl n-propyl ether	$CH_2CH_2CH_3$	−27
Polyvinyl n-butyl ether	$CH_2CH_2CH_2CH_3$	−32

increases with main-chain polarity, and it is assumed that the associated reduction in main-chain mobility is due to the increase in intermolecular forces. In particular it is suggested that the higher values for the polychlor-acrylic esters arise from the increased valence forces associated with the chlorine molecules.

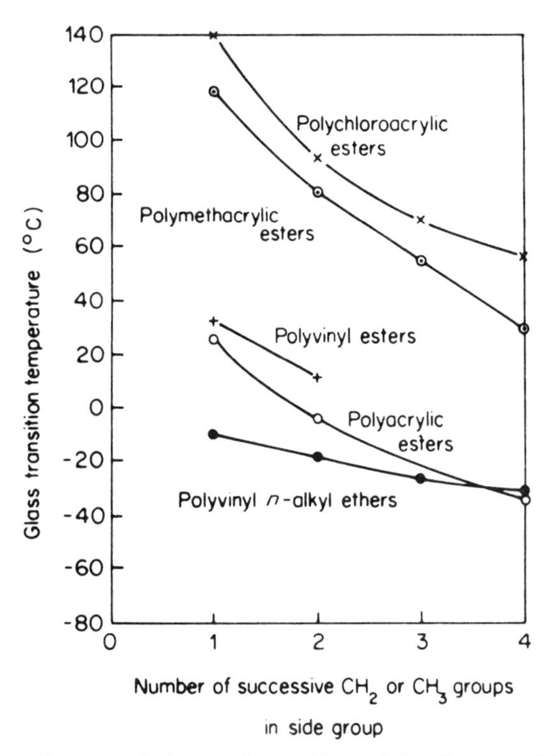

Figure 9.2 The effect of polarity on the position of the glass transition temperature for five polymer series. (Reproduced with permission from Vincent, in *The Physics of Plastics* (P. D. Ritchie, ed.), Iliffe 1965)

9.2.2 Effect of Molecular Mass and Cross-linking

The length of the main chain does not affect the dynamic mechanical properties of polymers in the glassy state, where molecular motions are of restricted extent, but the glass transition temperature is depressed at very low relative molecular masses as a consequence of the additional free volume introduced by the increased proportion of chain ends [7].

As already discussed (section 6.1.1 above) molecular mass has a large effect in the glass transition range, where viscous flow transforms to a plateau range of rubber-like behaviour due to entanglements between the longer molecular chains.

Chemical cross-links reduce the free volume by bringing adjacent chains close together and so raise the temperature of the glass transition, as is shown in Figure 9.3, for phenol-formaldehyde resin cross-linked with hexamethylene tetramine to different extents. The transition region is greatly broadened [8], so that in very highly cross-linked materials, where motions of extensive segments of the main chain are not possible, there is no glass transition.

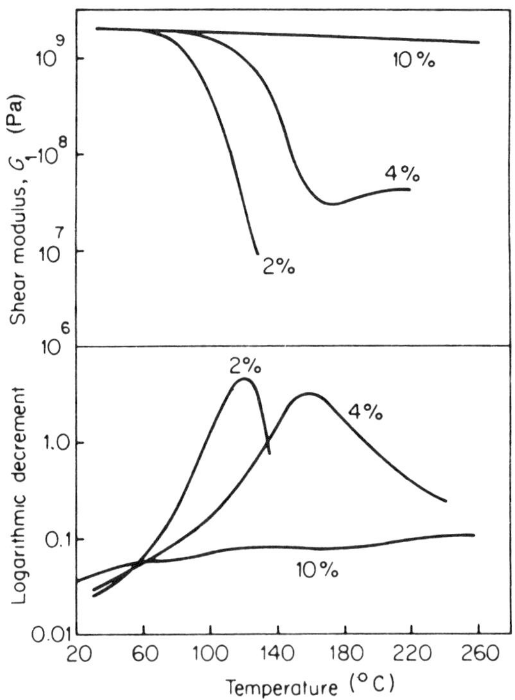

Figure 9.3 Shear modulus G_1 and logarithmic decrement of a phenol-formaldehyde resin cross-linked with hexamethylene tetramine at stated concentrations. (Reproduced with permission from Drumm, Dodge and Nielsen, *Ind. Eng. Chem.*, **48**, 76 (1956))

9.2.3 Blends, Grafts and Copolymers

The mechanical properties of blends and grafts polymers are determined primarily by the mutual solubility of the two homopolymers. For complete solubility the properties of the mixture are close to those of a random copolymer of the same composition, as is shown in Figure 9.4, which compares a 50:50 mixture of polyvinyl acetate and polymethyl acrylate with the equivalent copolymer [8]. Note that the damping peak occurs at 30 °C, compared with 15 °C for the polymethyl acrylate and 45 ° for polyvinyl acetate.

A theoretical interpretation of the glass transition temperature of a copolymer is based on the assumption that the transition occurs at a constant fraction of free volume. Gordon and Taylor [9] assume that in an ideal copolymer the partial specific volumes of the two components are constant and equal to the specific volumes of the two homopolymers. The specific volume–temperature coefficients for the two components in the rubbery and glassy states are assumed to remain the same in the copolymer as in the

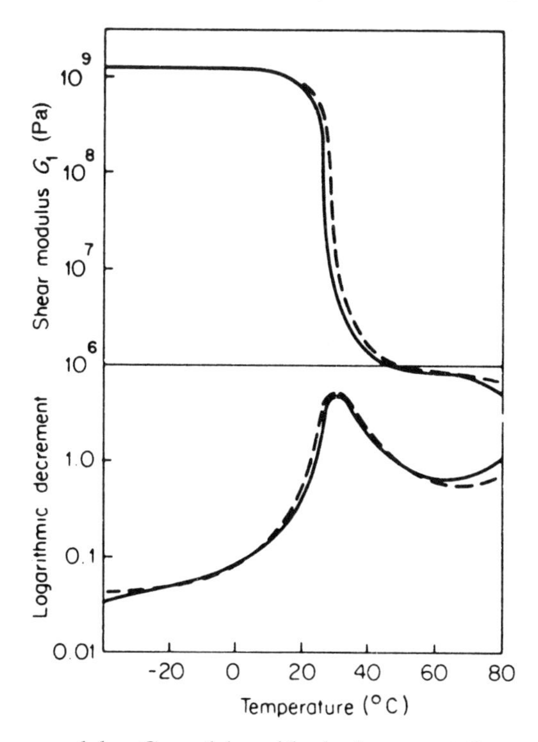

Figure 9.4 Shear modulus G_1 and logarithmic decrement for a miscible blend of polyvinyl acetate and polymethyl acrylate (– – –) and a copolymer of vinyl acetate and methyl acrylate (____). (Reproduced with permission from Neilsen, *Mechanical Properties of Polymers*, Van Nostrand-Reinhold, New York, 1962)

homopolymers, and to be independent of temperature. The glass transition temperature T_g for the copolymer is then given by [10]

$$\frac{1}{T_g} = \frac{1}{(w_1 + Bw_2)}\left[\frac{w_1}{T_{g1}} + \frac{Bw_2}{T_{g2}}\right]$$

where w_1 and w_2 are the mass fractions of the two monomers whose homopolymers have transition temperatures T_{g1} and T_{g2} respectively, and B is a constant close to unity.

Where the two polymers in a mixture are insoluble they exist as separate phases, so that two glass transitions are observed, as shown in Figure 9.5 for a polyblend of polystyrene and styrene-butadiene rubber [8]. The two loss peaks are very close to those for pure polystyrene and pure styrene-butadiene rubber.

9.2.4 Effects of Plasticizers

Plasticizers, which are relatively low molecular mass organic materials, added to soften rigid polymers, must be soluble in the polymer and usually

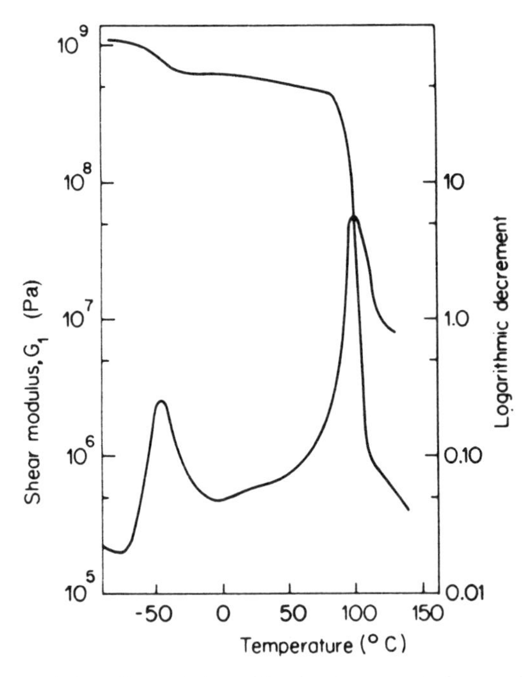

Figure 9.5 Shear modulus G_1 and logarithmic decrement for an immiscible polyblend of polystyrene and a styrene-butadiene copolymer (Reproduced with permission from Nielsen, *Mechanical Properties of Solids*, Van Nostrand-Reinhold, New York, 1962).

they dissolve it completely at high temperature. Figure 9.6 shows the change in the loss peak associated with the glass transition of polyvinyl chloride (PVC) when plasticized with varying concentrations of di(ethylhexyl) phthalate [11]. In this polymer plasticization is of major commercial importance: rigid PVC is used in applications such as replacement window frames, and the plasticized material supplies flexible sheeting and inexpensive footwear.

Plasticizers make it easier for changes in molecular conformation to occur, and so lower the temperature of the glass transition. They also broaden the loss peak, with the extent of broadening depending on the nature of the interactions between the polymer and the plasticizer. A broad damping peak is found where the plasticizer has a limited solubility in the polymer or tends to associate in its presence. The increased width of the damping peak as the plasticizer becomes a poorer solvent is shown in Figure 9.7 for plasticized PVC [8]. Diethyl phthalate is a relatively good solvent, dibutyl phthalate is intermediate, and n-dioctyl phthalate is a very poor solvent.

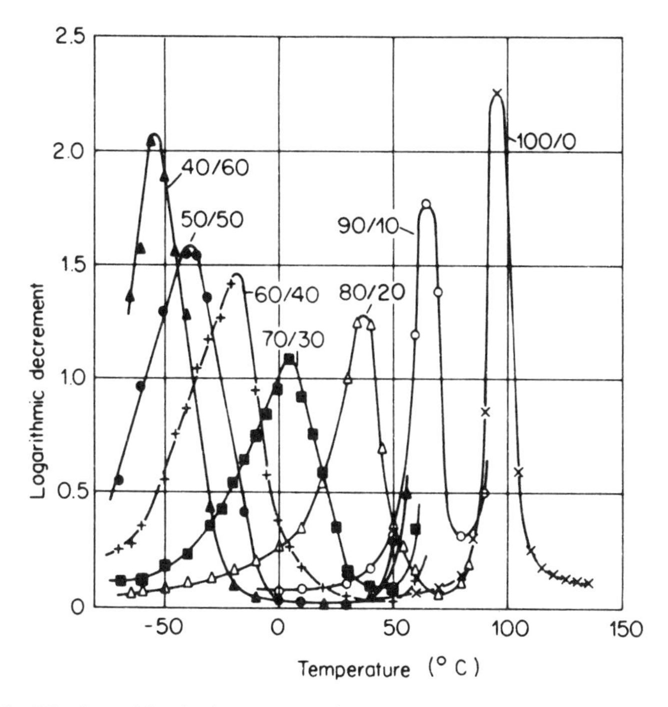

Figure 9.6 The logarithmic decrement of polyvinyl chloride plasticized with various amounts of di(ethylhexyl) phthalate. (Reproduced with permission from Wolf, *Kunststoffe*, **41**, 89 (1951))

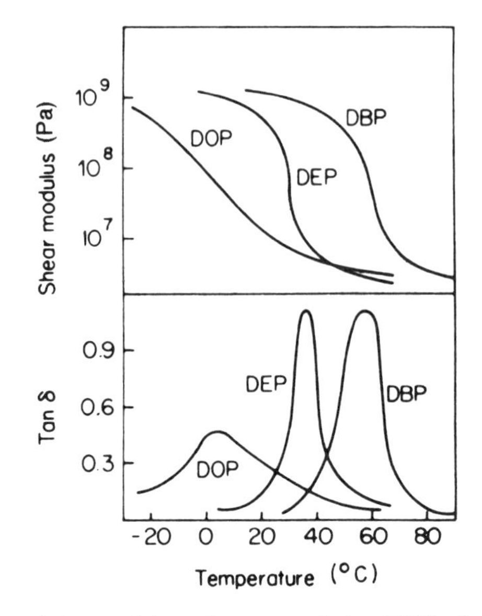

Figure 9.7 Shear modulus and loss factor tan δ for PVC plasticized with diethyl phthalate (DEP), dibutyl phthalate (DBP) and n-dioctyl phthalate (DOP). (Reproduced with permission from Neilsen, Buchdahl and Levreault, *J. Appl. Phys.*, **21** 607 (1950))

9.3 RELAXATION TRANSITIONS IN CRYSTALLINE POLYMERS

9.3.1 General Introduction

Semicrystalline polymers are less sensitive to wide variations of stiffness with temperature than those that are totally amorphous, but even so stiffnesses may vary by an order of magnitude over the useful working range of a given material. Oriented crystalline polymers may additionally show contrasts between extensional and shear deformations, and also angular dependent changes in relaxation strength.

Some polymers, notably low-density polyethylene, show clearly resolved α, β and γ processes. The high-temperature α relaxation is frequently related to the proportion of crystalline material present, the β process is related to a greatly broadened glass–rubber relaxation, and the γ relaxation has been associated, at least in part, with the amorphous phase. Other materials, an example to be discussed shortly being that of PET, show only two relaxation processes. In these cases the α relaxation is akin to the β process in polymers where all three relaxations are evident.

In earlier editions of this work it was noted that the interpretation of viscoelastic relaxations in crystalline polymers was at a very speculative stage but, as a working hypothesis, it was assumed that the tangent of the phase lag angle ($\tan \delta$), or its equivalent the logarithmic decrement (Λ) was an appropriate measure of the relaxation strength. Boyd, in two important review articles [12, 13] has demonstrated that the situation is more complex: for instance the apparent trend of relaxation strength with changing crystallinity can depend on whether $\tan \delta$, the real (in-phase) or imaginary (out-of-phase) components of modulus (G_1 and G_2 respectively) or compliance (J_1, J_2) are used to record the relaxations; and the interpretation of the data depends on the composite model used to determine the interaction between crystalline and amorphous phases.

We shall begin with a brief and simplified discussion of the main features of experimental observations, and proceed to consider the interpretation of these features, first for polymers of a relatively low level of crystallinity, taking PET as a particular example, and then for highly crystalline polymers such as polyethylene.

9.3.2 Relaxation in Low Crystallinity Polymers

The temperature variation of the complex modulus of PET as a function of crystallinity has been studied by Takayanagi [14] in extension at 138 Hz (Fig. 9.8) and by Illers and Breuer [15] in shear at ~1 Hz. At the lowest levels of crystallinity there is a sudden and severe drop in stiffness associated with the α process that is characteristic of amorphous polymers. With increasing crystallinity the α peak broadens as the change in stiffness is greatly reduced. This behaviour is consistent with that of a composite for which only one phase softens, with the broadening of the peak resulting from the restriction of long-range segmented motions in the amorphous phase by the remaining crystals. Illers and Breuer noted also that the temperature at which the loss peak (G_2) was a maximum increased up to 30% crystallinity and then decreased slightly at high crystallinities. Studies involving small angle X-ray scattering indicate that high crystallinity specimens have both thicker crystal layers and thicker amorphous layers than those of low crystallinity [16]. This latter feature will reduce the constraints imposed by crystal surfaces.

In contrast to the α relaxation both the shape and location of the subglass β process are insensitive to the degree of crystallization. Dielectric studies [17] yield the same conclusion. The process is therefore consistent with localized molecular motions, in contrast with the restrained long-range segmental motions involved in the glass–rubber α relaxation.

Boyd [13] analyses the behaviour in terms of a composite model by plotting relaxation strength, determined from the shear modulus, as a

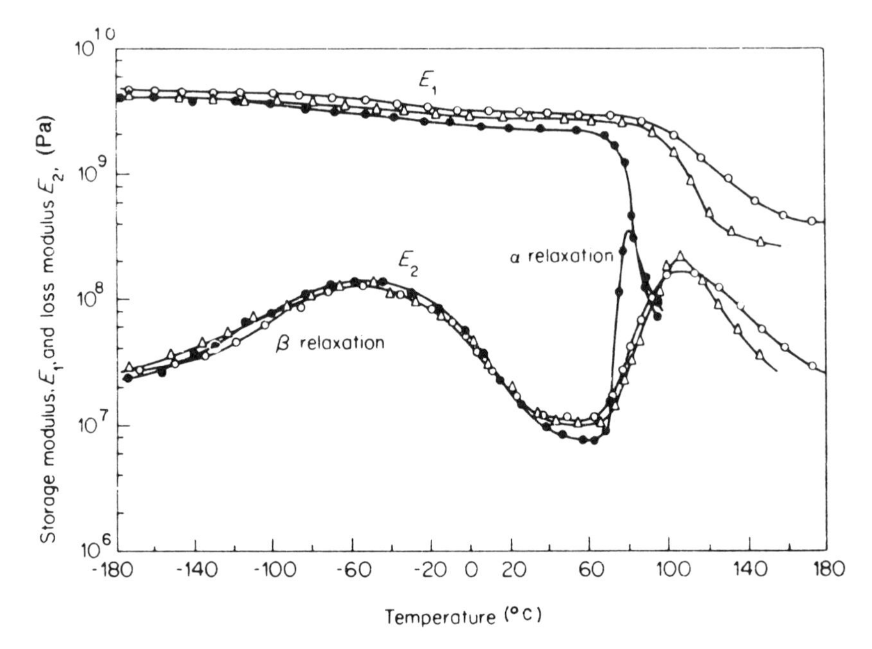

Figure 9.8 Storage modulus E_1 and loss modulus E_2 as a function of temperature at 138 Hz for PET samples of differing degrees of crystallinity (●, 5%; △, 34%; ○, 50%). (Reproduced with permission from Takayanagi, *Mem, Fac. Eng., Kyushu Univ.*, **23**, 1 (1963))

function of crystallinity. For the β process, the shear modulus for the amorphous phase lies between the upper (Voigt) and lower (Reuss) bounds in both relaxed and unrelaxed states. The α process, however, behaves in a lower bound like manner. This model of PET in terms of separate but linked crystalline and amorphous phases contrasts with the single-phase aggregate model that Ward has found applicable in other studies of the mechanical behaviour of the polymer.

9.3.3 Relaxation Processes in Polyethylene

Polyethylene is the obvious choice for investigating relaxations in the more highly crystalline polymers. Its structure has been studied in great detail, and the material is readily obtainable in two forms. Low-density polyethylene (LDPE) typically contains about three short side branches per 100 carbon atoms, together with about one longer branch per molecule. High-density polyethylene (HDPE) is much closer to the pure $(CH_2)n$ polymer, and the proportion of side branches is often less than five per 1000 carbon atoms. The main features of the temperature dependence of tan δ for each material are indicated schematically in Figure 9.9. The LDPE shows clearly resolved α, β and γ loss peaks. In HDPE the low temperature γ peak is very similar to that in LDPE, but the β relaxation is hardly resolved and the α relaxation is often considerably modified, appearing to consist of at least two processes (α and α') with different activation energies. The high-temperature behaviour is also dependent on whether loss angle or loss modulus is the quantity being measured. At one time some investigators questioned the existence of the β relaxation in HDPE, but further work involving a range of polymers intermediate between the extremes indicated in Figure 9.9 has established its presence.

The composite nature of the α process in HDPE is thought to be a consequence of specimen inhomogeneity [18], as was demonstrated by McCrum and Morris in torsional pendulum experiments which compared two slowly cooled specimens, one of which had its lateral surface removed by milling (Figure 9.10). The α' process was not evident in the milled specimen due to the removal of an oriented surface layer with the preferred orientation for maximum energy loss.

We must conclude that α, β and γ relaxations occur in all forms of polyethylene. The first stage in the process of analysis is to determine whether a given relaxation is related to the crystalline or the amorphous component, or to an interaction involving both phases. Next one deduces a process that is able to account for each relaxation. Finally an attempt must

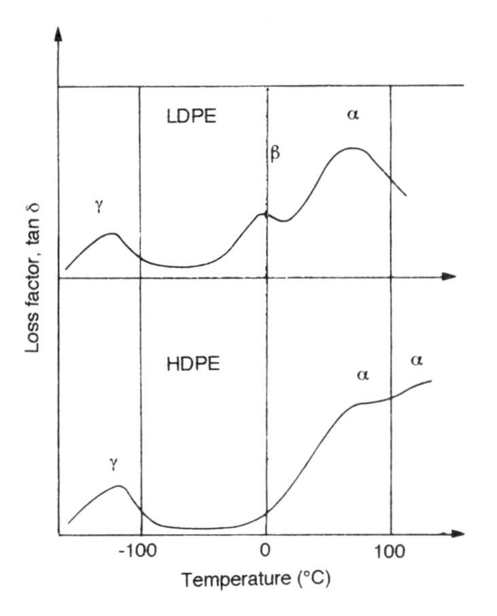

Figure 9.9 Schematic diagram showing α, α', β and γ relaxation processes in LDPE and HDPE

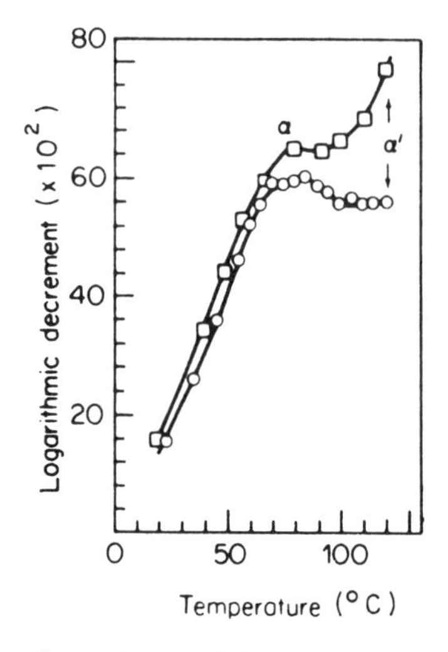

Figure 9.10 Temperature dependence of logarithmic decrement at 0.67 Hz for a polyethylene specimen crystallized by slow cooling from the melt (□). The other points (○) are for the same specimen with the oriented surface layers removed by milling. (Reproduced with permission from McCrum and Morris, *Proc. Roy. Soc.*, **A292**, 506 (1966))

be made to model a molecular mechanism that can cause the process to occur.

In general, we shall follow the arguments deduced by Boyd in his reviews [12, 13] which are essential reading for those who intend working in this field. His warnings on pitfalls in interpreting experimental data are extremely useful. We will, however, also take into account research by other groups, especially that of Ward and co-workers, which is not entirely in agreement with all of Boyd's conclusions. For example, Boyd suggests that the behaviour of HDPE shifts smoothly to that characteristic of LDPE with reducing crystallinity, and considers that specimens of intermediate crystallinity can be modelled as composites of linear HDPE with more disordered material. Experiments by Ward and his colleagues, to be discussed shortly, suggest that such an assumption may not be correct.

Measurements of the mechanical relaxations can usefully be supplemented by the equivalent dielectric relaxation experiments, in which the information obtained relates solely to the crystalline component [19, 20]. Pure polyethylene shows no dielectric response, so experiments are made on specimens that have been lightly decorated with dipoles by means of chlorination or oxidation. The proportion of dipoles is so small that the overall properties of the material are unlikely to be altered by a significant extent.

Dielectric studies indicate a much narrower relaxation width than in mechanical measurements, and it has been suggested that in the latter case interaction occurs between crystalline material and an amorphous component that is constrained by the adjacent well-ordered polymer. An interesting feature both of mechanical and dielectric experiments is that the location of the α relaxation depends on the crystal lamellar thickness. The mechanical β relaxation is considered to be equivalent to the glass–rubber transition in amorphous polymers. Its insignificance in highly crystalline HDPE is a consequence of the low proportion of non-crytalline material, coupled with the high value of the unrelaxed modulus at the upper temperature limit of the process.

The γ relaxation is very small in material that approaches 100% crystallinity. Crissman [21] found no loss peaks in the polycrystalline long-chain paraffin $C_{94}H_{190}$, yet the onset of both α and γ relaxations was observed in specimens of very low molecular mass LDPE in which the measured X-ray long period coincided with the chain length. There could be no chain folding, and so no amorphous surface layer, but a small amount of non-crystalline material was present because of polydispersity. The size of the γ relaxation increased as soon as the chain length was sufficient for chain folding to occur. As the temperature of this relaxation is below the glass temperature, only small-scale motions are possible within the non-crystalline component.

Mansfield and Boyd [22] have proposed that the dielectric α process can be represented by the torsional movement of a segment of chain about $12CH_2$ units in length. This motion, which will cause the short twisted mismatch region to move through the crystal one carbon atom at a time, is consistent with the dependence of activation energy on crystal thickness (Figure 9.11)

In the mechanical situation the translational component of the crystal process can lead to reorganization of the crystal surface and hence modify the connections of amorphous chains to the crystal surface. An example is shown in Fig. 9.12, where translational motion of a decorated dipole

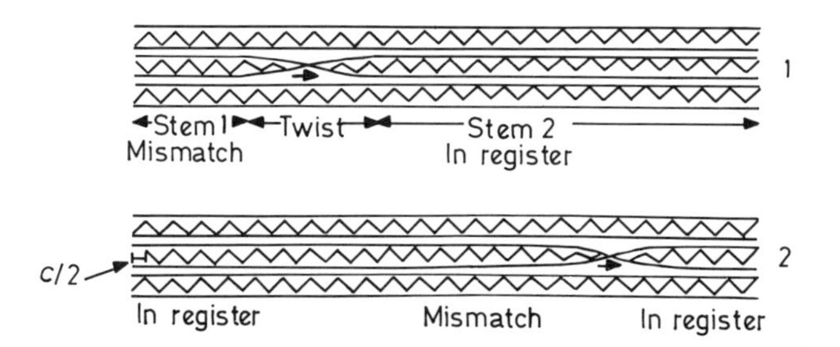

Figure 9.11 Propagation of a localized smooth twist along the chain. As the twist starts (1) it leaves behind it a translational mismatch. As the twist proceeds (2) the mismatch becomes attenuated at large distances from the twist by elastic distortion of the stem valence angles and bond length. (Reproduced with permission from Mansfield and Boyd, *J Polym Sci, Phys Ed.*, **13**, 1407 (1975)

(indicated by a horizontal arrow) permits lengthening of the tight tie chain in (a), and so enables further deformation of the amorphous fraction.

The above model is consistent with the results of Stachurski and Ward [23] for sheets of HDPE that were cold-drawn uniaxially and then annealed. Figure 9.13 shows the variation of tan δ with temperature at different angles to the draw direction (note that in these specimens the β relaxation is difficult to detect). The explanation depends on proposals that the crystal lamellae make an acute angle of about 40° with the initial draw direction [24]. Applying the stress along the initial draw direction then gives the maximum resolved shear stress parallel to the lamellar planes, so that the α relaxation in HDPE is primarily an interlamellar shear process.

The β relaxation is very broad compared with that in completely amorphous polymers due to the immobilizing effect of the crystals on the amorphous fraction. Boyd speculates that the shortest relaxation times may be associated with motions of very loose folds and relatively non-extended tie chains; conversely tight folds are unable to relax. The relative prominence of the β-relaxation in LDPE compared with HDPE is enhanced by the lower value of the relaxed β process modulus in LDPE, which will increase the relative intensity of the β and decrease that of the α. On a molecular basis the branching of LDPE gives a more loosely organized amorphous component, capable of relaxing to a lower limiting rubbery modulus.

Caution must be exercised, however, in assuming that the behaviour of LDPE is basically similar to that of HDPE. Experiments by Stachurski and Ward [25] on (a) cold-drawn, and (b) cold-drawn and subsequently annealed sheets of LDPE are illustrated in Figure 9.14. For the cold-drawn sheet the maximum loss in the 0 °C region occurs at 45° to the draw direction, which is

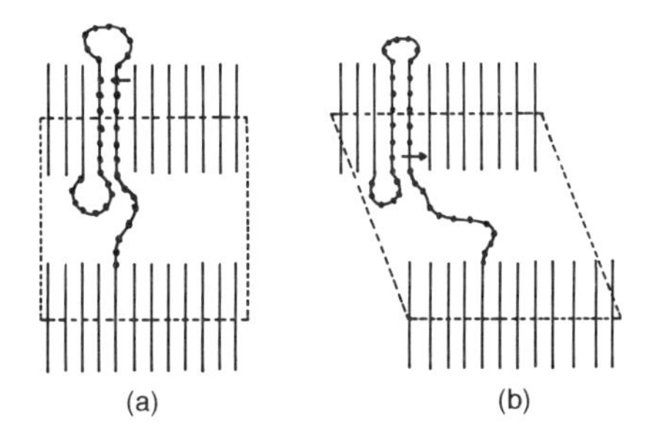

Figure 9.12 Further relaxation of the amorphous fraction resulting from translational mobility in the crystal (the latter acquired in the α process). Illustrated in this case is reorganization of the interface in (a) through shortening of two loops that in (b) permits lengthening of a tight tie chain, which in turn permits more deformation of the amorphous fraction. Also shown is a decorating dipole in the crystal which moves through a number of translational, rotational steps. One such step suffices for dielectric activity. (Boyd, *Polymer*, **28**, 1123 (1985))

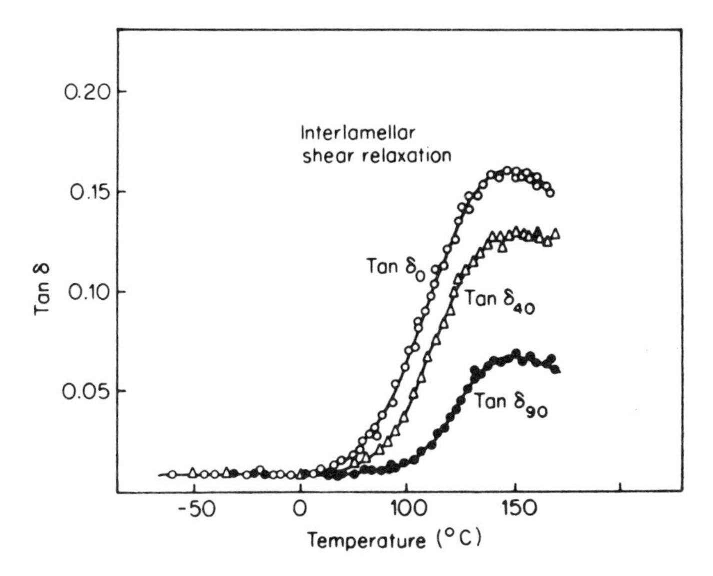

Figure 9.13 Temperature dependence of tan δ in cold-drawn and annealed HDPE sheet in different directions at 50 Hz. (Reproduced with permission from Stachurski and Ward, *J. Macromol. Sci.* B, **3**, 445 (1969))

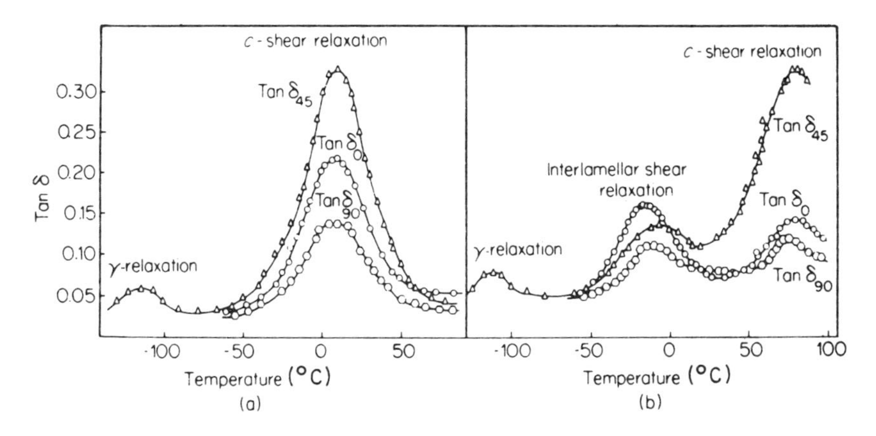

Figure 9.14 Temperature dependence of tan δ in three directions in (a) cold-drawn, and (b) cold-drawn and annealed LDPE sheets at approximately 500 Hz. (Reproduced with permission from Stachurski and Ward, *J. Polymer Sci. A-2*, **6**, 1817 (1986))

the anisotropy appropriate for a relaxation which involves shear parallel to the draw direction in a plane containing the draw direction. For the cold-drawn and annealed sheet this relaxation has moved to about 70 °C. These results, taken together with other measurements on specially oriented sheets, suggest that the chain or c-axis orientation of the crystalline regions is the governing factor. As the relaxation involves shear in the c-axis direction in planes containing the c axis it has been termed the 'c-shear relaxation'. The annealed LDPE sheet also shows a β relaxation below 0 °C with an anisotropy similar to that discussed for the α relaxation in HDPE, and which therefore must be associated with a process involving interlamellar shear. This unexpected feature does not imply that both these relaxations are associated with identical molecular processes. Boyd's model of the α relaxation is dependent on the mobility of fold surfaces, but it is likely that the β relaxation in LDPE requires mobility of chains close to branch points. It appears that interlamellar shear in the mechanical α-process in HDPE requires coupled motions of chains that run through the lamellae (c-shear) together with chains on the lamellar surface. In contrast, c-shear and interlamellar shear in LDPE seem to be two distinct mechanical relaxations, presumably due to the influence of branch points on the molecular motions.

As it occurs below the glass transition temperature the γ relaxation will involve simple conformational motions that are relatively short range in character. Such motions must leave the molecular stems adjacent to the bonds undergoing transition relatively undisturbed, must require only a modest activation energy, and the swept-out volume during the relaxation should be small. Following Willbourn's [26] suggestion that the γ relaxation in many amorphous and semicrystalline polymers can be attributed to a restricted motion of the main chain that involves at least four successive —CH_2 groups both Shatzki [27] and Boyer [28] have proposed that subglass relaxations can be modelled in terms of so-called 'crankshaft' mechanism (Fig. 9.15). Shatzki's five-bond mechanism involves the simultaneous rotation about bonds 1 and 7 such that the intervening carbon bonds move as a crankshaft. Boyer's proposal involves only three intermediate carbon bonds. The mechanisms are, as Boyd [12] has pointed out, the simplest allowed moves of a tetrahedrally bonded chain on a diamond lattice that leave the adjacent stem bonds in place. The internal energetics of the

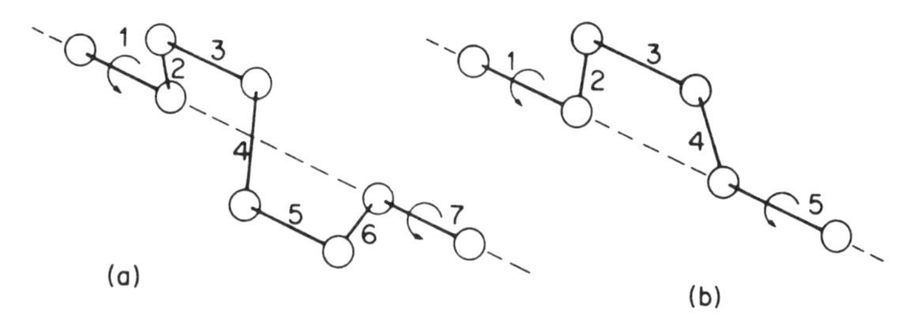

Figure 9.15 The crankshaft mechanisms of (a) Shatzki and (b) Boyer. (Reproduced with permission from McCrum, Read and Williams, *Anelastic and Dielectric Effects in Polymer Solids*, Wiley, London, 1967)

five-bond model are modest, but the swept-out volume is large in the context of a motion in a glassy matrix. For the three-bond transition a double energy barrier system with an intermediate energy minimum is implied, and the motion associated with one of these barriers requires a significant free volume, and so is inhibited by the matrix. Despite these drawbacks a crankshaft mechanism has been proposed as relevant to the γ-relaxation in polyethylene.

Boyd [12] discusses a motion related to the three-bond mechanism that can accomplish the appropriate shape change without encountering problems associated with free volume. It involves the conformational sequence GTG′ occurring in an otherwise all *trans* chain (G and G′ represent alternative *gauche* transformations). From Fig. 9.16 it can be seen that this sequence, known as a kink [29], has the effect of displacing the separated *trans* components of a planar zigzag yet leaving them parallel to one another. Interchanging the senses of the *gauche* bonds

$$\ldots \text{TTGTG'TT} \ldots \rightarrow \ldots \text{TTG'TGTT} \ldots$$

converts the kink into a mirror image of itself. The kink inversion process, shown in Figure 9.17, involves only a small swept-out volume and requires a

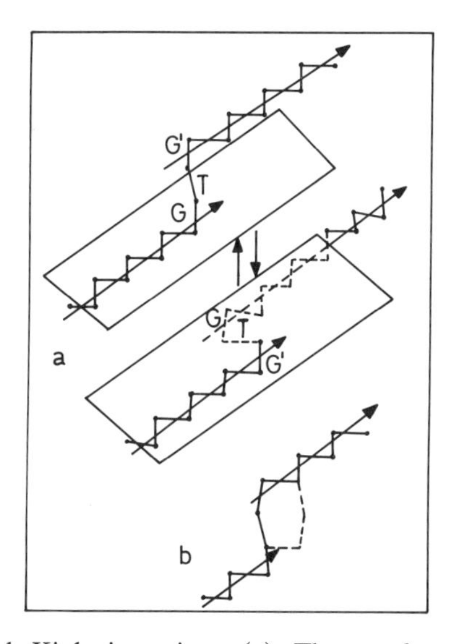

Figure 9.16 Kinks and Kink inversion. (a) The conformational sequence ... TTTGTG′TTT ... has parallel offset planar zigzag stems (indicated by arrows) on either side of the GTG′ portion. The transition TGTG′ → TG′TGT (called here kink inversion) creats a mirror image of the kink about the displaced stems. (b) A three-bond crankshaft move is shown at a kink site (as dashed line). This move results in advancing the kink along the chain by $2CH_2$ units. (Reproduced with permission from Boyd, *Polymer*, **26**, 1123 (1985)

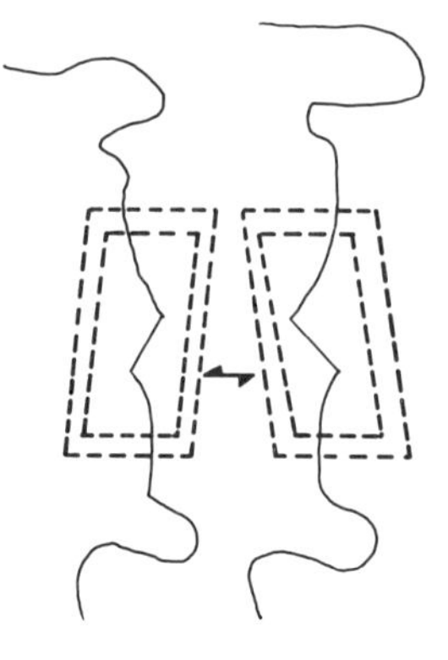

Figure 9.17 Strain fields set up by stem displacement accompanying kink inversion. (Reproduced with permission from Boyd, *Polymer,* **26**, 1123 (1985))

modest activation energy. The stem displacement causes a localized shape change that can propagate through the specimen as a shear strain. The kink inversion process is therefore a possible candidate on which to base a molecular model of the γ relaxation, but it must be emphasized that there is no direct evidence which demonstrates that it is appropriate for modelling the behaviour of polyethylene.

9.4 CONCLUSIONS

We have seen that there is a general understanding of the main factors that can modify the relaxation behaviour of non-crystalline polymers. With semicrystalline polymers it is frequently possible to attribute each relaxation to either the crystalline component or the amorphous component, or to an interaction whereby the crystalline component constrains motions in the less-well-ordered material. For polyethylene considerable progress has been made in unravelling the complex relaxation processes, although it is far from clear whether measurements indicate that two or more mechanisms can operate simultaneously, with the dominant mechanism being dependent on structural features such as the density of branch points. Progress has also been made towards understanding relaxation mechanisms in other polymers, which we have no space to discuss. For these materials a smaller amount of structural information is available than is the case for polyethylene, so we cannot hope for a complete picture of relaxation behaviour. Nevertheless the methods used for the elucidation of the relaxations in polyethylene can provide guide-lines for future advancement.

REFERENCES

1. K. Deutsch, E. A. Hoff and W. Reddish, *J. Polymer Sci.*, **13**, 365 (1954).
2. J. G. Powles, B. I. Hunt and D. J. H. Sandiford, *Polymer*, **5**, 505 (1964).
3. K. M. Sinnot, *J. Polymer Sci.*, **42**, 3 (1960).
4. J. Heijboer, *Physics of Non-Crystalline Solids*, North-Holland, Amsterdam, 1965, p. 231
5. N. G. McCrum, B. E. Read and G. Williams, *Anelastic and Dielectric Effects in Polymeric Solids*, Wiley London, 1967.
6. P. I. Vincent, *Physics of Plastics* (ed. P. D. Ritchie), Iliffe, London 1965.
7. T. G. Fox and P. J. Flory, *J. Appl. Phys.*, **21**, 581 (1950); *J. Polymer Sci.* **14**, 315 (1954).
8. L. E. Nielsen, *Mechanical Properties of Polymers,* Van Nostrand-Reinhold, New York, 1962.
9. M. Gordon and J. S. Taylor, *J. Appl. Chem.*, **2**, 493 (1952).
10. L. Mandelkern, G. M. Martin and F. A. Quinn, *J. Res. Natl. Bur. Stand.*, **58**, 137 (1959); T. G. Fox and S. Loshaek, *J. Polymer Sci.*, **15**, 371 (1955).
11. K. Wolf, *Kunstoffe*, **41**, 89 (1951).
12. R. H. Boyd, *Polymer*, **26**, 1123 (1985).
13. R. H. Boyd, *Polymer*, **26**, 323 (1985).
14. M. Takayanagi, *Mem. Fac. Eng, Kyushu Univ.*, **23**, 1 (1963).
15. K. H. Illers and H. Breuer, *J. Colloid Sci.*, **18**, 1 (1963).
16. H. G. Kilian, H. Halboth and E. Jenckel, *Kolloid Z.*, **176**, 166 (1960).
17. J. C. Coburn, PhD Dissertation, Univ. of Utah, 1984.
18. N. G. McCrum and E. L. Morris, *Proc. Roy. Soc.* **A292**, 506 (1966).
19. G. R. Davies and I. M. Ward, *J. Polymer Sci.*, **137**, 353 (1969).
20. C. R. Ashcraft and R. H. Boyd, *J. Polymer Sci., Phys. Edn*, **14**, 2153 (1976).
21. J. M. Crissman, *J. Polymer Sci., Phys Edn*, **13**, 1407 (1975).
22. M. Mansfield and R. H. Boyd, *J. Polymer Sci., Phys. Edn*, **16**, 1227 (1978).
23. Z. H. Stachurski and I. M. Ward, *J.Macromol. Sci (Phys)*, **B3**, 445 (1969).
24. I. L. Hay and A. Keller, *J. Materials Sci.*, **2**, 538 (1967); T. Seto and T. Hara, *Rept. Prog. Polymer Phys. (Japan)*, **7**, 63 (1967).
25. Z. H. Stachurski and I. M. Ward, *J. Polymer Sci. A-2*, **6**, 1817 (1968).
26. A. H. Willbourn, *Trans. Faraday Soc.*, **54**, 717 (1958).
27. T. F. Shatzki, *J. Polymer Sci.*, **57**, 496 (1962).
28. R. F. Boyer, *Rubber Rev.*, **34**, 1303 (1963).
29. W. Pechold, S. Blasenbrey and S. Woerner, *Kolloid-Z, Z. Polym.*, **189**, 14 (1963).

Chapter 10

Creep, Stress Relaxation and Non-linear Viscoelasticity

In the course of the extensive studies of the creep and recovery behaviour of textile fibres already referred to, Leaderman [1] became one of the first to appreciate that the simple assumptions of linear viscoelasticity might not hold even at small strains. For nylon and cellulosic fibres he discovered that although the creep and recovery curves may be coincident at a given level of stress, the creep compliance plots indicated a softening of the material as stress increased, except at the shortest times (Figure 10.1). For materials such as the polypropylene filament illustrated in Figure 10.2 the non-linearity is even more pronounced: at a given stress the instantaneous recovery is

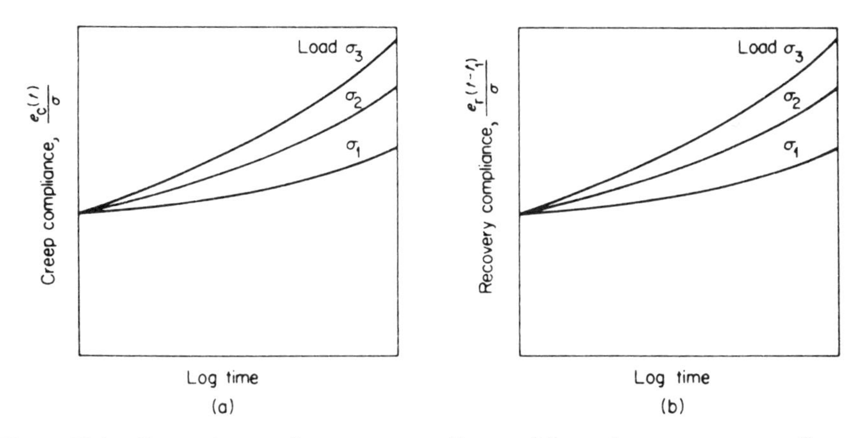

Figure 10.1 Comparison of creep compliance (a) and recovery compliance (b) at three load levels σ_1, σ_2, σ_3 for a non-linear viscoelastic material obeying Leaderman's modified Boltzmann superposition principle. Note that the creep and recovery curves for a given load level are identical

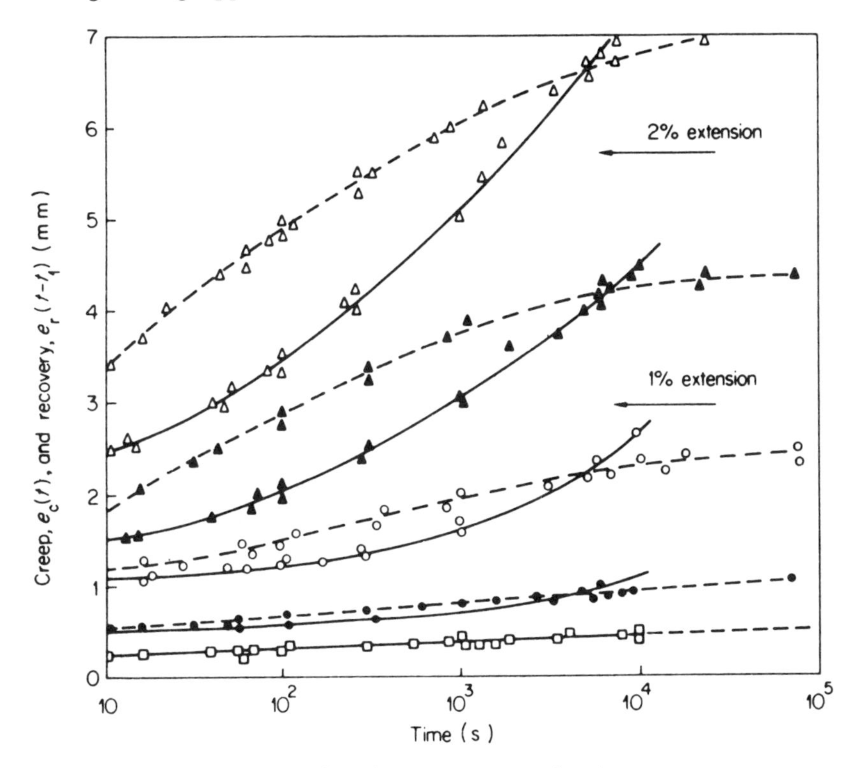

Figure 10.2 Succesive creep (——) and recovery (– – –) for an oriented monofilament of polypropylene of total length 302 mm. The load levels are 587 g (△), 401.8 g (▲), 281 g (●) and 67.7 g (□). (Reproduced with permission from Ward and Onat, *J. Mech. Phys. Solids*, **11** 217 (1963)

greater than the instantaneous elastic deformation, although the delayed recovery proceeds at a slower rate than the preceding creep.

Leaderman's approach was to modify the basic Boltzmann superposition principle of linear viscoelasticity, so that the strain was given by

$$e(t) = \frac{\sigma}{E} + \int_{-\infty}^{t} \frac{\mathrm{d}f(\sigma)}{\mathrm{d}\tau} J(t-\tau)\,\mathrm{d}\tau, \qquad (10.1)$$

where $f(\sigma)$ is an empirical function of stress that depends arbitrarily on the test fibres. This form of empiricism is inadequate to describe the behaviour in loading programmes of greater complexity than creep and recovery, and emphasizes that no general treatment is known to cope with the problems of non-linear viscoelasticity.

In this chapter we draw attention to some of the problems and outline a few methods of approach, which are amplified in the advanced text by Ward

[2]. A theoretical survey, using sophisticated mathematical methods, is given in the text by Lockett [3], and the contrasting approach of the practical engineer is typified in the work of Turner [4].

10.1 THE ENGINEERING APPROACH

10.1.1 Isochronous Stress–Strain Curves

The aim here is to predict the behaviour for a proposed application, using the minimum of experimental data. Empirical relations between stress, strain and time are obtained, which may have no physical significance, and their use may be restricted to very specific stress or strain programmes. In the general case creep curves need to cover the complete range of stresses over as long a period of time as feasible. However, Turner [4] has shown that when stress and time dependence are approximately separable it may be possible to interpolate creep curves at intermediate stresses from a knowledge of two creep curves, combined with a knowledge of the stress–strain curve for a fixed time (say 100 s): so-called isochronal

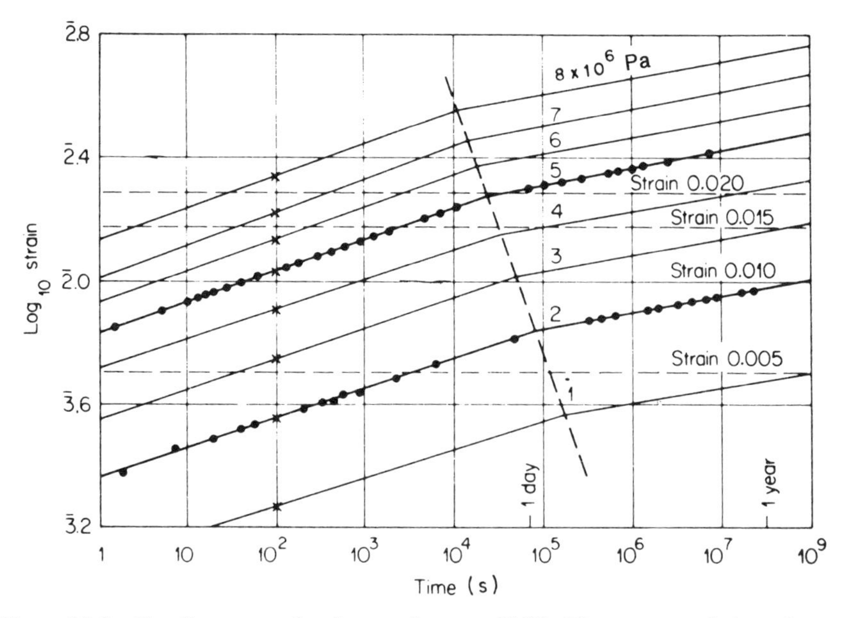

Figure 10.3 Tensile creep of polypropylene at 60 °C. The stress and time dependence are approximately separable and therefore creep curves at intermediate stresses can be interpolated from a knowledge of two creep curves (●) and the isochronous stress–strain relationship (x). (Reproduced with permission from Turner, *Polymer Eng. Sci.*, **6** 306 (1966))

stress–strain curves represented by the vertical line in Figure 10.3. These data cannot, however, be used to predict the response in multistep loading situations for non-linear viscoelastic materials, because the Boltzmann superposition principle does not then hold. In a number of cases it is claimed that simplifying regularities become evident if data are replotted in terms of *fractional recovery*, defined as the strain recovered as a fraction of the maximum creep strain, and *reduced time*, which is the ratio of recovery time to creep time.

10.1.2 Power Laws

An alternative approach has been the application to polymers of power law and other analytical relationships. Pao and Martin [5] have extended to polymers the approach adopted for metals by Martin [6], where the total strain in creep e is considered to consist of three independent components: an elastic strain e_1, a transient recoverable viscoelastic strain e_2 and a permanent non-recoverable plastic strain e_3. At a fixed stress σ, $e_1 = \sigma/E$, where E is Young's modulus, e_2 is similar in form to the creep of a Kelvin element (section 4.2.3.1) but with the added complication of relating to σ^n rather than σ and e_3 is found by integrating the condition that the plastic strain rate is a function of stress only. For simplicity the functions of stress for both e_2 and e_3 are assumed to be simple power laws, and usually the same power law is adopted for each component. Under a constant stress σ the total creep strain is then

$$e = \frac{\sigma}{E} + K\sigma^n[1 - \exp(-qt)] + B\sigma^n t, \tag{10.2}$$

where K, n, p and B are constants for a given material. By using the concept of equivalent stress and equivalent strain the approach can be extended to three-dimensional stress situations, for example it can be extended to creep in bending and torsion [5].

Another approach, exemplified by the work of Findley and Khosla [7] is based on analytical relations originally derived for metals [8].

Under some circumstances creep strain e_c and time t can be related through

$$e_c(t) = e_0 + mt^n, \tag{10.3}$$

where e_0 and m are functions of stress for a given material and n is a material constant. A more general relation for single-step loading tests is

$$e_c(\sigma, t) = e'_0 \sinh\frac{\sigma}{\sigma_e} + m' t^n \sinh\frac{\sigma}{\sigma_m} \tag{10.4}$$

where m', σ_e and σ_m are constants for a given specimen.

The Andrade creep law for metals [9] is a simplified form of equation (10.3)

$$e_c(t) = e_0 + \beta' t^{1/3}, \tag{10.5}$$

where β' is a constant, and has been quite widely used for the creep of amorphous glassy polymers, notably by Plazek [10] and Van Holde [11].

The approaches outlined above can be useful to the design engineer for applications relating to loading under a constant stress, but in no case is a general representation achievable for creep and recovery under complicated loading programmes, and creep data cannot be related directly to stress relaxation and dynamic mechanical experiments.

10.2 THE RHEOLOGICAL APPROACH

There is no general method of dealing with non-linear viscoelastic behaviour, and here we summarize three of the approaches that have been applied with some success in specific instances. The first approach takes linear viscoelasticity as the starting-point; then we consider a functional representation in which the separability of stress, strain and time is preserved, so leading to a single-integral representation that is consistent with rigorous continuum mechanics; finally we consider a development based on a multiple-integral method. In all cases the approach is phenomenological, with constitutive equations being found that describe, on the basis of continuum mechanics, the viscoelastic behaviour of a particular set of specimens. As a consequence the relationships derived cannot be interpreted directly at the molecular level.

10.2.1 Large-strain Behaviour of Elastomers

T. L. Smith [12] has described the large-strain behaviour of elastomers by taking as his starting-point the (linear) Maxwell element (section 4.2.3.2 above) for which stress σ and strain e are related through

$$\frac{\mathrm{d}e}{\mathrm{d}t} = \frac{\sigma}{\eta} + \frac{1}{E}\frac{\mathrm{d}\sigma}{\mathrm{d}t}$$

The retardation time $\tau = \eta/E$, where η represents the viscous component and E the elastic component.

At a constant strain rate the model can be generalized for a continuous distribution of relaxation times to yield the 'constant strain-rate modulus'

$$F(t) = \frac{1}{t}\int_{-\infty}^{\infty} \tau H(\tau)\left[1 - \exp\left(-\frac{t}{\tau}\right)\right] \mathrm{d}\ln\tau + E_e, \tag{10.6}$$

where E_e is the equilibrium modulus.

Smith assumed that at large strains

$$F(t) = \frac{g(e)\sigma(e, t)}{e} \tag{10.7}$$

where *g(e)* is a function of strain that approaches unity as the strain approaches zero. Hence,

$$\log F(t) = \log\left\{\frac{g(e)}{e}\right\} + \log(e, t). \tag{10.8}$$

As Figure 10.4 shows, the approach was very successful empirically. The stress–strain curves at each strain rate are analysed by constructing plots of log (nominal stress) $v \log t$, which are parallel straight lines as predicted by equation (10.8). The displacements between the lines are $\log\{g(e)/e\}$, which implies that this factor is independent of time. At all but the lowest temperatures $g(e)/e$ was independent of temperature, as might have been anticipated from earlier discussions of time–temperature equivalence.

At strains up to 100% Smith found that true stress (i.e. the stress corrected for changes in area) was proportional to strain, although at larger strains an empirical relation

$$\sigma = E\,\frac{e}{\lambda^2}\exp A\left(\lambda - \frac{1}{\lambda}\right) \tag{10.9}$$

where λ is the extension ratio and A a constant, was employed [13]. These results imply that the non-linearity is attributable to large strains, and the time and strain dependence remain separable.

10.2.2 More Complicated Single-integral Representations

In the work described here the successful separation of the strain and time functions achieved by Smith and Leaderman is extended, and combined with the condition that the stress–strain relationship at finite strains should, at very long times when equilibrium is attained, be similar to that for the elastic behaviour of rubbers.

The elastic response of many rubbers as a function of extension ratio λ has the form indicated in Figure 10.5, where a plot of $\sigma/(\lambda^2 - \lambda^{-1})$ against $1/\lambda$ shows a linear relationship at low strains (high $1/\lambda$) and a minimum at larger strains. The stress relaxation behaviour of several elastomers, and also plasticized polyvinyl chloride, presented in terms of an isochronal stress–strain relationship, takes a similar form [14, 15]. Based on this similarity Bernstein, Kearsley and Zapas [14] developed a theoretical model

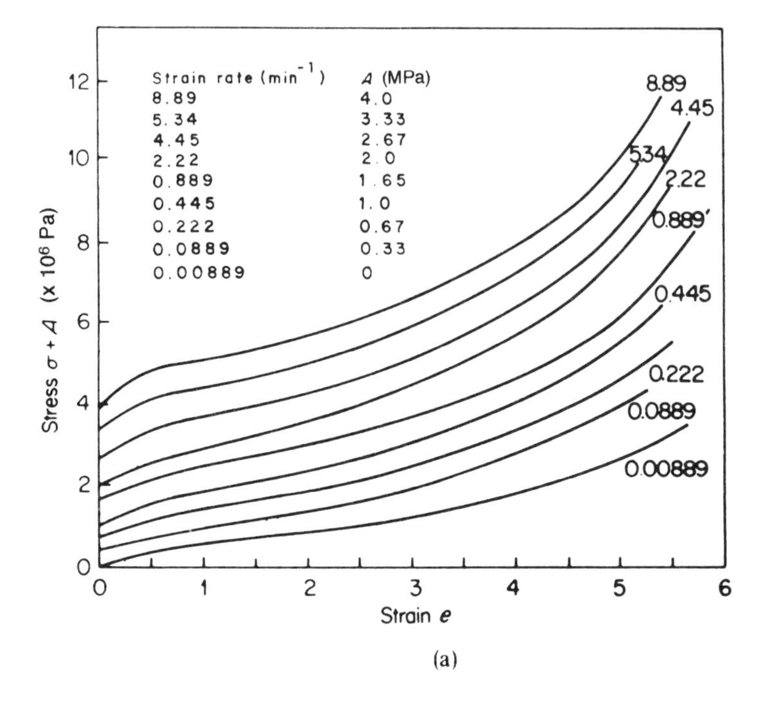

Strain rate (min⁻¹) A (MPa)
8.89 4.0
5.34 3.33
4.45 2.67
2.22 2.0
0.889 1.65
0.445 1.0
0.222 0.67
0.0889 0.33
0.00889 0
8.89
4.45
5.34
2.22
0.889
0.445
0.222
0.0889
0.00889
Stress σ + A (x 10⁶ Pa)
Strain e

(a)

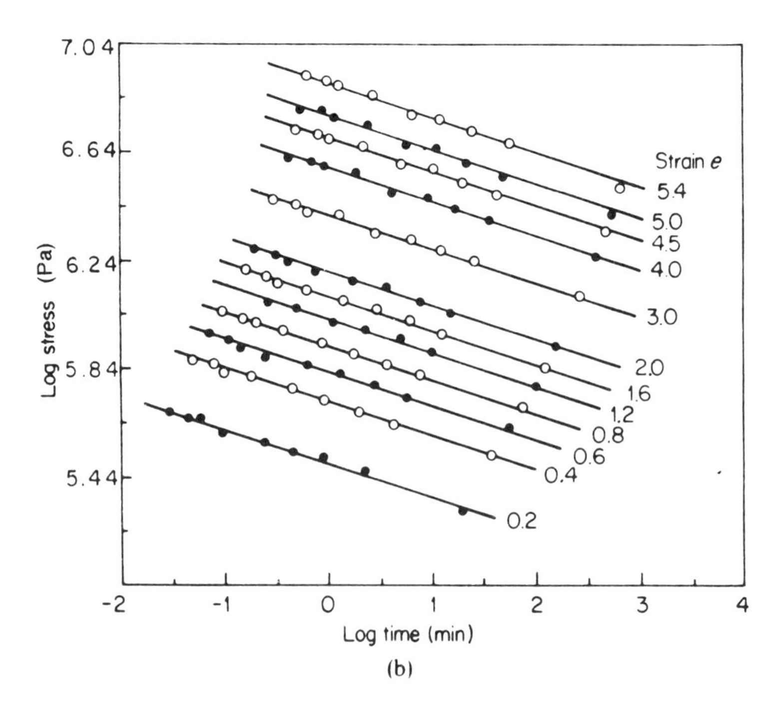

Strain e
5.4
5.0
4.5
4.0
3.0
2.0
1.6
1.2
0.8
0.6
0.4
0.2
Log stress (Pa)
Log time (min)

(b)

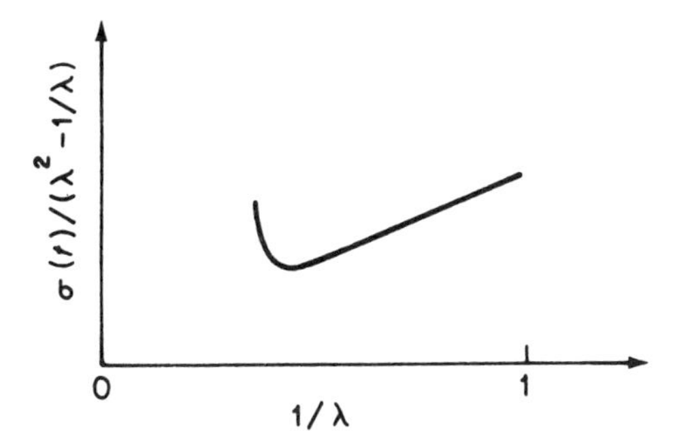

Figure 10.5 Schematic diagram of the equilibrium response for a typical elastomer. (Reproduced with permission from de Hoff, Lianis and Goldberg, *Trans. Soc. Rheol.*, **10**, 385 (1966))

(known as BKZ theory) in which the stress-relaxation behaviour is represented by single integral terms of the general form

$$\int_{-\infty}^{t} A(t-\tau) f[E(\tau)]\,\mathrm{d}\tau,$$

where $E(\tau)$ is a generalized measure of strain.

It was then shown that the stress relaxation in simple extension appropriate for an elastomer at short times (or low temperature) can be expressed by

$$\frac{\sigma(t)}{\lambda^2 - \lambda^{-1}} = (\lambda^2 - 1)\left[\frac{1}{2} A_1(t) + A_2(t)\right] + \frac{1}{\lambda}[A_1(t) + A_2(t)] + A_3 - A_1(t), \tag{10.10}$$

where A_1 and A_2 represent limiting values of time-dependent terms having the form of independent single-integral stress relaxation functions, and A_3 is a constant. When dealing with the stress relaxation behaviour at long times (and high temperatures) the constant A_3 is replaced by a single integral stress relaxation function $A_3(t)$. Zapas and Craft [16] later showed that this modified form of BKZ theory was appropriate for multistage stress relaxation, creep and recovery and constant strain-rate experiments.

Figure 10.4 Tensile stress–strain curves of SBR vulcanized rubber at −34.4 °C and at strain rates between 8.89×10^{-3} and 8.89 min^{-1}. (a) The stress ordinates are displaced by an amount A to enable distinction between curves for different strain rates. (b) Variation of log stress with log time at different strain values for SBR vulcanized rubber. The data were obtained from analysis of the curves shown in (a) and the strain values are indicated for each case. (Reproduced with permission from Smith, *Trans. Soc. Rheol.*, **6**, 61 (1962))

10.3 THE MULTIPLE INTEGRAL REPRESENTATION

The representations of non-linear viscoelasticity previously mentioned incorporate the assumption that strain (or stress) and time dependence are separable. The inadequacy of such representations except in the region of rubber-like response was indicated by a study of oriented polypropylene fibres [17] summarized in Figure 10.2. Creep and recovery coincide only at the lowest stress level, which suggests that a linear viscoelastic region exists only at very low stresses; at higher stresses the 'instantaneous' recovery is always greater than the 'instantaneous' creep, but the subsequent rate of recovery is always slower than the equivalent creep rate.

Replotting the data as isochronous curves of compliance versus stress at different times (Figure 10.6), we see that the creep compliance curves are approximately independent of stress at low stresses, but the recovery curves never coincide. Furthermore, multistep loading results in an additional creep greater than that predicted by the Boltzmann superposition principle, the magnitude of which depends on the time at which the additional stress is applied.

Despite the above complexity the data can be treated by adopting a multiple integral representation of a form earlier proposed by Green and Rivlin [18]. For a *linear* system subjected to incremental stresses $\Delta\sigma_1(\tau_1)$, $\Delta\sigma_2(\tau_2)$ . . . we have seen that the deformation at time t is given by

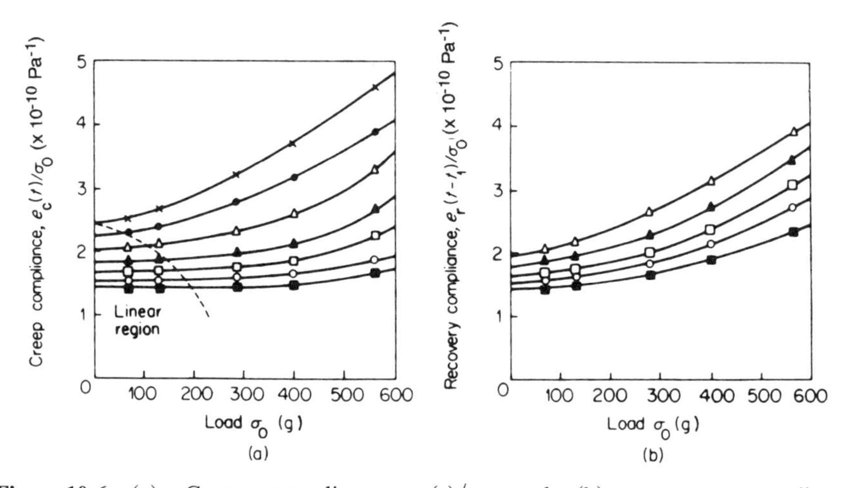

Figure 10.6 (a) Creep compliance $e_c(t)/\sigma_0$ and (b) recovery compliance $e_r(t - t_1)/\sigma_0$ as a function of applied load σ_0 for an oriented polypropylene monofilament. The time of loading for the recovery test was 9.3×10^3 s. The times in seconds are given for both creep and recovery by the following key: (X), 9300; (●), 3000; (△), 1000; (▲) 300; (□), 100; (○), 40; (■) 15. (Reproduced with permission from Ward and Onat, *J. Mech. Phys. Solids*, **11** 217 (1963)

$$e(t) = \Delta\sigma_1 J_1(t - \tau_1) + \Delta\sigma_2 J_2(t - \tau_2) + \ldots,$$

where *J(t)* is the creep compliance function. Terms are now added to account for the *joint* contributions of the various loading steps to the final deformation. These terms are of the form

$$+ \Delta\sigma_1\Delta\sigma_2 J_2(t - \tau_1, t - \tau_2) + \Delta\sigma_1\Delta\sigma_2\Delta\sigma_3 J_3(t - \tau_1, t - \tau_2, t - \tau_3), \text{etc.}$$

where J_2, J_3, etc. are functions of the differences in time between the instant at which the deformation is measured and the instant at which a given increment of stress is applied.

For a continuous load history, the deformation is given by

$$\begin{aligned} e(t) = & \int_{-\infty}^{t} J_1(t - \tau) \frac{\mathrm{d}\sigma(\tau)}{\mathrm{d}\tau} \mathrm{d}\tau \\ & + \int_{-\infty}^{t}\int_{-\infty}^{t} J_2(t - \tau_1, t - \tau_2) \frac{\mathrm{d}\sigma(\tau_1)}{\mathrm{d}\tau_2} \frac{\mathrm{d}\sigma(\tau_2)}{\mathrm{d}\tau_2} \mathrm{d}\tau_1 \mathrm{d}\tau_2 \\ & + \ldots + \int_{-\infty}^{t} \ldots \int_{-\infty}^{t} J_{\mathrm{N}}(t - \tau_1, \ldots, t - \tau_N) \frac{\mathrm{d}\sigma(\tau_1)}{\mathrm{d}\tau_1} \\ & \ldots \frac{\mathrm{d}\sigma(\tau_N)}{\mathrm{d}\tau_N} \mathrm{d}\tau_{\tau_1} \ldots \mathrm{d}\tau_N. \end{aligned} \tag{10.11}$$

The first term is linear, and represents the Boltzmann superposition principle.

This representation appears extremely complicated, but in practical applications it is found that for a specific experiment with a particular material only a limited number of multiple integral terms need to be considered, and the behaviour under complex loading programmes can be predicted from a few simple experiments.

For creep under a constant stress σ_0 the multiple integral representation gives

$$e_{\mathrm{c}}(t) = J_1(t)\sigma_0 + J_2(t, t)\sigma_0^2 \ldots + J_N(t, \ldots t)\sigma_0^N. \tag{10.12}$$

The experimental data of Ward and Onat were compared with the above relation by plotting the isochronous compliance surves of Figure 10.6. A linear viscoelastic solid would yield a series of straight lines parallel to the load axis, and a small region indicated does approximate to such behaviour, but in general the curves are parabolic, which suggests that

$$\frac{e_{\mathrm{c}}(t)}{\sigma_0} = A + B\sigma_0^2, \tag{10.13}$$

where $A = J_1(t)$ and $B = J_3(t, t, t)$ are functions of time only. It appears therefore that the non-linear behaviour of the particular specimen may be represented by retaining only the first- and third-order terms, in which case the recovery response in a creep and recovery test where $\sigma = 0$, $\tau < 0$; $\sigma = \sigma_0$, $0 < \tau < t_1$; $\sigma = 0$, $\tau > t_1$, is given by

$$e_r(t - t_1) = J_1(t - t_1)\sigma_0 + J_3(t - t_1, t - t_1, t - t_1)\sigma_0^3 + 3[J_3(t, t, t - t_1) - J_3(t, t - t_1, t - t_1)]\sigma_0^3. \quad (10.14)$$

Recovery is predicted to take the form

$$\frac{e_r(t - t_1)}{\sigma_0} = A' + B'\sigma_0^2, \quad (10.15)$$

where $A' = J_1(t - t_1)$ and

$$B' = J_3(t - t_1, t - t_1, t - t_1) + 3[J_3(t, t, t - t_1) - J_3(t, t - t_1, t - t_1)].$$

The results, shown in Figure 10.6(b), confirm that the prediction is correct. A two-step loading test, in which the stress was increased to $2\sigma_0$ after time t_1 provided additional confirmation that for the particular specimen only first- and third-order terms need to be retained. However, a later study [19] of a wide range of polypropylene fibres indicated that although only a few terms of the multiple integral representation were required to describe the non-linearity of a given specimen, the actual terms involved changed in a consistent manner with changes in fibre morphology and properties, particularly the molecular orientation.

There have been few studies of the temperature dependence of non-linear viscoelastic materials, although such behaviour might be expected to assist in an explanation in structural or molecular terms. Morgan and Ward [20] carried out creep, recovery and load superposition experiments over the temperature range 28–60 °C on an oriented polypropylene filament similar to that studied by Ward and Onat. Some features of linear viscoelasticity survived: at a given stress both creep and recovery curves could be superposed by means of a simple horizontal shift, with the shifts being identical for all stress levels (Figure 10.7). For similar material Hadley (unpublished) obtained plots of the retardation function at several stress levels, which indicated that as the stress was increased the peak of the spectrum shifted towards shorter times (Figure 10.8).

Although it is of considerable interest to know that the multiple integral representation can be applied in a simplified form to a particular polymer specimen, it must be stressed that it is not possible to predict in advance how an unknown sample will respond to testing, because the simplifications that are applicable must be derived experimentally from an initial test programme.

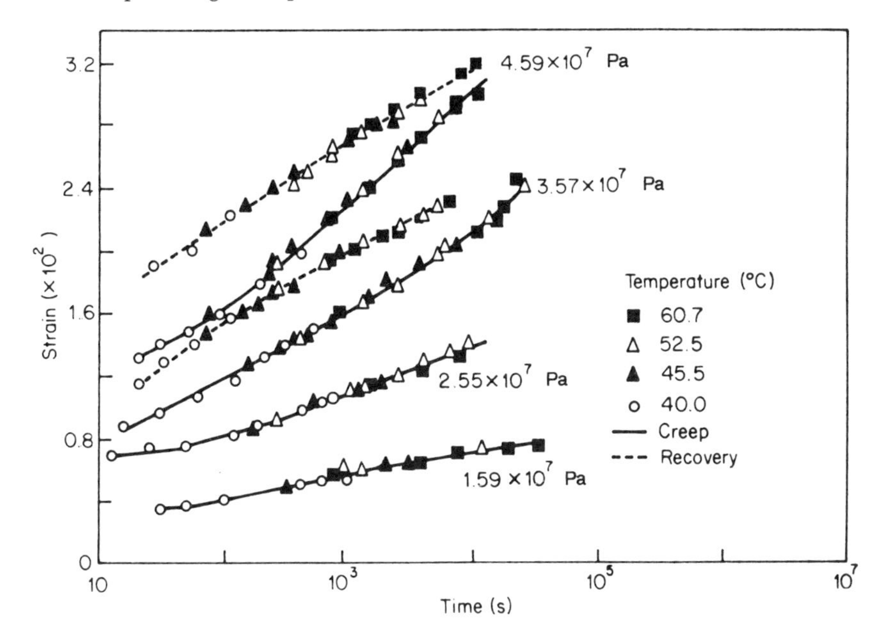

Figure 10.7 Master creep and recovery curves for an oriented polypropylene monofilament reduced to 40 °C (Reproduced with permission from Morgan and Ward, *J. Mech. Phys. Solids*, **19**, 165 (1971))

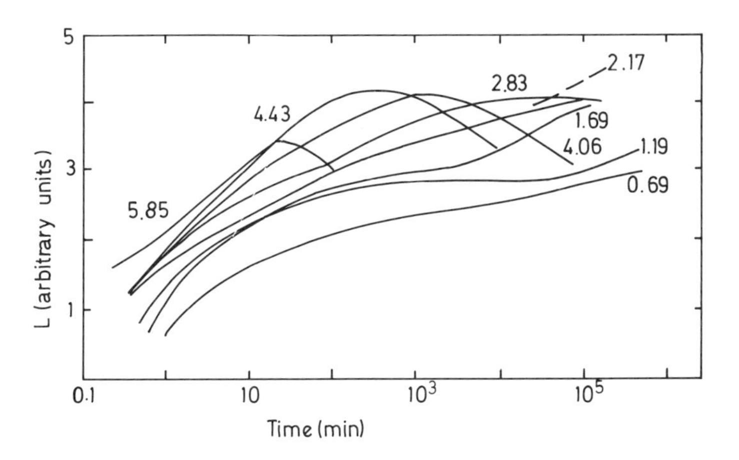

Figure 10.8 Retardation spectrum for a polypropylene fibre. Numbers indicate stress in pascals × 10^7

10.4 CREEP AND STRESS RELAXATION AS THERMALLY ACTIVATED PROCESSES

We have shown (section 4.2.2.3) that the standard linear solid, a three-component spring and dashpot model, provides to a first approximation a description of linear viscoelastic behaviour. Eyring and his colleagues [21] assumed that the deformation of a polymer was a thermally activated rate process involving the motion of segments of chain molecules over potential barriers, and modified the standard linear solid so that the movement of the dashpot was governed by the activated process. The model, which now represents non-linear viscoelastic behaviour, is useful because its parameters include an activation energy and an activation volume that may give an indication of the underlying molecular mechanisms. The activated rate process may also provide a common basis for the discussion of creep and yield behaviour.

10.4.1 The Eyring Equation

Macroscopic deformation is assumed to be the result of basic processes that are either intermolecular (e.g. chain-sliding) or intramolecular (e.g. a change in the conformation of the chain), whose frequency ν depends on the ease with which a chain segment can surmount a potential energy barrier of height ΔH.

In the absence of stress, dynamic equilibrium exists, so that an equal number of chain segments move in each direction over the potential barrier at a frequency given by

$$\nu = \nu_0 \exp\left(-\frac{\Delta H}{RT}\right) \tag{10.16}$$

The equation above is identical to equation (6.16) describing the frequency of a molecular event.

An applied stress σ is assumed to produce linear shifts $\beta\sigma$ of the energy barriers in a symmetrical way (Figure 10.9), where β has the dimensions of volume. The flow in the direction of the applied stress is then given by

$$\nu_1 = \nu_0 \exp\left[-\frac{(\Delta H - \beta\sigma)}{RT}\right]$$

compared with a smaller flow in the backward direction of

$$\nu_2 = \nu_0 \exp\left[-\frac{(\Delta H + \beta\sigma)}{RT}\right]$$

The net flow in the forward direction is then

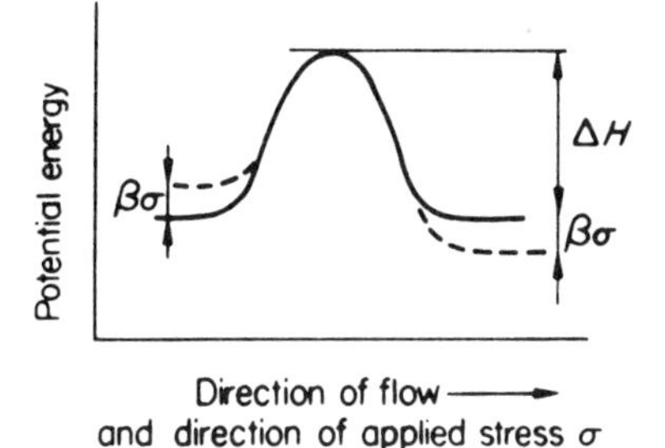

Figure 10.9 The Eyring model for creep

$$v' = v_1 - v_2 = v_0 \exp\left(-\frac{\Delta H}{RT}\right)\left\{\exp\left(\frac{\beta\sigma}{RT}\right) - \exp\left(-\frac{\beta\sigma}{RT}\right)\right\} \quad (10.17)$$

The resemblance of the large bracket to the sinh function should be noted.

Assuming that the net flow in the forward direction is directly related to the rate of change of strain, we have

$$\frac{\mathrm{d}e}{\mathrm{d}t} = \dot{e} = \dot{e}_0 \exp\left(-\frac{\Delta H}{RT}\right)\sinh\left(\frac{\upsilon\sigma}{RT}\right) \quad (10.18)$$

where $\dot{e}_0$ is a constant pre-exponential factor and υ, which replaces β, is termed the activation volume for the molecular event.

The rate of strain equation (10.18) defines an 'activated' viscosity, which is then incorporated in the dashpot of the standard linear solid model, and leads to a more complicated relationship between stress and strain than that for the linear model. The activated dashpot model was tested against Leaderman's data for several fibres [1], and, by a suitable choice of model parameters, gave a good fit, at a given level of stress, over the four decades of time observed.

Subsequently the limitations of simple viscoelastic models have been recognized, and it is accepted that exact fitting of data requires a retardation or relaxation time spectrum, so we must consider why the activated dashpot model was so successful. Although creep curves are sigmoidal when plotted on a logarithmic time-scale, over a long intermediate time region they are, to good approximation, linear. The model predicts creep of the form $e = a' + b' \log t$, which is appropriate to this central region.

10.4.2 More Recent Experiments

Sherby and Dorn [22] investigated the creep under constant stress of glassy PMMA at different temperatures by applying step temperature changes, and constructed plots of creep rate versus total creep strain at a given stress level (Figure 10.10). The data were then superposed by assuming that the temperature dependence at each stress level followed an activated process, to give a relation between strain rate and strain (Figure 10.11). The temperature shifts were interpreted in terms of an activated process where the activation energy fell in a linear manner with increasing stress, to give a creep rate of the form

$$\dot{e} \exp[(\Delta H - v)/RT] = F(e), \tag{10.19}$$

which is the high stress approximation of the Eyring equation, where $\sinh x \approx \frac{1}{2}\exp x$ and v is the activation volume.

Mindel and Brown [23] performed a Sherby–Dorn type analysis on data for the compressive creep of polycarbonate. Superposition was achieved using an equation of the form of (10.19) with an activation volume of $5.7\,\text{nm}^3$, which was very close to the values of the activation volume obtained from measurements of the strain rate dependence of the yield stress (section 11.5.1).

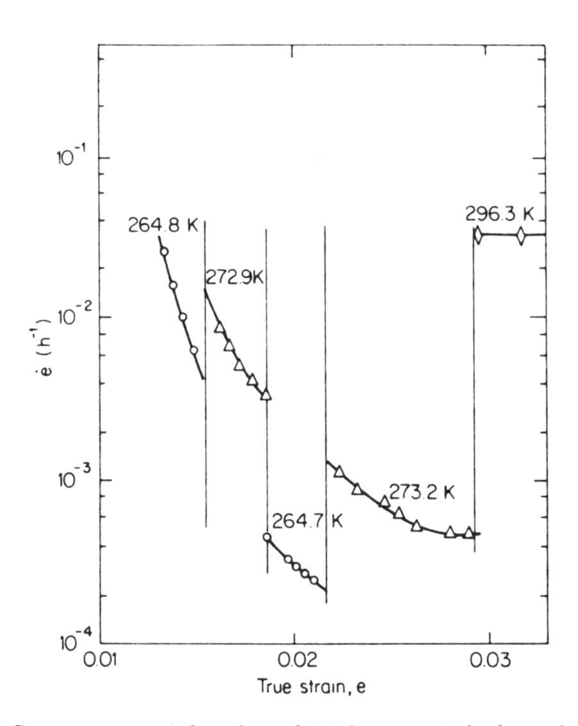

Figure 10.10 Creep rates as a function of total creep strain for polymethyl methacrylate at indicated temperatures for a stress level of 5.6×10^7 Pa. (Reproduced with permission from Sherby and Dorn, *J. Mech. Phys. Solids*, **6**, 145 (1958))

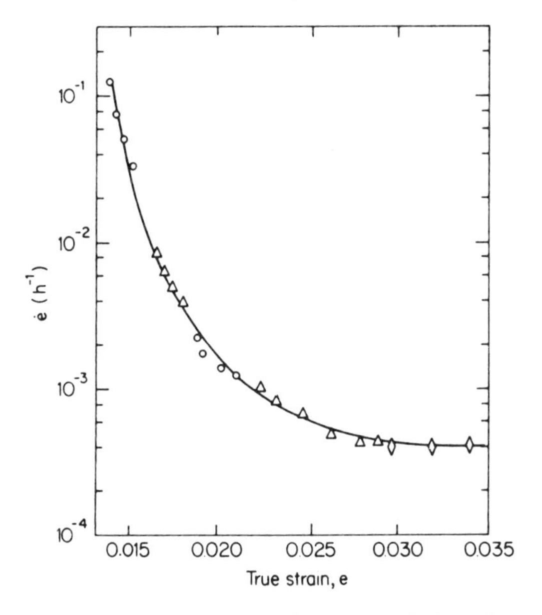

Figure 10.11 Superposition of creep data for polymethyl methacrylate at different temperatures at a stress level of 5.6×10^7 Pa. (Reproduced with permission from Sherby and Dorn, *J. Mech. Phys. Solids,* **6**, 145 (1985))

The results suggest that creep rate can be represented by a general equation of the form

$$\dot{e} = f_1(T)f_2(\sigma/T)f_3(e), \tag{10.20}$$

where $f_1(T)$, $f_2(\sigma/T)$ and $f_3(e)$ are separate functions of the variables T, σ and e. Although $f_1(T)$ has the exponential form expected for a thermally activated process, the exponential for $f_2(\sigma/T)$ is modified to take into account the hydrostatic component of stress, giving different activation volumes for tensile, shear and pressure measurements. Mindel and Brown also proposed that in the region where the creep rate is falling rapidly with increasing strain $f_3(e)$ has the form

$$f_3(e) = \text{const}\exp(-ce_{\text{R}}),$$

where e_{R} is the recoverable component of the creep strain and c is a constant. We then have

$$\dot{e} = \dot{e}_0 \exp[-(\Delta H - \tau V + p\Omega)/RT]\exp(-ce_{\text{R}}), \tag{10.21}$$

where τ and p are the shear and hydrostatic components of stress and V and Ω the shear and pressure activation volumes. The equation may be rewritten as

$$\dot{e} = \dot{e}_0 \exp[-(\Delta H - \{\tau - \tau_i\}V + p\Omega)/RT], \tag{10.22}$$

where $ce_R = \tau_i V/RT$; τ_i, which has the character of an internal stress, increases with strain and is proportional to absolute temperature as would be expected for the stress in a rubber-like network.

Wilding and Ward [24] have used the Eyring rate process to model the creep of ultrahigh modulus polyethylene, and show that at high strains, which correspond to long creep times, the creep rate reaches a constant value called the plateau (or equilibrium) creep rate (Figure 10.12). For polymers of low relative molecular mass the stress and temperature dependence of the final creep rate can be modelled by a single activated process with an activation volume of 0.08 nm^3. In molecular terms this volume is that swept out by a single molecular chain moving through the lattice by a discrete distance.

For polymers of a higher molecular mass, and for copolymers, the permanent flow process was activated only at high stress levels, which suggested that there are two Eyring processes coupled in parallel (Figure 10.13). This suggestion is akin to the representation proposed to describe the strain rate dependence of the yield stress in polymers [25–27]. Process A has the smaller tensile activation volume ($\sim$0.05 nm^3) and larger pre-exponential factor, and is activated only at high stress levels. Process B has the larger tensile activation volume ($\sim$1 nm^3) and a smaller pre-exponential

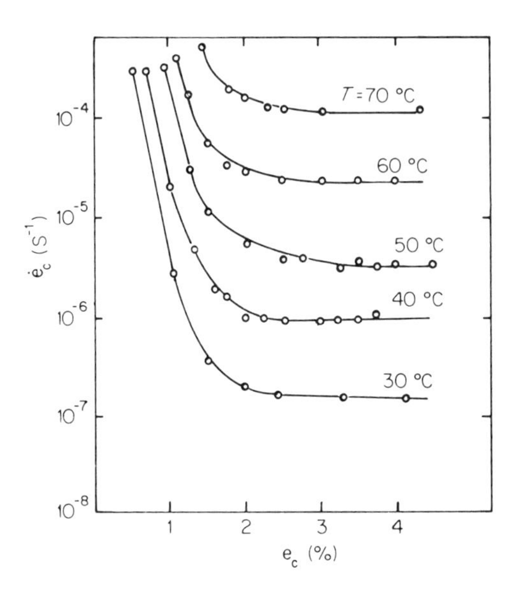

Figure 10.12 Sherby–Dorn plots of creep of ultrahigh modulus polyethylene at different temperatures. (Reproduced with permission from Wilding and Ward, *Plastics and Rubber Processing and Applications*, **1**, 167 (1981))

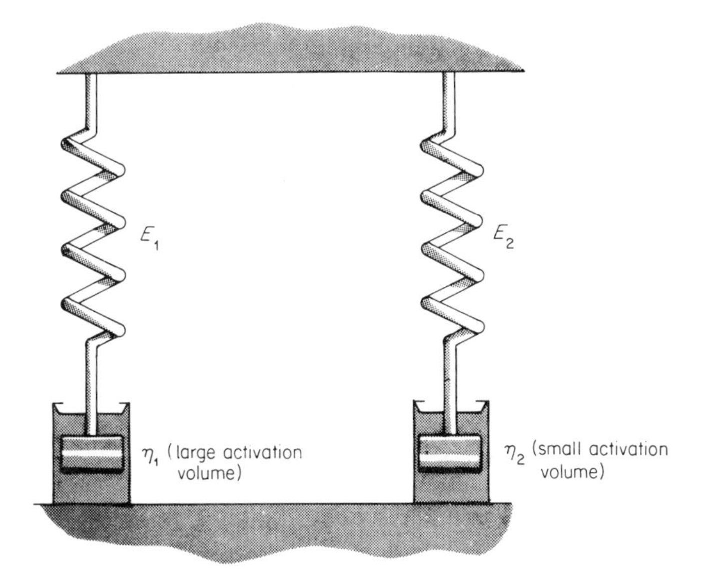

Figure 10.13 The two-process model for permanent flow creep. (Reproduced with permission from Wilding and Ward, *Plastics and Rubber Processing and Applications*, **1**, 167 (1981)

factor, and is operative at low stress levels. At low stresses there will be little permanent flow because B carries almost the whole load. Although the overall creep and recovery behaviour can be represented by a model containing two activated dashpots, a spectrum of relaxation times would be required to give an accurate fit to experimental data over the complete time and strain scales.

In summary, it must be admitted that none of the approaches considered is entirely satisfactory: the constitutive relations of the theoretician tend to be of so general a nature that the experimentalist cannot appreciate their relevance to a particular problem; on the other hand the empiricism of the experimentalist is severely restricted in its capacity for prediction and physical insight.

REFERENCES

1. H. Leaderman, *Elastic and Creep Properties of Filamentous Materials and Other High Polymers*, Textile Foundation, Washington, DC, 1943.
2. I. M. Ward, *Mechanical Properties of Solid Polymers*, J. Wiley, Chichester, 1983.
3. F. J. Lockett, *Non-Linear Viscoelastic Solids*, Academic Press, London, 1972.
4. S. Turner, *Polymer Eng Sci.*, **6**, 306 (1966).

5. Y. H. Pao and J. Marin, *J. Appl. Mech.,* **19**, 478 (1952); **20**, 245 (1953).
6. J. Marin, *Trans. ASME*, **59**, A21 (1937).
7. W. N. Findley and G. Khosla, *J. Appl. Phys.,* **26**, 821 (1955).
8. P. G. Nutting, *J. Franklin Inst.*, **235**, 513 (1943).
9. E. N. Da C. Andrade, *Proc Roy Soc.,* **A84**, 1 (1910).
10. D. J. Plazek, *J. Colloid Sci.*, **15**, 50 (1960).
11. K. Van Holde., *J. Polymer Sci.,* **24**, 417 (1957).
12. T. L. Smith, *Trans. Soc. Rheol.,* **6**, 61 (1962).
13. G. M. Martin, F. L. Roth and R. D. Stiehler, *Trans. Inst. Rubber Ind.,* **32**, 189 (1956).
14. B. Bernstein, E. A. Kearsley and L. P. Zapas, *Trans. Soc. Rheol.,* **7**, 391 (1963).
15. P. H. de Hoff, G. Lianis and W. Goldberg, *Trans. Soc. Rheol.,* **10**, 385 (1966).
16. L. J. Zapas and T. Craft, *J. Res. Natl. Bur. Stand A.*, **69**, 541 (1965).
17. I. M. Ward and E. T. Onat, *J. Mech. Phys. Solids.,* **11**, 217 (1963).
18 A. E. Green and R. S. Rivlin, *Arch. Rat. Mech. Analysis,* **1**, 1 (1957); **4**, 387 (1960).
19. D. W. Hadley and I. M. Ward, *J. Mech. Phys. Solids,* **13**, 397 (1965).
20. C. J. Morgan and I. M. Ward, *J. Mech. Phys. Solids,* **19**, 165 (1971).
21. G. Halsey, H. J. White and H. Eyring, *Text Res. J.,* **15**, 295 (1945).
22. O. D. Sherby and J. E. Dorn., *J. Mech. Phys. Solids,* **6**, 145 (1958).
23. M. J. Mindel and N. Brown, *J. Mater Sci.,* **8**, 863 (1973).
24. M. A. Wilding and I. M. Ward, *Polymer*, **19**, 969 (1978); **22**, 870 (1981).
25. J. A. Roetling, *Polymer*, **6**, 311 (1965).
26. C. Bauwens-Crowet, J. C. Bauwens and G. Homès, *J. Polymer Sci. A2,* **7**, 735 (1969).
27. R. E. Robertson, *J. Appl. Polymer Sci.,* **7**, 443 (1963).

Chapter 11

The Yield Behaviour of Polymers

Until comparatively recently, the yield behaviour of polymers has not received much attention, mainly because it was not thought profitable to treat it as a distinct mode of mechanical behaviour, different in kind from either the viscous flow processes which occur at high temperatures or the large extensions observed in the temperature range above the glass transition. The yield process in a polymer was often considered to be a softening due to a local rise in temperature and was referred to as a localized 'melting'.

A number of different factors contribute to the present interest in the yield behaviour. First, it has been recognized that the classical concepts of plasticity are relevant to forming, rolling and drawing processes in polymers. Secondly, there have been a number of striking experimental studies of 'slip bands' and 'kink bands' in polymers which suggest that deformation processes in polymers might be similar to those in crystalline materials such as metals and ceramics. Finally, it is evident that distinct yield points are observed and there is much interest in understanding these in the context of other ideas in polymer science.

Our first task in this chapter is to discuss the relevance of classical ideas of plasticity to the yielding of polymers. Although the yield behaviour is temperature and strain rate dependent it will be shown that provided that the test conditions are chosen suitably, yield stresses can be measured which satisfy conventional yield criteria.

The temperature and time dependence often obscure some generalities of the yield behaviour. For example, it might be concluded that some polymers show necking and cold-drawing, whereas others are brittle and fail catastrophically. Yet another type of polymer (a rubber) extends homogeneously to rupture. A salient point to recognize is that polymers in general

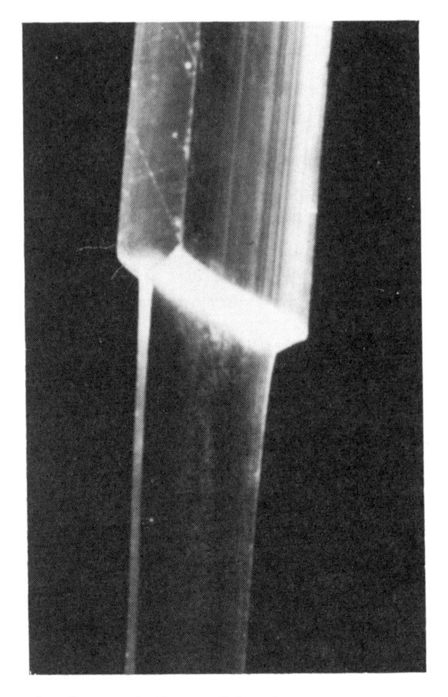

Figure 11.1 Photograph of a neck formed in the redrawing of oriented polyethylene

can show all these types of behaviour depending on the exact conditions of test (Figure 12.1), quite irrespective of their chemical nature and physical structure. Thus explanations of yield behaviour which involve, for example, cleavage of crystallites or lamellar slip or amorphous mobility are only relevant to specific cases. As in the case of linear viscoelastic behaviour or rubber elasticity we must first seek an understanding of the relevant phenomenological features, decide on suitable measurable quantities and then provide a molecular interpretation of the subsequent constitutive relations.

11.1 DISCUSSION OF THE LOAD–ELONGATION CURVES IN TENSILE TESTING

The most dramatic manifestation of yield is seen in a tensile test when a neck or deformation band occurs, as in Figure 11.1, with the plastic deformation concentrated either entirely or primarily in a small region of the specimen. The precise nature of the plastic deformation depends both on the geometry of the specimen and on the form of the applied stresses, and will be discussed more fully later.

The characteristic necking and cold-drawing behaviour is as follows. On the initial elongation of the specimen, homogeneous deformation occurs and the conventional load–extension curve shows a steady increase in load with increasing elongation (AB in Figure 11.2). At the point B the specimen thins to a smaller cross-section at some point, i.e. a neck is formed. Further elongation brings a fall in load. Continuing extension is achieved by causing the shoulders of the neck to travel along the specimen as it thins from the initial cross-section to the drawn cross-section. The existence of a finite or natural draw ratio is an important aspect of polymer deformation and is discussed in section 11.6.1 below. Ductile behaviour in polymers does not always give a stabilized neck, so that the requirements for necking and cold-drawing must now be considered in some detail.

11.1.1 Necking and the Ultimate Stress

It is important to distinguish between the nominal stress, which is the load at any time during deformation divided by the initial cross-sectional area, and the true stress, which is the load divided by the actual cross-section at any time. The cross-section of the sample decreases with increasing extension, so that the true stress may be increasing when the apparent or conventional stress or load may be remaining constant or even decreasing. This has been very well discussed by Nadai [1] and Orowan [2] and their argument will be followed here.

Consider the conventional stress–strain curve or the load–elongation curve for a ductile material (Figure 11.3). The ordinate is equal to the nominal stress σ_a obtained be dividing the load P by the original cross-sectional area A_0:

$$\sigma_a = P/A_0.$$

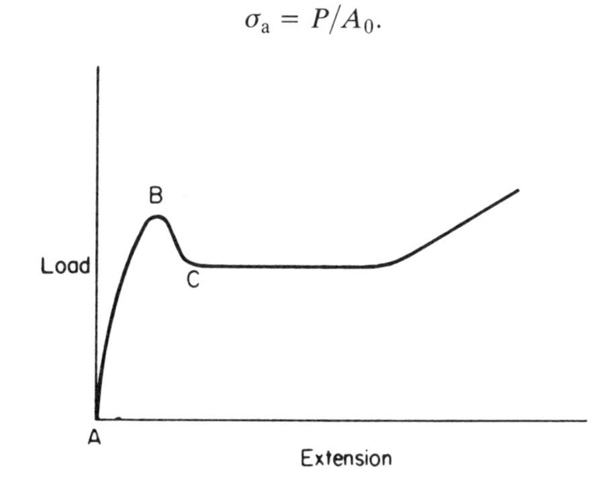

Figure 11.2 Typical load–extension curve for a cold-drawing polymer

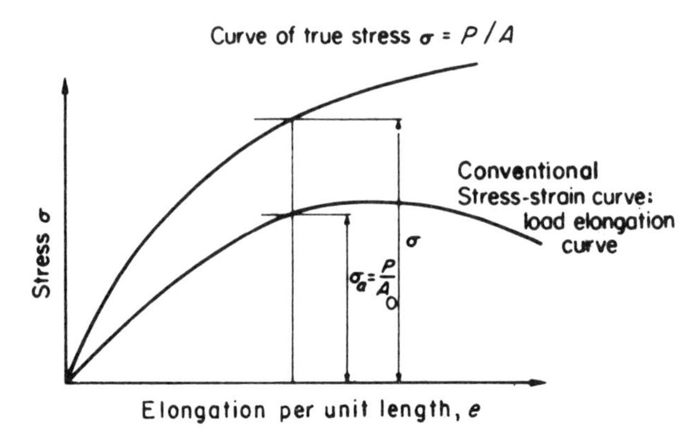

Figure 11.3 Comparison of load–elongation curve and true stress elongation curve

This gives a stress–strain curve of the form shown. The load reaches its maximum value at the instant the uniform extension of the sample stops. At this elongation the specimen begins to neck and consequently the load falls as shown by the last part of the stress–strain curve. Finally the sample fractures at the narrowest point of the neck.

It is more instructive to plot the true tensile stress at any elongation rather than the nominal stress σ_a.

The true stress is $\sigma = P/A$, where A is the actual cross-section at any time. We now assume, as is usual for plastic deformation, that the deformation takes place at constant volume. Then $Al = A_0 l_0$, and if we put $l/l_0 = 1 + e$ where e is the elongation per unit length,

$$A = \frac{A_0 l_0}{l} = \frac{A_0}{1 + e}$$

The true stress is given by

$$\sigma = \frac{P}{A} = \frac{(1 + e)P}{A_0} = (1 + e)\sigma_a.$$

From which the load

$$P = \frac{A_0 \sigma}{1 + e}$$

Thus if we know σ, the true stress, as a function of e, i.e. the true stress–strain curve, P can be computed for any elongation. In particular P_{max}, the maximum load, is defined by the condition $dP/de = 0$, i.e.

$$\frac{dP}{de} = \frac{A_0}{(1+e)^2}\left[(1+e)\frac{d\sigma}{de} - \sigma\right] = 0$$

or

$$\frac{d\sigma}{de} = \frac{\sigma}{1+e}$$

The measured ultimate stress can be obtained from the true stress–strain curve by the simple construction due to Considère shown in Figure 11.4. The ultimate stress is obtained when the tangent to the true stress–strain surve $d\sigma/de$ is given by the line from the point -1 on the elongation axis. The angle α in Figure 11.4 is defined by

$$\tan\alpha = \frac{d\sigma}{de} = \frac{\sigma}{1+e} = \frac{\sigma}{\lambda}$$

where λ is the draw ratio. The ultimate stress has a much greater significance than is the case for metals as it is a determining factor in deciding whether a polymer will neck and cold-draw.

The significance of the argument at this stage relates to the failure of plastics in the ductile state. Orowan [2] first pointed out that for ductile materials the ultimate stress is entirely determined by the stress–strain curve, that is, by the plastic behaviour of the material, without any reference to its fracture properties, provided that fracture does not occur before the load maximum corresponding to $d\sigma/de = \sigma/(1+e)$ is reached. Yield stress is thus an important property in many plastics, and defines the practical limit of behaviour much more than the ultimate fracture, unless the plastic fails by brittle fracture.

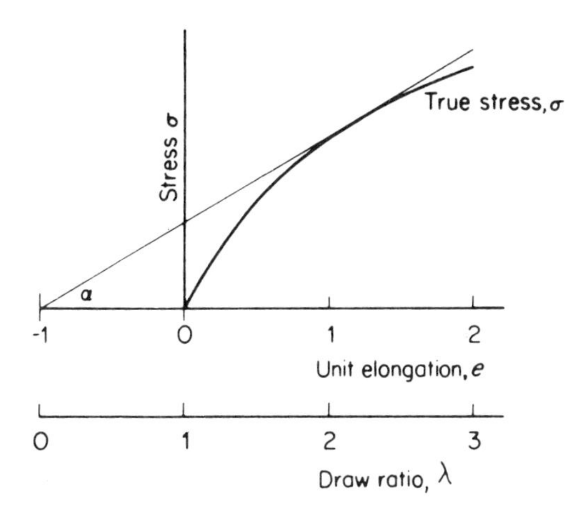

Figure 11.4 The Considère construction

11.1.2 Necking and Cold-drawing: A Phenomenological Discussion

Figure 11.5 shows that there are three distinct regions on the true stress–strain curve of a typical cold-drawing polymer.

1. Initially the stress rises in an approximately linear manner as the applied strain increases.
2. At the yield point there is a fall in true stress and a region where the stress rises less steeply with the strain. In some earlier discussions [3] it was not recognized that a fall in true stress can occur, and it was proposed only that the true stress rises less steeply with increasing strain (dotted line in Figure 11.5). This region was attributed to 'strain softening'.
3. Finally at large extensions the slope of the true stress–strain curve increases again, i.e. a 'strain-hardening' effect occurs.

There are two ways in which a neck may be initiated. First, if the cross-section of the sample is not uniform, the element with the smallest effective cross-section will be subjected to the highest true stress, and so will reach the yield point before any other element in the sample. Secondly, a fluctuation in material properties may cause a localized reduction of the yield stress in a given element so that this element reaches the yield point at a lower applied load. When a particular element has reached its yield point it is easier to continue deformation entirely within this element because it has a lower flow stress stiffness than the surrounding material. Hence further deformation of the sample is accompanied by straining in only one region and a 'neck' is formed.

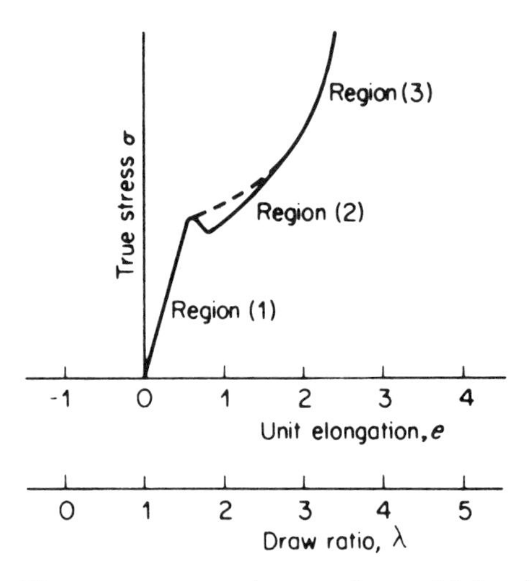

Figure 11.5 The true stress–strain curve for a cold-drawing polymer

This localized deformation will continue until strain hardening increases the effective stiffness of the element, i.e. it reaches the third part (3) of the true stress–strain curve in Figure 11.5. At this point the deformation will stabilize in the highly strained element and in order to accommodate further extension of the sample as a whole, new elements will be brought to their yield point. In this way a neck propagates along the length of a sample as successive elements are brought to a degree of strain hardening such that the flow stress is greater than the yield stress of the undeformed material. This process of necking and cold-drawing results in a non-uniform distribution of stress and strain along the length of the test specimen.

11.1.3 Use of the Considère Construction

In Figure 11.6 two tangent lines have been drawn to the true stress–strain curve from the point $e = -1$ or $\lambda = 0$. In terms of the extension ratio or draw ratio λ, Considère's construction gives the conventional stress

$$\frac{P}{A_0} = \frac{\sigma}{1 + e} = \frac{\sigma}{\lambda}$$

Thus a line from the point $\lambda = 0$ to a point on the true stress–strain curve has slope σ/λ and gives us the conventional stress, i.e. the applied tension at that point.

The first tangent line has been drawn from O to D, i.e. to the yield point. At this point $\mathrm{d}\sigma/\mathrm{d}\lambda = \sigma/\lambda$ and hence the conventional stress is a maximum. As further deformation takes place the slope of $\mathrm{d}\sigma/\mathrm{d}\lambda$ decreases continuously as the true stress–strain curve is traced, and then starts to increase again until the point E is reached, where strain hardening occurs. The final tension settles down to a value represented by the slope of the line OE, the second tangent line, (i.e. $\mathrm{d}\sigma/\mathrm{d}\lambda = \sigma/\lambda$ again) and drawing takes place by

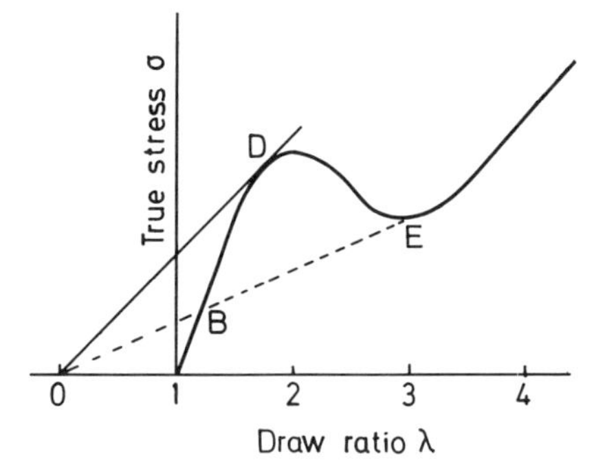

Figure 11.6 A true stress–strain curve for a cold-drawing polymer and the Considère construction

deforming successive elements by an amount corresponding to this extension ratio. After the entire sample has been drawn, further deformation may take place along the steeper part of the stress–strain curve until fracture occurs.

It should be noted that these arguments do not satisfactorily explain how the sample draws at a constant tension throughout its length (along the line BE in the diagram) and at the same time apparently follows the true stress–strain curve along BDE. This anomaly has been attributed by Vincent [3] to our ignorance regarding the actual stress conditions in the neck, where from corresponding work in metals [4] it can be implied that there is a complex combination of tensile and hydrostatic stresses.

The Considère construction can be used as a criterion to decide whether a polymer will neck, or will neck and cold-draw. There are three possible situations, shown in Figure 11.7:

1. $d\sigma/d\lambda$ is always greater than σ/λ, and it can be seen that there is no tangent line which can be drawn to the true stress–strain curve from the point $\lambda = 0$. The polymer therefore extends uniformly with increasing load, and no neck is formed.
2. $d\sigma/d\lambda = \sigma/\lambda$ at one point. The polymer extends uniformly up to the point where $d\sigma/d\lambda = \sigma/\lambda$ and then necks. The neck gets steadily thinner and then the measured load decreases until fracture occurs. This case has been discussed previously, where the treatment of Nadai and Orowan [1,2] was presented (section 11.1.1).
3. $d\sigma/d\lambda = \sigma/\lambda$ at two points. This is the case where necking and cold-drawing occurs, and so the requirement of two tangents becomes a criterion for necking and cold-drawing.

Note that these arguments do not take into account the strain rate dependence of the yield or flow stress σ. As will be discussed in detail

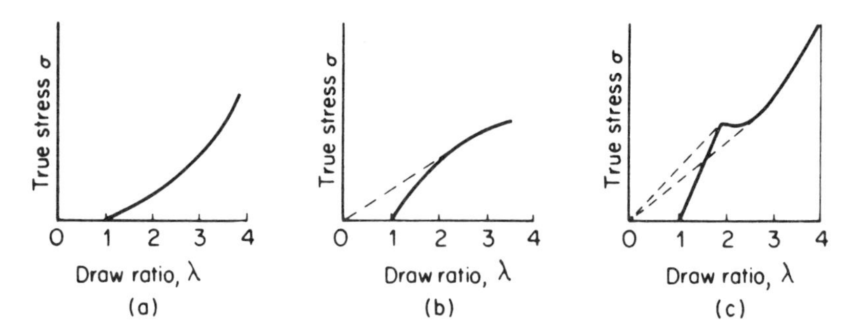

Figure 11.7 The three types of true–strain curves for polymers (a) $d\sigma/d\lambda > \sigma/\lambda$; (b) $d\sigma/d\lambda = \sigma/\lambda$ at one point, and (c) $d\sigma/d\lambda = \sigma/\lambda$ at two points. (Reproduced with permission from Vincent, *Polymer*, **1**, 7 (1960). (C) IPC Business Press Ltd)

below, the flow stress increases with increasing strain rate. Hence in the neck where the cross-sectional area is decreasing rapidly, the strain rate is correspondingly increasing, and so leads to an increase in the flow stress which tends to stabilize the neck.

11.1.4 Definition of Yield Stress

Yield stress may most simply be regarded as the minimum stress at which permanent strain is produced when the stress is subsequently removed. Although this deformation is satisfactory for metals, where there is a clear distinction between elastic recoverable definition and plastic irrecoverable deformation, in polymers the distinction is not so straightforward. In many cases, such as the tensile tests discussed above, yield coincides with the observation of a maximum load in the load–elongation curve. The yield stress can then be defined as the true stress at the maximum observed load (Figure 11.8(a)). Because this stress is achieved at a comparatively low elongation of the sample it is often adequate to use the engineering definition of the yield stress as the maximum observed load divided by the initial cross-sectional area.

In some cases there is no observed load drop and another definition of yield stress is required. One approach is to determine the stress where the two tangents to the initial and final parts of the load–elongation curve intersect (Figure 11.8(b)).

An alternative is to attempt to define an initial linear slope on the stress–strain curve and then to draw a line parallel to this which is offset by a specified strain, say 2%. The interception of this line with the stress–strain curve then defines the offset or proof stress, which is considered to be the yield stress (Figure 11.8(c)).

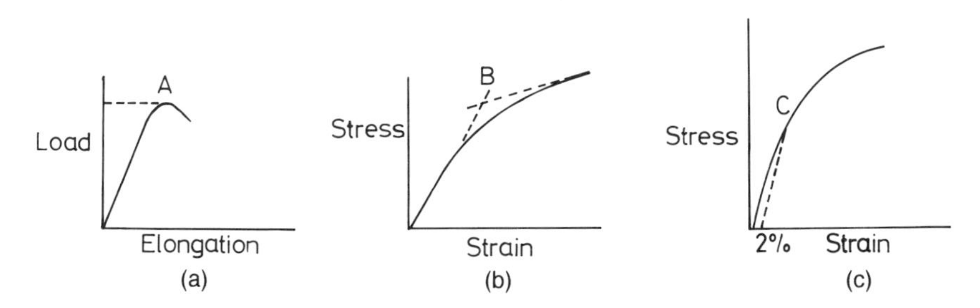

Figure 11.8 (a) The yield stress is defined as load divided by the cross-sectional area at the point A. (b) The yield stress is defined as the stress at the point B. (c) The yield stress is defined as the stress at the point C

11.2 IDEAL PLASTIC BEHAVIOUR

11.2.1 The Yield Criterion: General Considerations

The simplest theories of plasticity exclude time as a variable and ignore any feature of the behaviour which takes place below the yield point. In other words we assume a rigid plastic material whose stress–strain relationship in tension is shown in Figure 11.9. For stresses below the yield stress σ_y there is no deformation. For stresses equal to or above the yield stress the deformation is determined by the measurement of the applied loads, yield can be produced by a wide range of stress states, not just simple tension. In general, it must therefore be assumed that the yield condition depends on all six components of stress, i.e. we can construct a *yield criterion* which can be written as

$$f(\sigma_{xx}, \sigma_{yy}, \sigma_{zz}, \sigma_{xy}, \sigma_{yz}, \sigma_{zx}) = \text{constant}$$

In the first instance we will simplify matters considerably by restricting our considerations to isotropic materials only. This means that we can refer the stresses to the principal axes of stress, so that the stress tensor

$$\begin{bmatrix} \sigma_{xx} & \sigma_{xy} & \sigma_{xz} \\ \sigma_{xy} & \sigma_{yy} & \sigma_{yz} \\ \sigma_{xz} & \sigma_{yz} & \sigma_{zz} \end{bmatrix}$$

becomes

$$\begin{bmatrix} \sigma_1 & 0 & 0 \\ 0 & \sigma_2 & 0 \\ 0 & 0 & \sigma_3 \end{bmatrix}$$

This simplification enables the yield criterion to be expressed in terms of only three variables (by eliminating the three shear-stress components) but at the expense of defining the reference axes (which requires three direction variables).

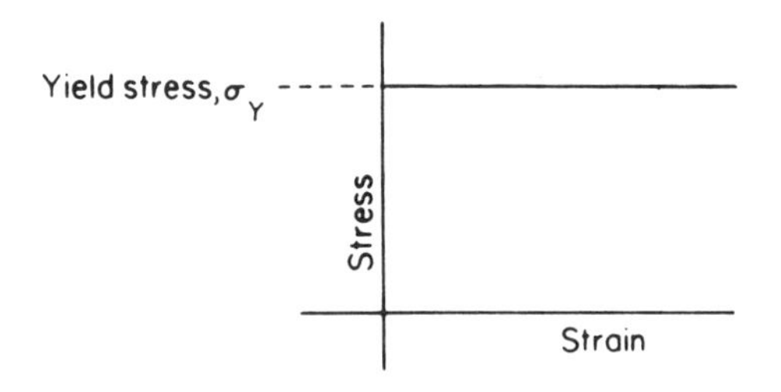

Figure 11.9 Stress–strain relationship for an ideal rigid-plastic material

11.2.2 The Tresca Yield Criterion

The earliest yield criterion to be suggested for metals was Tresca's proposal that yield occurs when the maximum shear stress reaches a critical value [5], i.e.

$$\sigma_1 - \sigma_3 = \text{constant}$$

with $\sigma_1 > \sigma_2 > \sigma_3$.

Of a similar nature is the Schmid critical resolved shear-stress law for the yield of metal single crystals [6].

11.2.3 The Coulomb Yield Criterion

The Tresca yield criterion assumes that the critical shear stress is independent of the normal pressure on the plane on which yield is occurring. Although this assumption is valid for metals, it is more appropriate in polymers to consider the possible applicability of the Coulomb yield criterion [7] which states that the critical shear stress τ for yielding to occur in any plane varies linerarly with the stress normal to this plane i.e.

$$\tau = \tau_c - \mu\sigma_N \tag{11.1}$$

The Coulomb criterion was originally conceived for the failure of soils and τ_C was termed the 'cohesion' and μ the coefficient of internal friction. For a compressive stress, σ_N has a negative sign so that the critical shear stress τ for yielding to occur on any plane increases linearly with the pressure applied normal to this plane.

The Coulomb criterion is often written as

$$\tau = \tau_c - \tan\phi\sigma_N, \tag{11.2}$$

where μ has been written as $\tan\phi$, for reasons which will now become apparent.

Consider uniaxial compression under a compressive stess σ_1 where yield occurs on a plane whose normal makes an angle θ with the direction of σ_1 (Figure 11.10).

The shear stress is $\tau_1 = \sigma\sin\theta\cos\theta$ and the normal stress $\sigma_N = -\sigma_1\cos^2\theta$. Yield occurs when

$$\sigma_1\sin\theta\cos\theta = \tau_c + \sigma_1\tan\phi\cos^2\theta,$$

i.e. when

$$\sigma_1(\cos\theta\sin\theta - \tan\phi\cos^2\theta) = \tau_c.$$

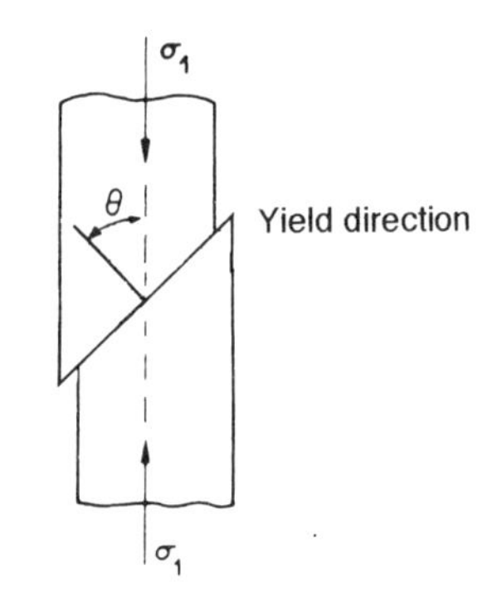

Figure 11.10 The yield direction under a compressive stress σ_1 for a material obeying the Coulomb criterion

For yield to occur at the lowest possible value of σ_1, $(\cos\theta\sin\theta - \tan\phi\cos^2\theta)$ must be a maximum, which gives

$$\tan\phi\tan 2\theta = -1 \quad \text{or} \quad \theta = \frac{\pi}{4} + \frac{\phi}{2} \tag{11.3}$$

Thus $\tan\phi$ determines the direction of yield and conversely the direction of yielding can be used to define ϕ, where $\tan\phi$ is the coefficient of friction. If the stress σ_1 is *tensile* the angle θ is given by

$$\theta = \frac{\pi}{4} - \frac{\phi}{2}$$

We see that the Coulomb yield criterion therefore defines both the stress condition required for yielding to occur and the directions in which the material will deform. Where a deformation band forms, its direction is that which is neither rotated nor distorted by the plastic deformation, because its orientation marks that direction which establishes material continuity between the deformed material in the deformation band and the undistorted material in the rest of the specimen. If volume is conserved, the band direction therefore denotes the direction of shear in a simple shear (by the definition of a shear strain). Thus for a Coulomb yield criterion the band direction is defined by equation (11.3).

11.2.4 The von Mises Yield Criterion

The von Mises yield criterion [8] assumes that the yield behaviour is independent of hydrostatic pressure and that the yield stresses in simple tension and compression are equal. It is expressed most simply in terms of the principal components of stress so that

$$(\sigma_1 - \sigma_2)^2 + (\sigma_2 - \sigma_3)^2 + (\sigma_3 - \sigma_1)^2 = \text{constant.}$$

In rather more sophisticated terms the von Mises criterion assumes that the yield criterion depends only on the components of the deviatoric stress tensor obtained by subtracting the hydrostatic components of stress from the total stress tensor. In terms of principal components of stress the deviatoric stress tensor is

$$\begin{bmatrix} \sigma_1' & 0 & 0 \\ 0 & \sigma_2' & 0 \\ 0 & 0 & \sigma_3' \end{bmatrix} = \begin{bmatrix} \sigma_1 - p & 0 & 0 \\ 0 & \sigma_2 - p & 0 \\ 0 & 0 & \sigma_3 - p \end{bmatrix}$$

where $p = -\frac{1}{3}(\sigma_1 + \sigma_2 + \sigma_3)$ is the hydrostatic presure. The von Mises yield criterion can then be written as

$$(\sigma_1 - p)^2 + (\sigma_2 - p)^2 + (\sigma_3 - p)^2 = \text{constant}. \tag{11.4}$$

The von Mises yield criterion is often written in terms of the so-called octahedral shear stress τ_{oct}, where

$$\tau_{\text{oct}} = \tfrac{1}{3}\{(\sigma_1 - \sigma_2)^2 + (\sigma_2 - \sigma_3)^2 + (\sigma_3 - \sigma_1)^2\}^{1/2}$$

giving the yield criterion as $\tau_{\text{oct}} = \text{constant}$.

We have seen that the Coulomb yield criterion defines both the stresses required for yield and also the directions in which the material deforms. In the case of the von Mises yield criterion we require a further development of the theory to predict the directions in which plastic deformation starts.

It is important to appreciate that plasticity is different in kind from elasticity, where there is a unique relationship between stress and strain defined by a modulus or stiffness constant. Once we achieve the combination of stresses required to produce yield in an idealized rigid plastic material, deformation can proceed without altering stresses, and is determined by the movements of the external constraints, e.g. the displacement of the jaws of the tensometer in a tensile test. This means that there is no unique relationship between the stresses and the total plastic deformation. Instead, the relationships which do exist relate the stresses and the incremental plastic deformation, as was first recognized by St Venant, who proposed that for an isotropic material the principal axes of the strain increment are parallel to the principal axes of stress.

If the material is assumed to remain isotropic after yield there is no dependence on the deformation or stress history. Furthermore, if we assume that the yield behaviour is independent of the hydrostatic component of stress then the principal axes of the strain increment are parallel to the principal axes of the deviatoric stress tensor.

Levy [9] and von Mises [8] independently proposed that the principal components of the strain-increment tensor

$$\begin{bmatrix} de_1 & 0 & 0 \\ 0 & de_2 & 0 \\ 0 & 0 & de_3 \end{bmatrix}$$

and the deviatoric stress tensor

$$\begin{bmatrix} \sigma_1' & 0 & 0 \\ 0 & \sigma_2' & 0 \\ 0 & 0 & \sigma_3' \end{bmatrix}$$

are proportional, i.e.

$$\frac{de_1}{\sigma_1'} = \frac{de_2}{\sigma_2'} = \frac{de_3}{\sigma_3'} = d\lambda, \tag{11.5}$$

where $d\lambda$ is not a material constant but is determined by our choice of the extent of deformation of the material, e.g. by the displacement of the jaws of the tensometer. Because

$$p = -\tfrac{1}{3}(\sigma_1 + \sigma_2 + \sigma_3), \qquad \sigma_1' + \sigma_2' + \sigma_3' = 0$$

and

$$d\lambda = \frac{de_1 + de_2 + de_3}{\sigma_1' + \sigma_2' + \sigma_3'}$$

it follows that $de_1 + de_2 + de_3 = 0$, i.e. that deformation takes place at constant volume.

If the stress–strain relations are referred to other than principal axes we have

$$de_{ij} = \sigma_{ij}' \, d\lambda,$$

i.e.

$$\frac{de_{xx}}{\sigma'_{xx}} = \frac{de_{yy}}{\sigma'_{yy}} = \frac{de_{zz}}{\sigma'_{zz}} = \frac{de_{yz}}{\sigma'_{yz}} = \frac{de_{zx}}{\sigma'_{zx}} = \frac{de_{xy}}{\sigma'_{zy}} \tag{11.6}$$

These equations are called the Levy–Mises equations.

11.2.5 Geometrical Representations of the Tresca, von Mises and Coulomb Yield Criteria

The assumption of material isotropy which implies that σ_1, σ_2 and σ_3 are interchangeable, means that the Tresca and von Mises yield criteria take very simple analytical forms when expressed in terms of the principal stresses. Thus the yield criteria form surfaces in principal stress space, i.e. that space where the three rectangular Cartesian axes are parallel to the principal stress directions. Points lying closer to the origin than the yield surface represent combinations of stress where yield does not occur; points on or outside the surface represent combinations of stress where yield does occur.

Because the yield criterion is independent of the hydrostatic component of stress we can replace σ_1, σ_2 and σ_3 by $\sigma_1 + p$, $\sigma_2 + p$ and $\sigma_3 + p$

respectively. Thus if the point σ_1, σ_2, σ_3 lies on the yield surface, so does the point $\sigma_1 + p$, $\sigma_2 + p$, $\sigma_3 + p$. This shows that the yield surface must be parallel to the [111] direction in principal stress space and can be represented geometrically by its cross-section normal to this direction (Figure 11.11). The material isotropy implies equivalence of σ_1, σ_2 and σ_3 and hence that the section has a threefold symmetry about the [111] direction. The assumption that the behaviour is the same in tension and compression implies equivalence of σ_1 and $-\sigma_1$, and hence we have finally sixfold symmetry about the [111] direction as shown in Figure 11.11.

The cross-section normal to the [111] direction therefore consists of 12 equivalent parts (see Figure 11.11), and for the Tresca and von Mises yield criteria take the particularly simple forms shown in Figure 11.12. The Tresca criterion gives a regular hexagon and the von Mises criterion a circle.

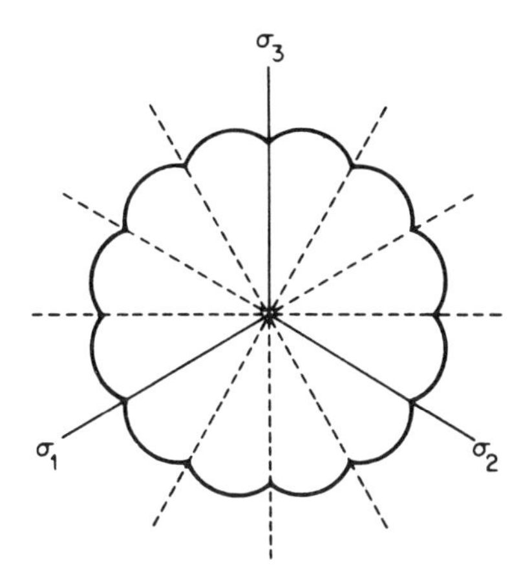

Figure 11.11 Cross-section of the yield surface normal to the [111] direction in principal stress space

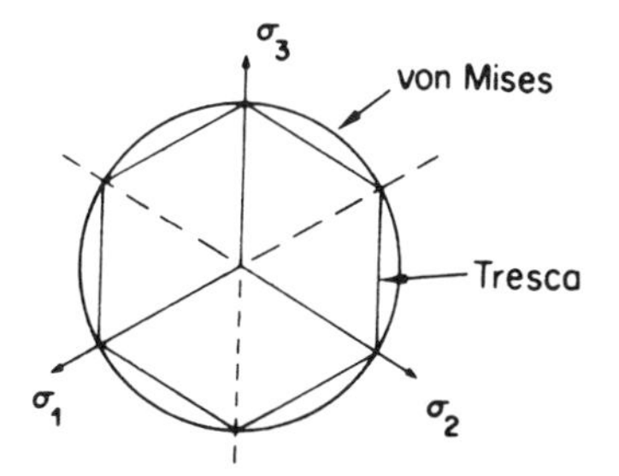

Figure 11.12 Cross-section of Tresca and von Mises yield surfaces normal to the [111] direction in principal stress space

11.2.6 Combined Stress States

For the analysis of combined stress in the two-dimensional situation the Mohr circle diagram (see Appendix A1.8) is of value. Normal stresses are represented along the x axis and shear stresses along the y axis, so that the Mohr circle thus represents a state of stress, with each point representing the stresses on a particular plane. The direction of the plane normal is given relative to the directions of the principal stresses by the rule that a rotation in real space of θ in a clockwise direction, corresponds to a rotation in Mohr circle space of 2θ in an anticlockwise direction. For further details the reader is referred to standard texts [10]. In Figure 11.13(a) two states of stress which produce yield with principal stresses σ_1 and σ_2, σ_3 and σ_4 respectively are represented by two circles of identical radius, tangential to the yield surface. The yield criterion in this case is assumed to be that of Tresca and the yield surface degenerates for the two-dimensional case to two lines parallel to the normal stress axis.

In Figure 11.13(b) two states of stress causing yield for a material which satisfies the Coulomb criterion are shown as σ_5 and σ_6, σ_7 and σ_8 respectively. In this case the yield stress depends on the magnitude of the (negative) normal stress, and so the diameters of the Mohr circles will vary with applied stress, increasing as we move to a more compressive stress field. The tangents to the Mohr circles represent the Coulomb yield surface, the critical shear yield stress for yield decreasing as the normal stress becomes more tensile. It can be shown that these tangents make an angle ϕ

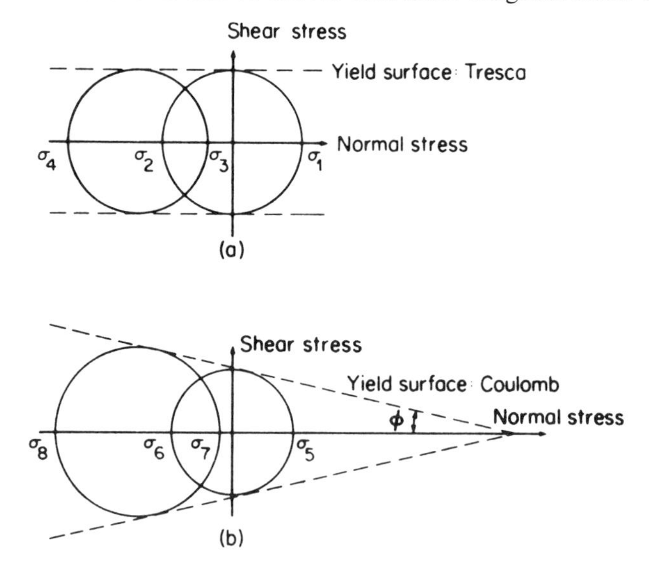

Figure 11.13 Mohr circle diagram for two states of stress which produce yield in a material satisfying (a) the Tresca yield criterion, and (b) the Coulomb yield criterion

with the normal stress axis, where $\tan \phi$ is the coefficient of friction as defined in 11.2.3.

11.3 THE YIELD PROCESS

We have seen that yield is often associated with a load drop on the load–extension curve, and always involves a change in slope on the true stress–strain curve. This load drop has sometimes been attributed either to adiabatic heating of the specimen or to the geometrical reduction in cross-sectional area on the formation of a neck. It is necessary to examine these explanations in detail, determine their shortcomings and establish the test conditions under which true yield behaviour can be observed. With this information it is then possible to consider the relevence of experimental data on the yield behaviour of polymers to our discussions of an ideal rigid plastic material.

11.3.1 The Adiabatic Heating Explanation

Under conventional conditions of cold-drawing, where the specimen is extended at strain rates of the order of 10^{-2} s^{-1} or higher, a considerable rise of temperature occurs in the region of the neck. Marshall and Thompson [11], following Muller [12], proposed that cold-drawing involves a local temperature rise and that necking occurs because of strain softening produced by the consequent fall in flow stress with rising temperature. The stability of the drawing process was then attributed to the stability of a localized process of heat transfer through the shoulders of the neck, with extension taking place at constant tension throughout the neck.

Hookway [13] later attempted to explain the cold-drawing of nylon 6.6 on somewhat similar grounds, suggesting that there is a possibility of local melting in the neck due to a combination of hydrostatic tension and temperature. The idea of local melting raises an important point with regard to the structure of the cold-drawn polymer. Nylon is crystalline in the undrawn, unoriented state. If the crystals do not break down in the necking and drawing process, the morphological state of the undrawn polymer will be particularly important in determining the structure of the drawn polymer.

There is no doubt that an appreciable rise in temperature does occur at conventional drawing speeds, and the ideas of Marshall and Thompson are very relevant to an understanding of the complex situation of fibre drawing. Calorimetric measurements by Brauer and Muller [14] have, however, shown that at slow rates of extension the increase in temperature is quite small (~10 °C) and not sufficient to give an explanation for necking and cold-drawing and cold-drawing in terms of adiabatic heating. More recently,

Lazurkin [15] demonstrated that necking can still take place under quasi-static conditions, for elastomers below their glass transition temperature, cold drawn at very low speeds. A comparable result was shown by Vincent [3] for (semicrystalline) polyethylene which cold draws at very slow extension rates at room temperature.

The adiabatic heating explanation arose at least in part because the initial yield process was not regarded as distinct from the drawing process. It is now recognized that up to the yield point the deformation of the sample is homogeneous and generally quite small strains are involved, whereas once a neck forms, the deformation is inhomogeneous and large strains are involved in the neck. The work of plastic deformation can then lead to a large rise in temperature in the neck. For example, Figure 11.14 shows results for the cold-drawing of PET [16] where both the yield stress and the drawing stress were measured as a function of strain rate. It can be seen that the yield stress continues to rise with increasing strain rate, beyond the strain rate at which the drawing stress falls quite distinctly. It is argued that provided the drawing is carried out at a low strain rate, any heat which is generated will be conducted away from the neck sufficiently rapidly for no significant temperature rise to occur. As the strain rate is increased and the process becomes more nearly adiabatic, the effective temperature at which the drawing is taking place is increased. In particular, heat is conducted into the unyielded portion of the sample, and so lowers the yield stress of the undeformed material and reduces the force necessary to propagate the neck.

The observed temperature rise in the neck has been found to be in approximate agreement with that calculated from the work done in drawing, assuming that no heat is generated due to crystallization. In PET, X-ray

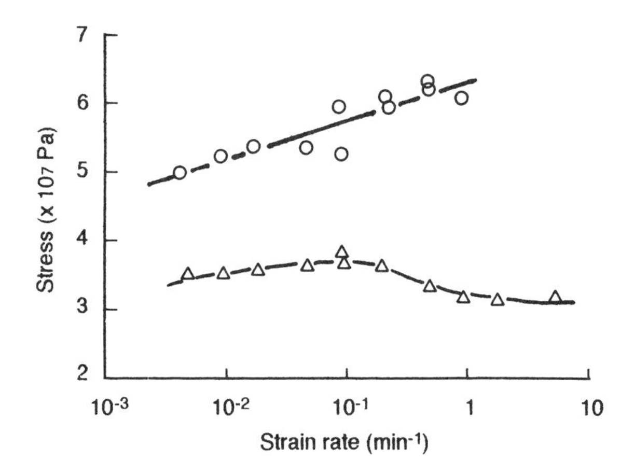

Figure 11.14 Comparison of yield stress (○) and drawing stress (△) as a function of strain rate for PET.

diffraction diagrams of cold-drawn fibres show that very little crystallization has occurred.

The work done per unit volume is given by $W = \sigma_D (\lambda_N - 1)$ where σ_D is the drawing stress and λ_N the natural draw ratio. From the results obtained $\sigma_D = 2.3 \times 10^7\ \mathrm{N\,m^{-2}}$ when $D_r = 3.6$, giving $W = 4.7\ \mathrm{MJ\,m^{-3}}$. For PET where the specific heat is $67\ \mathrm{J\,kg^{-1}\,K^{-1}}$ and the density is $1.38\ \mathrm{Mgm^{-3}}$, the calculated temperature rise is 57 °C, compared with the measured value of 42 °C.

11.3.2 The Isothermal Yield Process: the Nature of the Load Drop

There is no doubt that a temperature rise does occur in cold-drawing under many conditions of test. We have shown, however, that there is very good evidence to support the view that necking can still take place under quasistatic conditions where there is no appreciable temperature rise. Vincent [3] therefore proposed that the observed fall in load is a geometrical effect because the fall in cross-sectional area during stretching is not compensated for by an adequate degree of strain hardening. This effect, called strain softening, was attributed to the reduction in the slope of the stress–strain curve with increasing strain.

Contrary to this latter explanation of the load drop in terms of geometric softening, results reported by Whitney and Andrews [17] showed a yield drop in compression for polystyrene and PMMA where there are no geometrical complications. Brown and Ward [18] then made a detailed investigation of yield drops in PET, studying isotropic and oriented specimens, in tension, shear and compression. They concluded that in most cases there is clear evidence for the existence of an intrinsic yield drop, i.e. that a fall in true stress can occur in polymers, as in metals.

There is, however, a significant difference between polymers and many metals with regard to yield behaviour. For a polymer, as shown in Figure 11.2, only one maximum is observed on the load–extension curve, in contrast with metals (illustrated by mild steel in Figure 11.15), where often two maxima are observed on a typical load–extension curve. The first maximum (point A in Figure 11.15) called the upper yield point, represents a fall in true stress, an intrinsic load drop, and corresponds to a sudden increase in the amount of plastic strain which relaxes the stress. From B to C Lüders bands propagate throughout the specimen. Lüders bands have also been observed in polymers [19]. At C the specimen is homogeneously strained and the stress begins to rise as the material work hardens uniformly. A second maximum is observed at point D, and is always associated with the beginning of necking in the specimen. Necking occurs when the effects of strain hardening of the metal are overwhelmed by the geometrical softening due to the reduction in the cross-sectional area of the specimen as it is

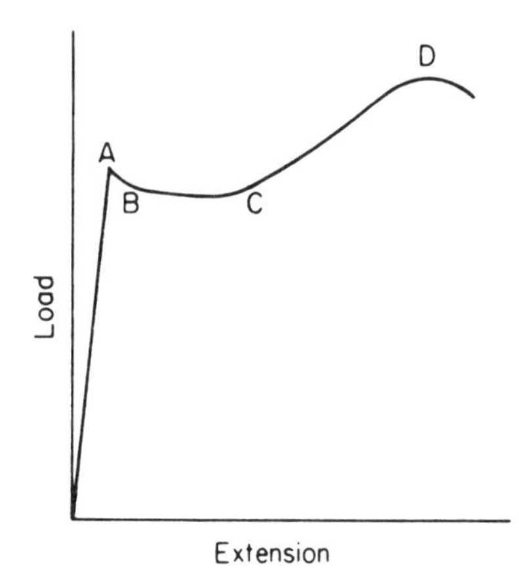

Figure 11.15 Load–extension curve in tension of mild steel

strained, i.e. the Orowan–Vincent explanation, discussed in section 11.1.1 above.

The second maximum, as we have seen previously, is not observed if the true stress–strain curve is plotted instead of the load–extension curve. The first maximum, on the other hand, would exist on the true stress–strain curve. It is called an intrinsic yield point, because it relates to the intrinsic behaviour of the material.

In polymers, as we have emphasized, only one maximum is observed in the load–extension curve. The investigations of Whitney and Andrews [17] and Brown and Ward [18], show that this maximum combines the effect of the geometrical changes and an intrinsic load drop, and cannot be attributed to the geometrical changes alone. In particular, the cold-drawing results are not accounted for by a decrease in the slope of the true stress–strain curve, as suggested in the explanation of Vincent. It is important to note that not every element of the material follows the same true stress–strain curve, since the stress for initiation is greater than for propagation of yielding, so confirming (as has already been noted in section 11.1.3) that it will not be possible to give a complete explanation of necking and cold-drawing in terms of the Considère construction on a true stress–strain curve.

11.4 EXPERIMENTAL EVIDENCE FOR YIELD CRITERIA IN POLYMERS

Many studies of the yield behaviour of polymers have bypassed the question of strain rate and temperature and sought to establish a yield criterion as discussed in section 11.2 above. In very general terms such studies divide into two categories: (1) those which attempt to define a yield criterion on the basis of determining yield for different stress states; (2) those which confine the experimental studies to an examination of the influence of hydrostatic pressure on the yield behaviour.

11.4.1 Application of Coulomb Yield Criterion to Yield Behaviour

From the early studies of yield behaviour of polymers, one example has been selected; the plane-strain compression tests on PMMA, carried out by Bowden and Jukes [20].

The experimental set-up is shown in Figure 11.16. A particular advantage of this technique is that yield behaviour can be observed in compression for materials which normally fracture in a tensile test. In this case PMMA was studied at room temperature, i.e. below its brittle–ductile transition temperature in tension.

The yield point in compression σ_1 was measured for various values of applied tensile stress σ_2. The results, shown in Figure 11.17, give $\sigma_1 = -11.1 + 1.365\sigma_2$, where both σ_1 and σ_2 are expressed as true stresses in units of pascals $\times 10^7$. The results therefore clearly do not fit the Tresca criterion, where $\sigma_1 - \sigma_2 =$ constant at yield; neither do they fit a von Mises yield criterion. They are, however, consistent with a Coulomb yield criterion with $\tau = 4.74 - 0.158\sigma_N$.

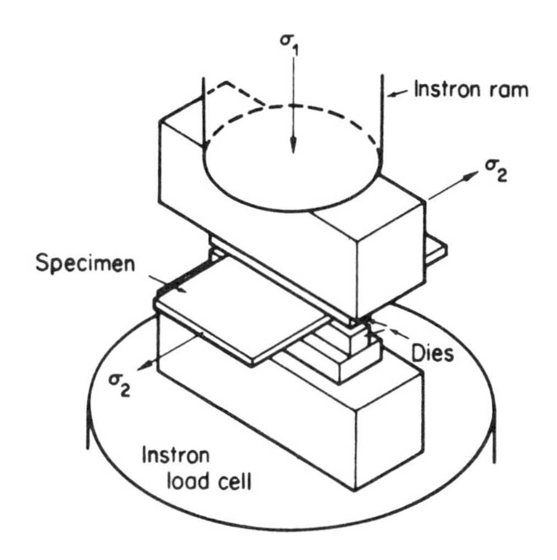

Figure 11.16 The plane-strain compression test (Reproduced with permission from Bowden and Jukes, *J. Mater. Sci.*, **3**, 183 (1968))

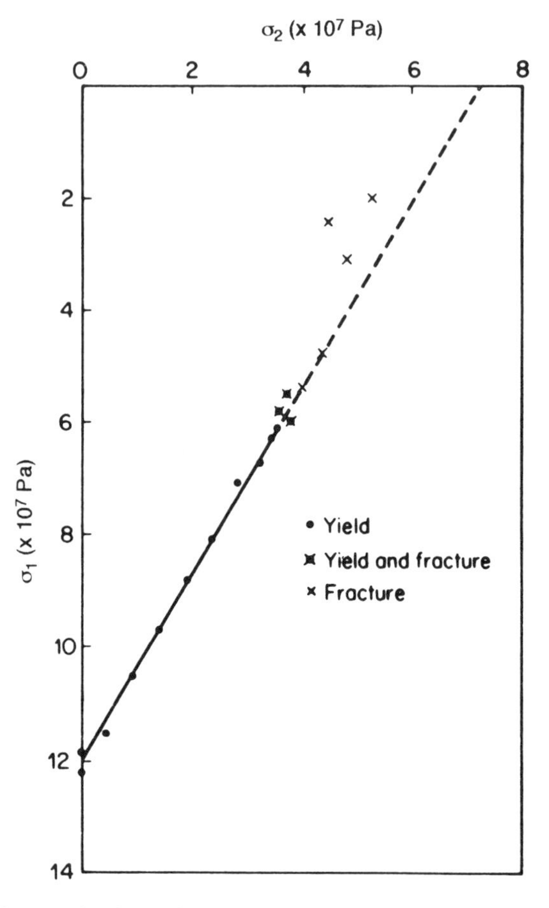

Figure 11.17 Measured values of the compressive yield stress σ_1 (true stress) plotted against applied tensile stress σ_2 (nominal stress). The full circles denote ductile yield, the crosses, brittle fracture, and the combined points, tests where ductile yielding occurred, followed immediately by brittle fracture, (Reproduced with permission from Bowden and Jukes, *J. Mater. Sci.*, **3**, 183 (1968))

11.4.2 Direct Evidence for the Influence of Hydrostatic Pressure on Yield Behaviour

There have been a number of detailed investigations of the influence of hydrostatic pressure on the yield behaviour of polymers. Because it illustrates clearly the relationship between a yield criterion which depends on hydrostatic pressure and the Coulomb yield criterion, an experiment will be discussed where Rabinowitz, Ward and Parry [21] determined the torsional stress–strain behaviour of isotropic PMMA under hydrostatic pressures up to 700 MPa. The results are shown in Figure 11.18.

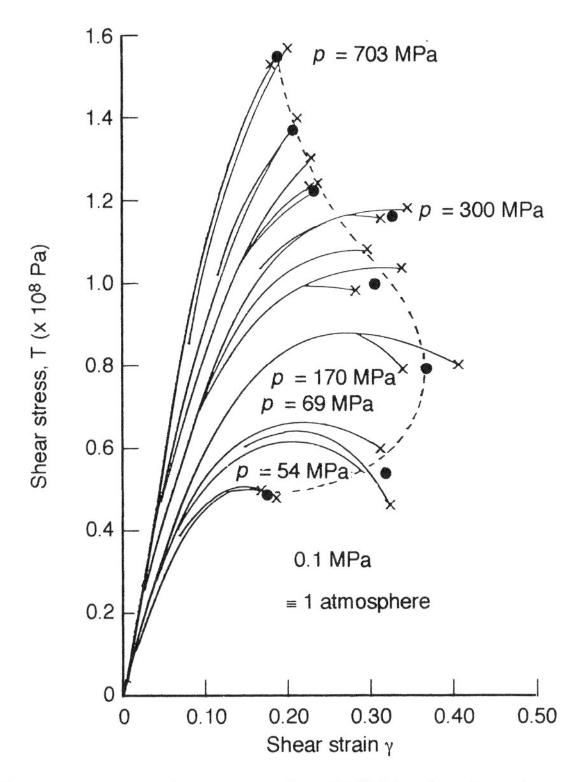

Figure 11.18 Shear stress–strain curves for PMMA showing fracture envelope. (Reproduced with permission from Rabinowitz, Ward and Parry, *J. Mater. Sci.*, **5**, 29 (1970))

There is a substantial increase in the shear yield stress up to a hydrostatic pressure of about 300 MPa. After this pressure brittle failure occurs, unless prevented by protecting the specimens from the hydraulic fluid [22] (e.g. by coating with a layer of solidified rubber solution). The strain at which yield occurs also increases with increasing pressure, similar to the results of other workers for tensile tests under pressure. The shear yield stress increases linearly with pressure to an excellent approximation (Figure 11.19).

There are two other ways in which these results can be represented. First, recalling section 11.2.6 and Figure 11.13, the Mohr circle diagram can be constructed from the data, as shown in Figure 11.20 where Bowden and Jukes's results appear as crossed points. This diagram leads naturally to a Coulomb yield criterion.

It is, however, equally reasonable to interpret Figure 11.18 directly in terms of the equation

$$\tau = \tau_0 + \alpha p, \tag{11.7}$$

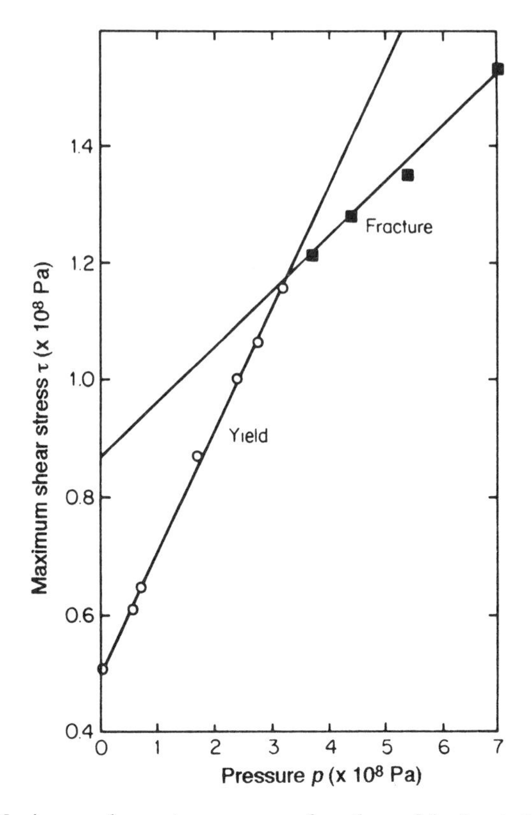

Figure 11.19 Maximum shear stress τ as a function of hydrostatic pressure p for PMMA. (○) Yield; (■) fracture. (Reproduced with permission from Rabinowitz, Ward and Parry, *J. Mater. Sci.*, **5**, 29 (1970))

where τ is the shear yield stress at pressure p, τ_0 is the shear yield stress at atmospheric pressure and α is the coefficient of increase of shear yield stress with hydrostatic pressure.

We will see that this simple form of pressure-dependent yield criterion is more satisfactory than the Coulomb criterion when a representation is developed which includes the effects of temperature and strain rate on the yield behaviour. In physical terms, the hydrostatic pressure can be seen as changing the state of the polymer by compressing the polymer significantly, unlike the situation in metals where the bulk moduli are much larger (~100 GPa compared with ~5 GPa). Although such experimental evidence as exists is not unequivocal in this respect, it seems likely that the flow rules for the polymer subjected to hydrostatic pressure are still given by equation (11.7), i.e. pressure increases the magnitude of the yield stresses but any volume changes are comparatively small.

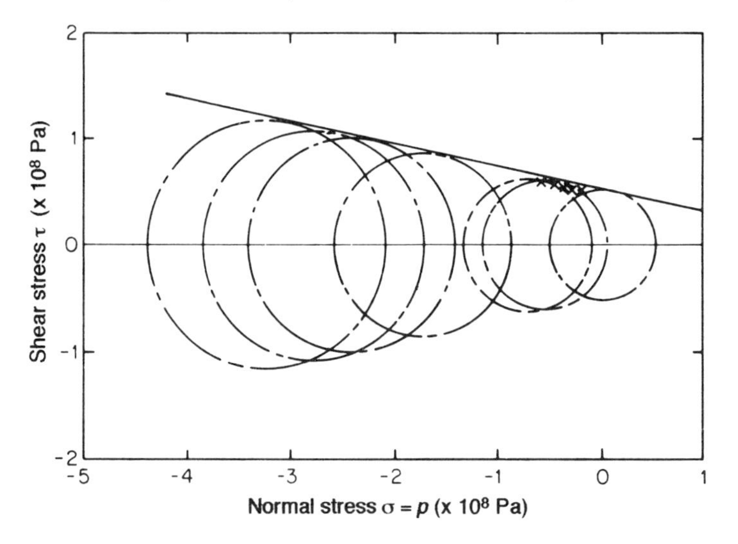

Figure 11.20 Mohr circles for yield behaviour of polymethyl methacrylate obtained from results of Rabinowitz, Ward and Parry. The crosses are the results of Bowden and Jukes. (Reproduced with permission from Rabinowitz, Ward and Parry, *J. Mater. Sci.*, **5**, 29 (1970))

Recent studies of yield behaviour, using a variety of multiaxial stressing experiments, can all be adequately described by a generalization of equation (11.7), i.e. a generalized von Mises equation where τ is replaced by the octahedral shear stress (see section 11.6.1 for a fuller development).

Finally, it can be noted that the coefficient α in equation (11.7) depends on the temperature of measurement and increases markedly near a viscoelastic transition. Briscoe and Tabor [23] have pointed out that α is equivalent to the coefficient of friction μ in sliding friction, and show that there is good numerical agreement between values of μ and the values of α obtained from yield stress/pressure measurements.

11.5 THE MOLECULAR INTERPRETATIONS OF YIELD AND COLD-DRAWING

11.5.1 Yield as an Activated Rate Process: The Eyring Equation

It is very apparent that polymers are not ideal plastic materials. However, we have seen that provided temperature and strain rate are maintained constant, and conditions chosen so that adiabatic heating does not occur, the yield behaviour can profitably be discussed in terms of theories of ideal plasticity.

It is now important to expand the discussion to embrace the effects of strain rate and temperature in order to seek some molecular understanding of the yield behaviour. Many workers [15, 24–31] have considered that the applied stress induces molecular flow much along the lines of the Eyring viscosity theory where internal viscosity falls to a value such that the applied strain rate is identical to the plastic strain rate $\dot{e}$ predicted by the Eyring equation.

Furthermore, the measurement of a yield stress in a constant strain rate test is analogous to the measurement of the creep rate at constant applied stress. We can therefore recall equation (10.18),

$$\dot{e} = \dot{e}_0 \exp - \left[\frac{\Delta H}{RT}\right] \sinh \left[\frac{\upsilon\sigma}{RT}\right] \tag{11.8}$$

where ΔH is the activation energy, σ is the applied tensile stress and υ the activation volume, which is considered to represent the volume of the polymer segment which has to move as a whole in order for plastic deformation to occur.

For high values of stress $\sinh \chi = \frac{1}{2} \exp \chi$ and

$$\dot{e} = \frac{\dot{e}_0}{2} \exp - \left(\frac{\Delta H - \upsilon\sigma}{RT}\right) \tag{11.9}$$

which gives the shear yield stress τ in terms of strain rate as

$$\frac{\sigma}{T} = \frac{\Delta H}{\upsilon T} + \frac{R}{\upsilon} \ln \frac{2\dot{e}}{\dot{e}_0}$$

i.e.

$$\frac{\sigma}{T} = \frac{R}{\upsilon} \left\{\frac{\Delta H}{RT} + \ln \frac{2\dot{e}}{\dot{e}_0}\right\} \tag{11.10}$$

Plots of (yield stress/T) against log (strain rate) for a series of temperatures should give a series of parallel straight lines. Figure 11.21 shows results obtained by Bauwens and co-workers [27] for the tensile yield stress of polycarbonate together with calculated lines based on equation (11.10) with constant values of ΔH and υ.

Haward and Thackray [30] have compared the Eyring activation volumes obtained from yield stress data with the volume of the 'statistical random link'. The latter was obtained from solution studies by assuming that the real chain can be represented by an equivalent chain with freely jointed links of a particular length. Table 11.1, based on data collated by Haward and Thackray, shows that the activation volumes are very large in molecular terms and range from about two to ten times that of the statistical random link. The result suggests that yield involves the cooperative movement of a

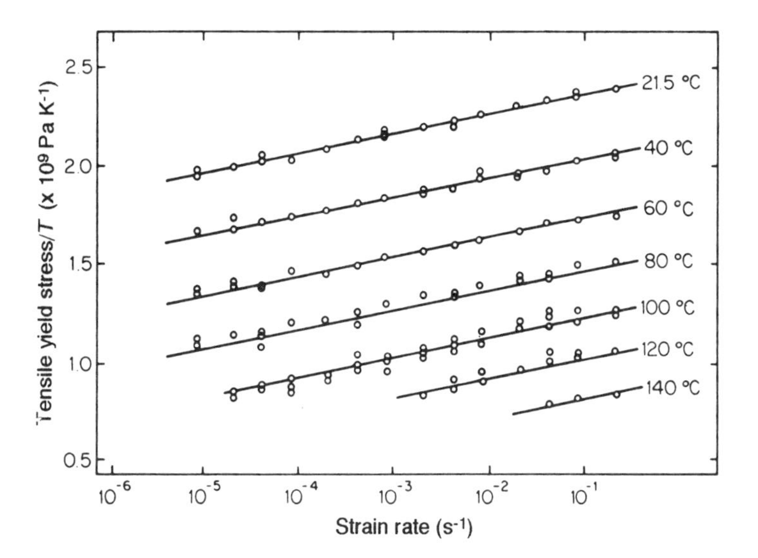

Figure 11.21 Measured ratio of yield stress to temperature as a function of the logarithm of strain rate for polycarbonate. The set of parallel straight lines is calculated from equation (11.10) (Reproduced with permission from Bauwens-Crowet, Bauwens and Homès, *J. Polymer. Sci.* **A2**, **7**, 735 (1969))

Table 11.1 A comparison of the statistical segment volume for a polymer measured in solution with the flow volumes derived from the Eyring theory (after Haward and Thackray [30])

Polymer	Volume of statistical link in solution (nm^3)	Eyring flow volume v (nm^3)
Polyvinyl chloride	0.38	8.6
Polycarbonate	0.48	6.4
Polymethyl methacrylate	0.91	4.6
Polystyrene	1.22	9.6
Cellulose trinitrate	2.62	6.1
Cellulose acetate (triacetate)	2.06	17.4

larger number of chain segments than would be required for a conformational change in dilute solution.

11.5.2 The Two-stage Eyring Process Representation

Extensive studies of the yield behaviour of PMMA and polycarbonate over very wide ranges of strain rate and temperature by Roetling and Bauwens have shown, that at low temperatures and high strain rates, the yield stresses increase more rapidly with increasing strain rate and decreasing temperature, than at high temperature and low strain rates. Following Ree and Eyring it has therefore been proposed that equation (11.10) should be extended by assuming that there is more than one activated rate process with all species of flow units moving at the same rate, the stresses being additive. For PMMA, PVC and polycarbonate it has been shown that the yield behaviour can be represented very satisfactorily by the addition of two activated processes. The equivalent equation to (11.10) above is then

$$\frac{\sigma}{T} = \frac{R}{v_1}\left\{\frac{\Delta H_1}{RT} + \ln\frac{2\dot{e}}{\dot{e}_{01}}\right\} + \frac{R}{v_2}\sinh^{-1}\left\{\frac{\dot{e}}{\dot{e}_{02}}\exp\frac{\Delta H_2}{RT}\right\} \tag{11.11}$$

where the two activated processes are denoted by the subscript symbols 1 and 2 respectively. At high temperatures and low strain rates, process 1 predominates; it has a comparatively low strain rate dependence (v_1 is large). We can therefore use the approximation $\sinh\chi = \frac{1}{2}\exp\chi$. Process 2, which becomes important only at low temperatures and high strain rates, shows a much higher strain rate dependence (v_2 is small compared to v_1). The sinh form is retained to cover the intermediate range where process 2 gives a smaller but significant contribution to the magnitude of the total yield stress. Figure 11.22 shows that experimental data for polyvinyl chloride [27] are well fitted by equation (11.11).

11.5.3 Pressure Dependence

We have seen that the effect of pressure on the shear yield stress of a polymer can be very well represented by equation (11.7), $\tau = \tau_0 + \alpha p$, which suggests that the Eyring equation may be very simply modified [32] to include the effect of the hydrostatic component of stress p, giving

$$\dot{e} = \frac{\dot{e}_0}{2}\exp - \left(\frac{\Delta H - \tau v + p\Omega}{RT}\right) \tag{11.12}$$

where v and Ω are the shear and pressure activation volumes respectively. This modification of the Eyring equation can be considered to arise from a linear increase in the activation energy with increasing pressure, and so is the simplest approach.

Equation (11.12) may be conveniently expressed in terms of the octahedral yield stress

$$\tau_{\text{oct}} = \tfrac{1}{3}\{(\sigma_1 - \sigma_2)^2 + (\sigma_2 - \sigma_3)^2 + (\sigma_3 - \sigma_1)^2\}^{1/2}$$

and the octahedral strain rate

$$\dot{\gamma}_{\text{oct}} = \tfrac{2}{3}\{(\dot{e}_1 - \dot{e}_2)^2 + (\dot{e}_2 - \dot{e}_3)^2 + (\dot{e}_3 - \dot{e}_1)^2\}^{1/2}$$

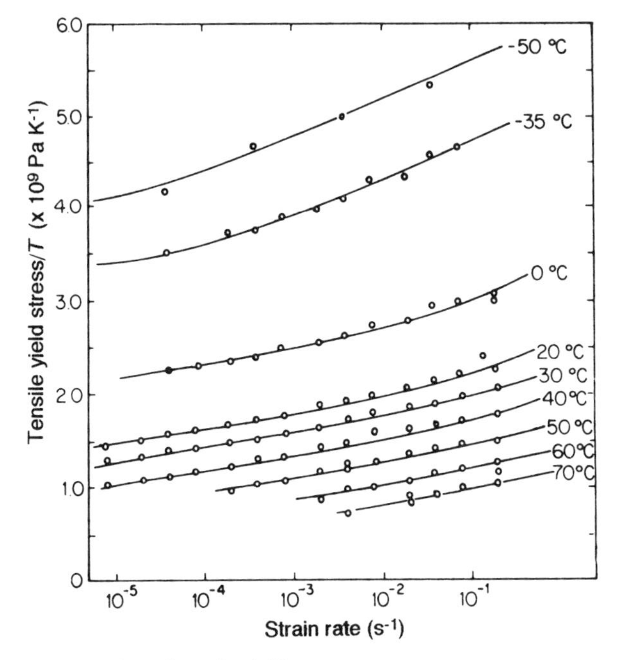

Figure 11.22 Measured ratio of yield stress to temperature as a function of logarithm of strain rate for PVC. The set of parallel curves is calculated from equation (11.11) (Reproduced with permission from Bauwens-Crowet, Bauwens and Homès, *J. Polymer. Sci.* **A2**, **7**, 735 (1969))

to give a generalized representation suitable for all stress fields

$$\dot{\gamma}_{\text{oct}} = \frac{\dot{\gamma}_0}{2}\exp\left[-\frac{(\Delta H - \tau_{\text{oct}}v + p\Omega)}{RT}\right] \tag{11.13}$$

For a constant strain rate test we therefore have

$$\Delta H - \tau_{\text{oct}}v + p\Omega = \text{constant},$$

from which an equation similar to equation (11.7) is obtained with

$$\tau_{\text{oct}} = (\tau_{\text{oct}})_0 + \alpha p,$$

where $\alpha = \Omega/v$. Figure 11.23 shows results for polycarbonate at atmospheric pressure [33] using data from torsion, tension and compression. It can be seen that on average the values of τ_{oct} lie in the order compression > torsion > tension. The differences are therefore consistent with the observed linear dependence of τ_{oct} on pressure shown by direct measurement of the yield stress in torsion over a range of hydrostatic pressures (section 11.4.1 above), and there is good numerical agreement between the two sets of measurements.

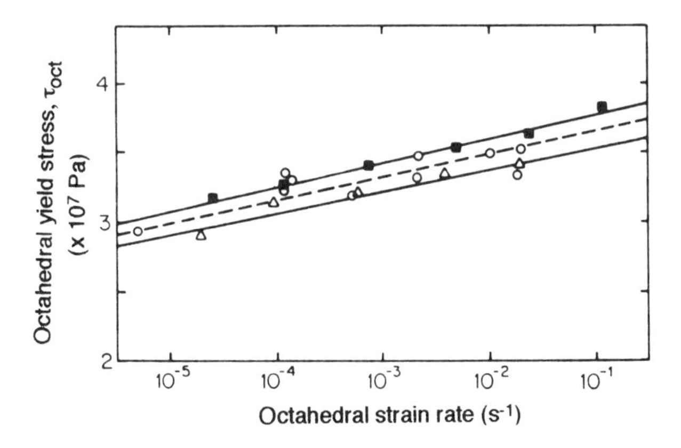

Figure 11.23 The strain rate dependence of the octahedral shear stress τ_{oct} at atmospheric pressure using data from torsion (○), tension (△), and compression (■). (Reproduced with permission from Duckett et al., *Brit. Polymer J.* **10**, 11 (1978))

11.6 COLD-DRAWING

11.6.1 General Considerations

The strain hardening, which is a necessary prerequisite for cold-drawing in polymers, has two possible sources.

1. Drawing causes molecular alignment so that the drawing stress (often called the flow stress) is increased. This is a general phenomenon, true for both crystalline and amorphous polymers.
2. Strain-induced crystallization may occur, similar to the crystallization observed in rubbers at high degrees of stretching.

In general, the morphological changes which occur in drawing are complex, and strain hardening is due to both molecular orientation and changes in morphology.

11.6.2 The Natural Draw Ratio, Maximum Draw Ratios and Molecular Networks

At room temperature, amorphous PET draws through a neck, so that the undrawn material is transformed to drawn material with a constant reduction in cross-section as it passes through the neck. If the drawing is uniaxial, as in the case of a fibre, the extension ratio λ is exactly similar to that defined for stretching a rubber (section 2.4.3 above), and is generally called the natural draw ratio. There is some evidence to suggest that this natural draw ratio arises because cold drawing extends the network of entangled molecular chains to its limiting extensibility.

Confirmation of this idea comes from the observation that the natural draw ratio observed for melt spun fibres is sensitive to the degree of molecular orientation introduced during the spinning process. It appears that the molecular network is formed as the polymer freezes from the melt, is subsequently stretched in the rubber-like state before the polymer cools below T_g and is eventually collected as a frozen stretched rubber. The amount of stretching in the threadline can be measured by shrinking these spun fibres back to state of zero strain, i.e. isotropy. These results can then be combined with measurements of the natural draw ratio to give the limiting extensibility of the network [34].

Consider the cold-drawing of a sample of length l_1. If the fibre were allowed to shrink back to its isotropic state, length l_0, the shrinkage s would be defined by

$$s = \frac{l_1 - l_0}{l_1} \tag{11.14}$$

Cold-drawing to a length l_2 gives a natural draw ratio

$$\lambda_N = \frac{l_2}{l_1} \tag{11.15}$$

Combining equations (11.14) and (11.15) we have

$$\frac{l_2}{l_0} = \frac{\lambda_N}{1 - s} \tag{11.16}$$

Table 11.2 shows collected results for a series of PET filaments. It can be seen that λ_N varies from 4.25 to 2.58 and s from 0.042 to 0.378, but the ratio l_2/l_0 calculated from equation (11.16) remains constant at a value of about 4.0.

Table 11.2 Value of $l_2/l_0 = \lambda_N/1 - s$ for samples of differing amounts of pre-orientation (polymer: polyethylene terephthalate, see [34])

Initial birefringence $\times 10^3$	Natural draw ratio λ_N	Shrinkage s	$\lambda_N/(1 - s)$
0.65	4.25	0.042	4.44
1.6	3.70	0.094	4.08
2.85	3.32	0.160	3.96
4.2	3.05	0.202	3.83
7.2	2.72	0.320	4.01
9.2	2.58	0.378	4.14

11.6.3 Crystalline Polymers

The plastic deformation of crystalline polymers, in particular polyethylene, has been studied intensively from the viewpoint of changes in morphology. Notable contributions to this area have been made by Keller and co-workers and by Peterlin, Geil and others [35–37]. It is now evident that very drastic reorganization occurs at the morphological level, with the structure changing from a spherulitic to a fibrillar type as the degree of plastic deformation increases. The molecular reorientation processes are very far from being affine or pseudo-affine and can also involve mechanical twinning in the crystallites. It is surprising that some of the continuum ideas for mechanical anisotropy are nevertheless still relevent, although they must be appropriately modified.

In a few highly crystalline polymers, notably high-density polyethylene, extremely large draws ratios, ~30 or more, have been achieved by optimizing the chemical composition of the polymers and the drawing conditions [38, 39]. These high draw ratios lead to oriented polymers with very high Young's moduli as discussed in section 10.7.3. In spite of the much more complex deformation processes in a crystalline polymer, it has been concluded [39] that the molecular topology and the deformation of a molecular network are still the overriding considerations in determining the strain-hardening behaviour and the ultimate draw ratio achievable. For high-molecular-weight, high-density polyethylene, the key network junction points are physical entanglements, as in amorphous polymers. For low-molecular-weight, high-density polyethylene, both physical entanglements and crystallites where more than one molecular chain is incorporated, can provide the network junction points. Junction points associated with the crystallites will be of a temporary nature. Very high draw ratios involve the breakdown of the crystalline structure and the unfolding of molecules, so that the simple ideas of a molecular network suggested for amorphous polymers have to be extended and modified.

REFERENCES

1. A. Nadai, *Theory of Flow and Fracture of Solids*, McGraw-Hill, New York, 1950.
2. E. Orowan, *Rept. Prog. Phys.,* **12**, 185 (1949).
3. P. I. Vincent, *Polymer*, **1**, 7 (1960).
4. P. W. Bridgeman, *Studies in Large Plastic Flow and Fracture*, McGraw-Hill, New York, 1952.
5. H. Tresca, *C. R. Acad. Sci. (Paris)*, **59**, 754 (1864); **64**, 809 (1867).
6. A. H. Cottrell, *Dislocations and Plastic Flow in Crystals*, Clarendon Press, Oxford, 1953.
7. C. A. Coulomb, *Mem. Math. Phys.,* **7**, 343 (1773).

8. R. von Mises. *Gottinger Nach. Math.-Phys. Kl.*, 582 (1913).
9. M. Levy, *C. R. Acad. Sci. (Paris)*, **70**, 1323 (1870).
10. S. Timoshenko and J. N. Goodier, *Theory of Elasticity*, McGraw-Hill, New York, 1951.
11. I. Marshall and A. B. Thompson, *Proc. Roy. Soc.*, **A221**, 541 (1954).
12. F. H. Müller, *Kolloidzeitschrift*, **114**, 59 (1949); **115**, 118 (1949), **126**, 65 (1952).
13. D. C. Hookway, *J. Text. Inst.*, **49**, 292 (1958).
14. P. Brauer and F. H. Müller, *Kolloidzeitschrift*, **135**, 65 (1954).
15. Y. S. Lazurkin, *J. Polymer Sci.*, **30**, 595 (1958).
16. S. W. Allison and I. M. Ward, *Brit. J. Appl. Phys.*, **18**, 1151 (1967).
17. W. Whitney and R. D. Andrews, *J. Polymer Sci. C,* **16**, 2981 (1967).
18. N. Brown and I. M. Ward, *J. Polymer Sci. A2,* **6**, 607 (1968).
19. J. Miklowitz, *J. Colloid Sci.*, **2**, 193 (1947).
20. P. B. Bowden and J. A. Jukes, *J. Mater. Sci.*, **3**, 183 (1968).
21. S. Rabinowitz, I. M. Ward and J. S. C. Parry, *J. Mater. Sci.*, **5**, 29 (1970).
22. J. S. Harris, I. M. Ward and J. S. C. Parry, *Mater. Sci.*, **6**, 110 (1971).
23. B. J. Briscoe and D. Tabor, in *Polymer Surfaces* (eds D. T. Clark and W. J. Feast), Wiley, New York, 1978, Ch. 1.
24. Y. S. Lazurkin and R. A. Fogelson, *Zhur. Tech. Fiz.*, **21**, 267 (1951).
25. R. E. Robertson, *J. Appl. Polymer Sci.*, **7**, 443 (1963).
26. C. Crowet and G. A. Homès, *Appl. Mater. Res.*, **3**, 1 (1964).
27. C. Bauwens-Crowet, J. A. Bauwens and G. Homès, *J. Polymer Sci. A2*, **7**, 735 (1969).
28. J. C. Bauwens, C. Bauwens-Crowet and G. Homès, *J. Polymer Sci. A2*, **7**, 1745 (1969).
29. J. A. Roetling, *Polymer*, **6**, 311 (1965).
30. R. N. Haward and G. Thackray, *Proc. Roy. Soc.*, **A302**, 453 (1968).
31. D. L. Holt, *J. Appl. Polymer Sci.*, **12**, 1653 (1968).
32. I. M. Ward, *J. Mater. Sci.*, **6**, 1397 (1971).
33. R. A. Duckett, B. C. Goswami, L. S. A. Smith, I. M. Ward and A. M. Zihlif, *Brit. Polymer J.*, **10**, 11 (1978).
34. S. W. Allison, P.R. Pinnock and I. M. Ward, *Polymer*, **7**, 66 (1966).
35. I. L. Hay and A. Keller, *Kolloidzeitschrift*, **204**, 43 (1965).
36. A. Peterlin, *J. Polymer Sci.*, **69**, 61 (1965).
37. P. H. Geil, *J. Polymer Sci. A*, **2**, 3835 (1964).
38. G. Capaccio and I. M. Ward, *Nature Phys. Sci.*, **243**, 43 (1973).
39. G. Capaccio and I. M. Ward, *Polymer*, **15**, 233 (1974).

Chapter 12

Breaking Phenomena

12.1 DEFINITION OF TOUGH AND BRITTLE BEHAVIOUR IN POLYMERS

The mechanical properties of polymers are greatly affected by temperature and strain rate, and the load–elongation curve at a constant strain rate changes with increasing temperature as shown schematically (not necessarily to scale) in Figure 12.1. At low temperatures the load rises approximately linearly with increasing elongation up to the breaking-point, when the

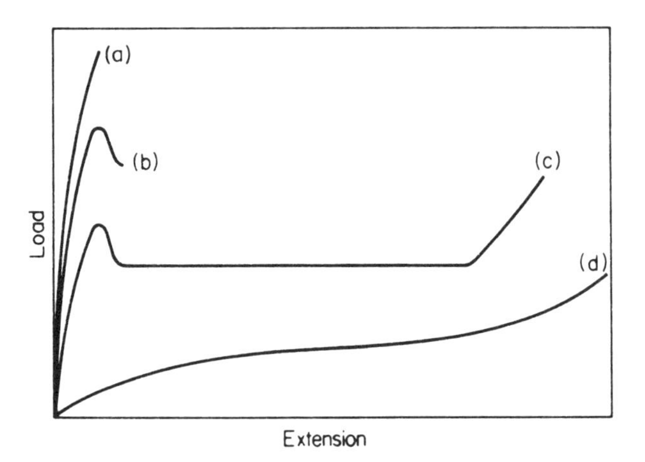

Figure 12.1 Load–extension curves for a typical polymer tested at four temperatures showing different regions of mechanical behaviour. (a) Brittle fracture, (b) ductile failure, (c) necking and cold-drawing and (d) homogeneous deformation (quasi-rubber-like behaviour)

polymer fractures in a brittle manner. At higher temperatures a yield point is observed and the load falls before failure, sometimes with the appearance of a neck: i.e. ductile failure, but still at quite low strains (typically 10–20%). At still higher temperatures, under certain conditions, strain hardening occurs, the neck stabilizes and cold-drawing ensues. The extensions in this case are generally very large, up to 1000%. Finally, at even higher temperatures, homogeneous deformation is observed, with a very large extension at break. In an amorphous polymer this rubber-like behaviour occurs above the glass transition temperature so the stress levels are very low.

For polymers the situation is clearly more complicated than that for the brittle–ductile transition in metals, as there are in general four regions of behaviour and not two. It is of considerable value to discuss the factors which influence the brittle–ductile transition, and then to consider further factors which are involved in the observation of necking and cold-drawing.

Ductile and brittle behaviour are most simply defined from the stress–strain curve. Brittle behaviour is designated when the specimen fails at its maximum load, at comparatively low strains (say $< 10\%$), whereas ductile behaviour shows a peak load followed by failure at a lower stress (Figure 12.1(a) and (b))

The distinction between brittle and ductile failure is also manifested in two other ways: (1) the energy dissipated in fracture, and (2) the nature of the fracture surface. The energy dissipated is an important consideration for practical applications, and forms the basis of the Charpy and Izod impact tests (discussed in section 12.8 below). At the testing speeds under which the practical impact tests are conducted it is difficult to determine the stress–strain curve, so that impact strengths are customarily quoted in terms of the fracture energy for a standard specimen.

The appearance of the fracture surface can also be an indication of the distinction between brittle and ductile failure, although the present state of knowledge concerning the crack propagation is not sufficiently extensive to make this distinction more than empirical.

12.2 PRINCIPLES OF BRITTLE FRACTURE OF POLYMERS

Modern understanding of the fracture behaviour of brittle materials stems from the seminal research of Griffith [1] on the brittle fracture of glass. The Griffith theory of fracture, which is the earliest statement of linear elastic fracture mechanics, has been applied extensively to the fracture of glass and metals, and more recently to polymers. Although it was initially conceived to describe the propagation of a crack in a perfectly elastic material at small elastic strains (hence linear elastic), subsequent work has shown that it is still applicable for situations including localized plastic deformation at the crack tip, which does not lead to general yielding in the specimen.

12.2.1 Griffith Fracture Theory

First, Griffith considered that fracture produces a new surface area and postulated that for fracture to occur the increase in energy required to produce the new surface must be balanced by a decrease in elastically stored energy.

Second, to explain the large discrepancy between the measured strength of materials and those based on theoretical considerations, he proposed that the elastically stored energy is not distributed uniformly throughout the specimen but is concentrated in the neighbourhood of small cracks. Fracture thus occurs due to the spreading of cracks which originate in pre-existing flaws.

In general the growth of a crack will be associated with an amount of work $\mathrm{d}W$ being done on the system by external forces and a change $\mathrm{d}U$ in the elastically stored energy U. The difference between these quantities, $\mathrm{d}W - \mathrm{d}U$, is the energy available for the formation of new surface. The condition for growth of a crack by a length $\mathrm{d}c$ is then

$$\frac{\mathrm{d}W}{\mathrm{d}c} - \frac{\mathrm{d}U}{\mathrm{d}c} \geqslant \gamma \frac{\mathrm{d}A}{\mathrm{d}c} \tag{12.1}$$

where γ is the surface free energy per unit area of surface and $\mathrm{d}A$ the associated increment of surface. If there is no change in the overall extension Δ when the crack propagates $\mathrm{d}W = 0$ and

$$-\left(\frac{\mathrm{d}U}{\mathrm{d}c}\right)_{\Delta} \geqslant \gamma \frac{\mathrm{d}A}{\mathrm{d}c} \tag{12.1a}$$

The elastically stored energy decreases and so $-(\mathrm{d}U/\mathrm{d}c)_{\Delta}$ is essentially a positive quantity.

Griffith calculated the change in elastically stored energy, using a solution obtained by Inglis [2] for the problem of a plate, pierced by a small elliptical crack, which is stressed at right angles to the major axis of the crack. Equation (12.1) then allows the fracture stress σ_{B} of the material to be defined in terms of the crack length $2c$ by the relationship

$$\sigma_{\mathrm{B}} = (2\gamma E^*/\pi c)^{1/2}, \tag{12.2}$$

where E^* is the 'reduced modulus', equal to Young's modulus E for a thin sheet in plane stress, and to $E/(1 - \nu^2)$, where ν is Poisson's ratio, for a thick sheet in plane strain.

12.2.2 The Irwin Model

An alternative formulation of the problem due to Irwin [3] considers the stress field near an idealized crack length $2c$ (Figure 12.2) In two-dimensional polar coordinates with the x axis as the crack axis and $r \ll c$,

$$
\begin{aligned}
\sigma_{xx} &= \frac{K_{\rm I}}{(2\pi r)^{1/2}} \cos(\theta/2)\,[1 - \sin(\theta/2)\sin(3\theta/2)], \\
\sigma_{yy} &= \frac{K_{\rm I}}{(2\pi r)^{1/2}} \cos(\theta/2)\,[1 + \sin(\theta/2)\sin(3\theta/2)], \\
\sigma_{zz} &= \nu(\sigma_{xx} + \sigma_{yy}) \text{ for plane strain}, \\
\sigma_{zz} &= 0 \text{ for plane stress}, \\
\sigma_{xy} &= \frac{K_{\rm I}}{(2\pi r)^{1/2}} \cos(\theta/2)\sin(\theta/2)\cos(3\theta/2)], \\
\sigma_{yz} &= \sigma_{zx} = 0.
\end{aligned}
\tag{12.3}
$$

In these equations θ is the angle between the axis of the crack and the radius vector.

The value of Irwin's approach is that the stress field around the crack is identical in form for all types of loading situation normal to the crack, with the magnitude of the stresses (i.e. their intensity) determined by $K_{\rm I}$ which is constant for given loads and geometry; $K_{\rm I}$ is called the stress intensity factor, the subscript I indicating loading normal to the crack. This crack opening mode I, is distinct from a sliding mode II, not considered here. As we approach the crack tip σ_{xx} and σ_{yy} clearly become infinite in magnitude and r tends to zero, but the products $\sigma_{xx}\sqrt{r}$ and $\sigma_{yy}\sqrt{r}$ and hence $K_{\rm I}$ remain finite.

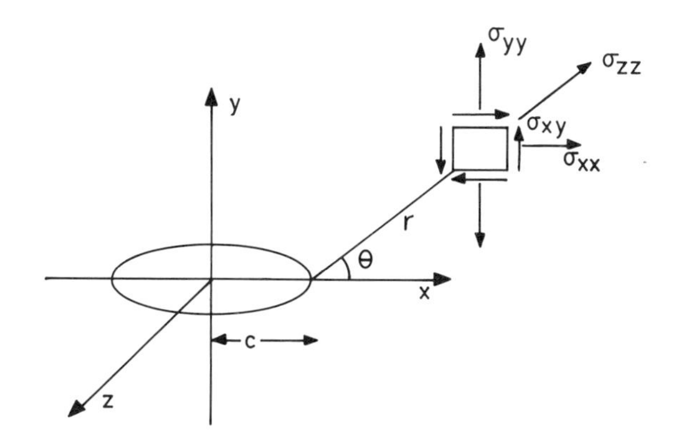

Figure 12.2 The stress field near an idealized crack of length $2c$

For an infinite sheet with a central crack of length $2c$ subjected to a uniform stress σ it was shown by Irwin that

$$K_{\mathrm{I}} = \sigma(\pi c)^{1/2}. \tag{12.4}$$

He postulated that when σ reaches the fracture stress σ_{B}, K_{I} has a critical value given by

$$K_{\mathrm{IC}} = \sigma_{\mathrm{B}}(\pi c)^{1/2}. \tag{12.5}$$

The fracture toughness of the material can then be defined by the value of K_{IC}, termed the critical stress intensity factor, which defines the stress field at fracture.

There is clearly a link with the earlier Griffith formulation in that equation (12.5) can be written as

$$\sigma_{\mathrm{B}} = (K_{\mathrm{IC}}^2/\pi c)^{1/2} \tag{12.6}$$

which is identical in form to equation (12.2).

12.2.3 The Strain Energy Release Rate

In linear elastic fracture mechanics it is useful also to consider the energy G available for unit increase in crack length, which is called the 'strain energy release rate'. Following equation (12.1) above G is

$$G = \frac{\mathrm{d}W}{\mathrm{d}A} - \frac{\mathrm{d}U}{\mathrm{d}A} = \frac{1}{B}\left[\frac{\mathrm{d}W}{\mathrm{d}c} - \frac{\mathrm{d}U}{\mathrm{d}c}\right] \tag{12.7}$$

where B is the thickness of the specimen. It is assumed that fracture occurs when G reaches a critical value G_{c}. The equivalent equation to (12.1) is then

$$G \geqslant G_{\mathrm{c}} \tag{12.8}$$

and G_{c} is equal to 2γ in the Griffith formulation, but is generalized to include all work of fracture, not just the surface energy.

Comparison of equations (12.2) and (12.6) shows

$$G_{\mathrm{IC}} = K_{\mathrm{IC}}^2/E^*. \tag{12.9}$$

Although the Griffith and Irwin formulations of the fracture problems are equivalent, most recent studies of polymers have followed Irwin. Before discussing results for polymers, it is useful to show how G_{c} can be calculated.

Consider a sheet of polymer with a crack of length $2c$ (Figure 12.3). We now define a quantity termed the compliance of the cracked sheet, C, which is the reciprocal of the slope of the linear load–extension curve from zero

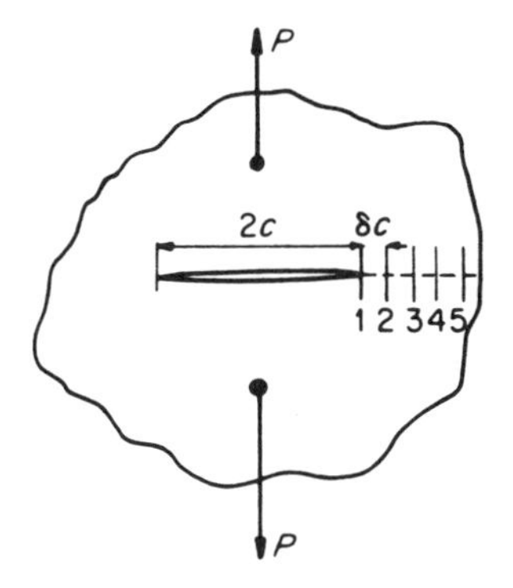

Figure 12.3 Schematic diagram of a specimen with a centre crack of length $2c$

load up to the point at which crack propagation begins. At the latter point the load is P and the extension is Δ, so $C = \Delta/P$.

This quantity C is not to be confused with an elastic stiffness constant as defined in section 7.1. The work done in an elemental step of crack propagation is illustrated by Figure 12.4. As the crack moves from 4 to 5 for example, the energy available for formation of new crack surface is the difference between the work done (45XY) and the increase in elastic stored energy (triangle 05Y – triangle 04X). This energy corresponds to the area of the shaded triangle in Figure 12.4, and for an increase of crack length dc is given by $\frac{1}{2}P^2\,\mathrm{d}C$. Hence

$$G_c = \frac{P^2}{2B}\frac{\mathrm{d}C}{\mathrm{d}c} \tag{12.10}$$

which is generally known as the Irwin–Kies relationship [4].

Here G_c can be determined directly by combining a load–extension plot from a tensile testing machine with determination of the movement of the crack across the specimen, noting the load P for given crack lengths (points 1, 2, 3, 4, 5 in Figure 12.4). Alternatively, test pieces of standard geometry can be used, for which the compliance is known as a function of crack length. For example, the relationship between the extension Δ (usually termed the deflection in this case) and the load P for a double cantilever beam specimen of thickness B (see Figure 12.5) is given by

$$\Delta = \frac{64c^3}{EBb^3}P$$

Hence

$$C = \frac{\Delta}{P} = \frac{64c^3}{EBb^3} \quad \text{and} \quad \frac{\mathrm{d}C}{\mathrm{d}c} = \frac{192c^2}{EBb^3} \tag{12.11}$$

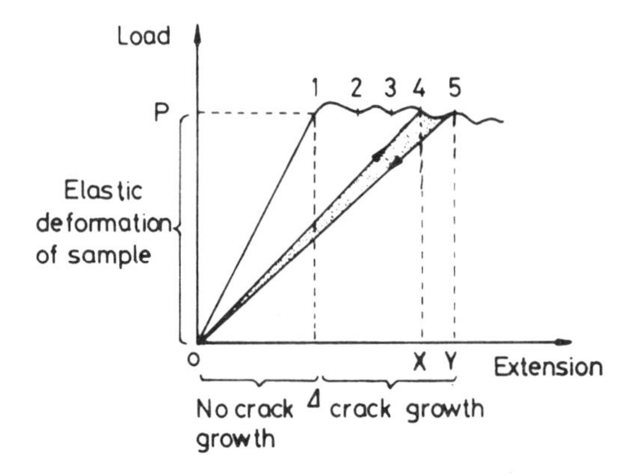

Figure 12.4 The load–extension curve for the specimen shown in Figure 12.3

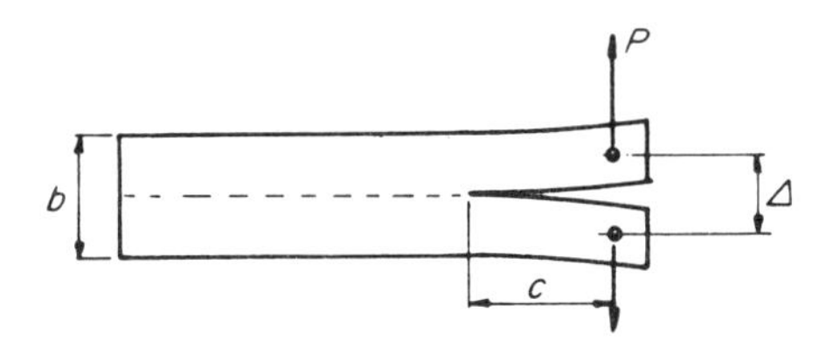

Figure 12.5 The double cantilever beam specimen

giving

$$G_c = \frac{P^2}{2B}\frac{dC}{dc} = \frac{P^2}{2B}\frac{192c^2}{EBb^3} \tag{12.12a}$$

or

$$G_c = \frac{3\Delta^2 b^3}{128c^4} E \tag{12.12b}$$

The critical strain-energy release rate (or, in the original Griffith terminology, the fracture surface energy γ) can therefore be obtained by measurements of either the load P or the deflection Δ for given crack lengths c.

The exact equivalent formulation in terms of the critical stress intensity factor can be obtained from equation (12.9) giving

$$K_{IC} = 4\sqrt{6}\,\frac{P^2}{Bb^{3/2}} \tag{12.13}$$

We have discussed only the calculation for a geometrically simple specimen, so that the principles involved are not obscured by complex stress analysis. For a comprehensive discussion of the calculation of the fracture toughness parameters G_c and K_c for specimens with different geometries, see standard texts [5–7]

12.3 CONTROLLED FRACTURE IN POLYMERS

In its simplest form, the Griffith theory and the linear elastic fracture mechanics which developed from it, ignore any contribution to the energy balance arising from the kinetic energy associated with movement of the crack. A basic study of the brittle fracture of polymers is likely to be most rewarding if the fracture takes place so slowly that a negligible amount of energy is dissipated in this way. Accordingly Benbow and Roesler [8] devised a method of fracture in which flat strips of polymer were cleaved lengthwise by gradually propagating a crack down the middle, as in the double cantilever beam of Figure 12.5.

Figure 12.6 shows a diagram of the Benbow and Roesler apparatus, which was used to test 6.35 mm thick specimens of commercial PMMA of a range of sizes. It was found empirically and subsequently explained theoretically

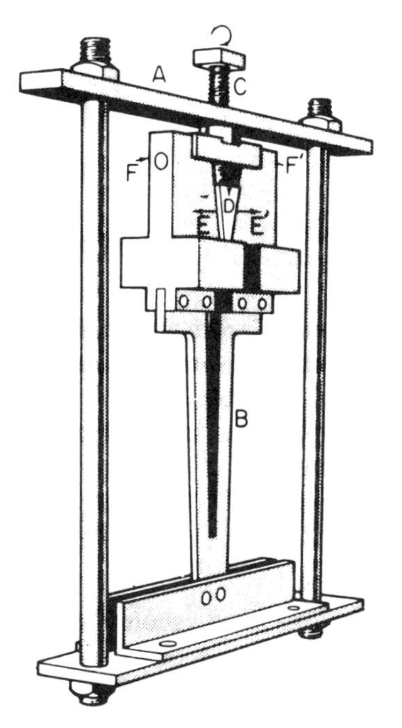

Figure 12.6 The cleavage apparatus of Benbow and Roesler. Flexure in the bar A compresses the sample B and stabilizes the crack direction. Turning the screw C moves the wedge D forward, forcing the clamps E, E′ apart. Rotation of the clamps, in the plane of the specimen, is prevented by the sliding bearings F, F′

by Cottrell [9], that the crack could be made to propagate straight along the strip by applying a preset lengthwise compression.

The increase in Griffith surface energy was equated to the decrease in elastically stored energy as the crack propagates to a greater length. In principle this treatment is identical to that outlined above for the double cantilever beam specimen where the crack length is very large. For any crack length the equations must take a similar dimensional form, and the form of the energy balance can be found from consideration of similarity. Essentially, the experiments involve determining the deflection Δ for given crack length c (symbols as in Figure 12.5). For small values of b/c, i.e. for large crack length, the situation is exactly that of the double cantilever beam, where from equation (12.12b) the surface energy γ is given by

$$\frac{2\gamma}{E} = \frac{G_c}{E} = \frac{3\Delta^2 b^3}{128c^4}$$

Knowing a value for the Young's modulus E, the surface energy can then be found.

Berry [10] using a simpler experimental procedure in which the crack direction was determined by routing an initial groove in the sample, adopted a slightly different approach and assumed that the force P required to bend the beam is given by an empirical formula $P = ac^{-n}\Delta$, where a is a constant. The energy balance equation was then written in terms of P, Δ, γ, c and n with n being determined by a subsidiary experiment. The surface energy can thus be obtained without a value for Young's modulus E, since this is effectively eliminated by measuring the force P. This procedure is clearly equivalent to the use of the Irwin–Kies relationship leading to equation (12.12a). Determining the compliance as a function of crack length has often been used in more recent work because it bypasses the problem as to whether fracture occurs under conditions of plane stress or plane strain.

Both Berry and Benbow and Roesler showed that the surface energy was independent of the sample dimensions, suggesting that it is a basic material property. Berry also examined the validity of equation (12.2) for the fracture of PMMA and polystyrene by measuring the tensile strength of samples containing deliberately introduced cracks of known magnitude. The relationship appeared to hold within the fairly large experimental errors experienced in fracture investigations, and gave values for the surface energy comparable with those obtained by cleavage tests. Berry's summary of his own results [10] and those of other workers are shown in Table 12.1.

An upper theoretical estimate of the surface energy is obtained if it is assumed that the energy required to form a new surface originates in the simultaneous breaking of chemical bonds only. Take the bond dissociation energy as 400 kJ and the concentration of molecular chains as 1 chain per 0.2 nm^2, giving 5×10^{18} molecular chains m^{-2}. To form 1 m^2 of new surface

Table 12.1 Fracture surface energies (in joules per square metre × 10^2)

	Polymers	
Method	Polymethyl methacrylate	Polystyrene
Cleavage (Benbow [11])	4.9 ± 0.5	25.5 ± 3
Cleavage (Svensson [12])	4.5	9.0
Cleavage (Berry [10])	1.4 ± 0.07	7.13 ± 0.36
Tensile (Berry [13])	2.1 ± 0.5	17 ± 6

requires about 1.5 J, which is two orders of magnitude less than that obtained from cleavage and tensile measurements.

12.4 SURFACE ENERGY AND CRAZING

The large discrepancy between experimental and theoretical values for the surface energy is comparable to that found for metals, where it was proposed by Orowan and others that the surface free energy may include a large term which arises from plastic work done in deforming the metal near the fracture surface as the crack propagates. Andrews [14] suggested that the quantity measured in the fracture of polymers should be described by $\mathfrak{J}$, the 'surface work parameter', to distinguish it from a true surface energy, and proposed a generalized theory of fracture which embraces viscoelastic as well as plastic deformation, both of which may be important in polymers.

On the basis of the results shown in Table 12.1, Berry concluded that the largest contribution to the surface energy of a glassy polymer comes from a viscous flow process which in PMMA, he suggested [15] was related to the interference bands observed on the fracture surfaces, as seen in Figure 12.7 He proposed that work was expended in the alignment of polymer chains ahead of the crack, the subsequent crack growth leaving a thin, highly oriented layer of polymeric material on the fracture surface. Following on from these ideas, Kambour [16–18] showed that a thin wedge of porous material termed a craze forms at a crack tip in a glassy polymer as shown schematically in Figure 12.8. The craze forms under plane strain conditions, so that the polymers is not free to contract laterally, and there is a consequent reduction in density. Several workers [19, 20] have attempted to determine the craze profile by examining the crack tip region in PMMA in an optical microscope. In reflected light, two sets of interference fringes were observed, which correspond to the crack and the craze respectively. It was found that the craze profile was very similar to the plastic zone model proposed by Dugdale [21] for metals, which will now be described.

Figure 12.7 Matching fracture surfaces of a cleavage sample of polymethyl methacrylate showing colour alternation (green filter). (Reproduced from Berry, in *Fracture 1959* (B.L. Averach et al., Eds), Wiley, New York, 1959, p. 263)

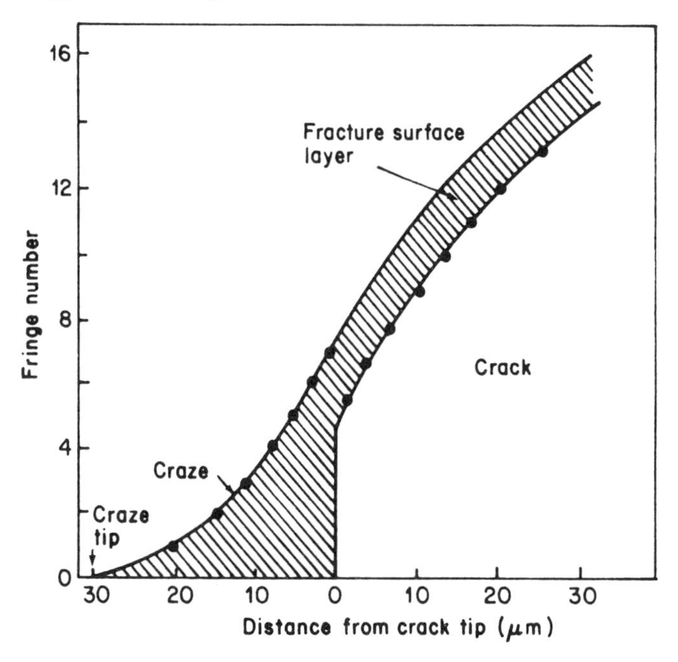

Figure 12.8 Schematic diagram of a craze (Reproduced with permission from Brown and Ward, *Polymer*, **14**, 469 (1973). (C) IPC Business Press Ltd)

Equations (12.3) imply that there is an infinite stress at the crack tip. In practice this clearly cannot be so, and there are two possibilities. First, there can be a zone where shear yielding of the polymer occurs. In principle this can occur in both thin sheets where conditions of plane stress pertain and in thick sheets where there is a plane strain. Secondly, for thick specimens under conditions of plane strain, the stress singularity at the crack tip can be released by the formation of a craze, which is a line zone, in contrast to the approximately oval (plane stress) or kidney-shaped (plane strain) shear yield zones. As indicated, its shape approximates very well to the idealized Dugdale plastic zone where the stress singularity at the crack tip is cancelled by the superposition of a second stress field in which the stresses are compressive along the length of the crack (Fig. 12.9). A constant compressive stress is assumed, and is identified with the craze stress. It is not the yield stress, and crazing and shear yielding are different in nature and respond differently to changes in the structure of the polymer.

Rice [22] has shown that the length of the craze for a loaded crack on the point of propagation is

$$R = \frac{\pi}{8}\frac{K_{\mathrm{IC}}^2}{\sigma_{\mathrm{c}}^2} \tag{12.15}$$

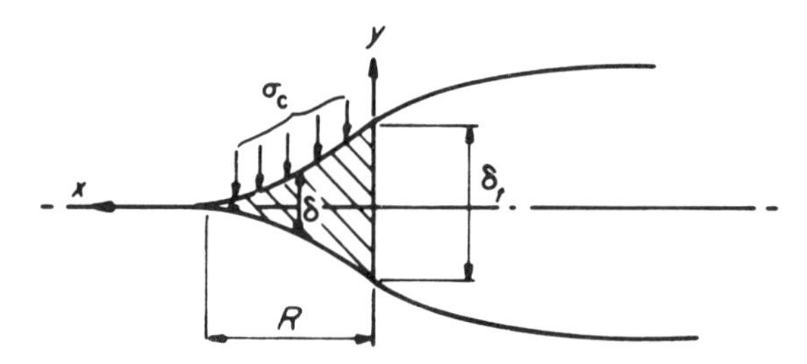

Figure 12.9 The Dugdale plastic zone model for a craze

and the corresponding separation distance δ between the upper and lower surfaces of the craze is

$$\delta = \frac{8}{\pi E^*}\,\sigma_c R\left[\zeta - \frac{x}{2R}\log\left(\frac{1+\zeta}{1-\zeta}\right)\right] \tag{12.16}$$

where $\zeta = (1-x/R)^{1/2}$.

The crack opening displacement (COD) δ_t is the value of the separation distance δ at the crack tip, where $x = 0$, and is therefore

$$\delta_t = 8\sigma_c R/\pi E^* = K_{IC}^2/\sigma_C E^*. \tag{12.17}$$

The fracture toughness of the polymer then relates to two parameters δ_t and σ_c, the craze stress, the product of which $\delta_t\sigma_c = G_{IC}$ the critical strain energy release rate. Direct measurements of craze shapes for several glassy polymers, including polystyrene, polyvinyl chloride and polycarbonate [20, 23] have confirmed the similarity to a Dugdale plastic zone. A result of some physical significance is that the crack opening displacement is often insensitive to temperature and strain rate for a given polymer, although it has been shown to depend on molecular mass. For constant COD, the true dependence of G_{IC} on strain rate and temperature is determined only by the sensitivity of the craze stress to these parameters. Since $G_{IC} = K_{IC}^2/E^*$ the fracture toughness K_{IC} will in addition be affected by E^*, which is also dependent on strain rate and temperature.

This approach offers a deeper understanding of the brittle–ductile transition in glassy polymers in terms of competition between crazing and yielding. Both are activated processes, in general with different temperature and strain rate sensitivities, and one will be favoured over the other for some conditions and vice versa for other conditions. An additional complexity can arise from the nature of the stress field which may favour one process rather than the other, but the latter consideration does not enter into our discussion of the craze at the crack tip. The line of travel of the crack is a line of zero shear stress within the plane but maximum triaxial stress. In later discussion we will see that such a stress field favours crazing and that, for long cracks where the stress field of the crack is the dominant factor, the craze length is determined solely by the requirement that the craze grows to cancel the stress singularity at the crack tip.

In several glassy polymers [23, 24] such as the polycarbonate shown in Figure 12.10, a complication occurs in that a thin line of material called a shear lip forms on the fracture surface where the polymer has yielded. Analogous to the behaviour of metals, it has been proposed that the overall strain energy release rate G_C^0 is the sum of the contribution from the craze and that from the shear lips. To a first approximation we would expect the latter to be proportional to the volume of yielded material. If the total width of the shear lip on the fracture surface is w, B is the specimen thickness, and the shear lip is triangular in cross-section, then

$$G_c^0 = G_{IC}\left(\frac{B - w}{B}\right) + \frac{\phi w^2}{2B} \tag{12.18}$$

where ϕ is the energy to fracture unit volume of shear lip. It has been shown that this relationship describes results for polycarbonate and polyether sulphone very well [23, 24] and that ϕ corresponds quite closely to the energy to fracture in a simple tensile extension experiment.

An alternative approach [25] assumes an additivity rule based on a plane strain K_{IC}, which pertains to fracture in the central part of the specimen and is designated K'_{IC}, and a plane stress K_{IC} which is effective for the two surface skins of depth $w/2$ and is designated K''_{IC}. For the overall specimen it is then proposed that

$$K_{IC} = \left(\frac{B - w}{B}\right) K'_{IC} + \left(\frac{w}{B}\right) K''_{IC}. \tag{12.18a}$$

Although equation (12.18a) is more empirically based than equation (12.18) and is not formally equivalent, it has been shown to model fracture results very well. Moreover, in this formulation w relates to the size of the so-called Irwin plastic zone r_y which can be simply defined on the basis of equation

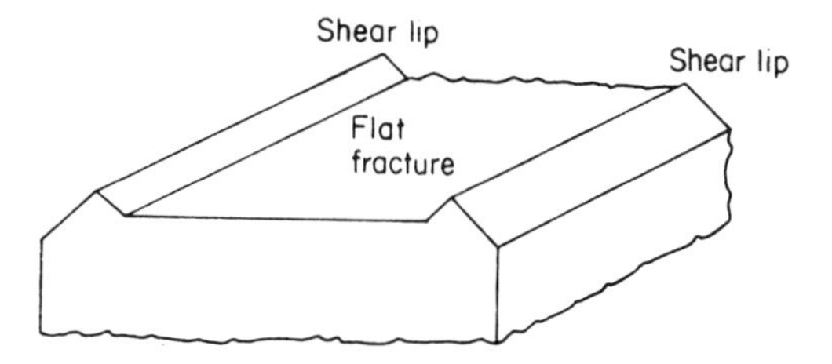

Figure 12.10 The shear lips in polycarbonate. (Reproduced with permission from Fraser and Ward, *Polymer*, **19**, 220 (1978))

(12.3) by assuming that at a point r_y, the stress reaches the yield stress σ_y. Hence

$$r_y = \frac{1}{2\pi}\left(\frac{K_{IC}}{\sigma_y}\right)^2$$

for plane stress and $w/2 = r_y$ in equation (12.18a).

For PMMA Berry showed that the surface energy was strongly dependent on polymer molecular mass [26]. His results (Figure 12.11) fitted an approximately linear dependence of the fracture surface energy on reciprocal molecular mass, such that $\gamma = A' - B'/\bar{M}_v$, where $\bar{M}_v$ is the viscosity average molecular mass. Many years previously Flory [27] had proposed that the brittle strength is related to the number average molecular mass.

More recently Weidmann and Döll [28] have shown that the craze dimensions decrease markedly in PMMA at low molecular masses. In a study of the molecular mass dependence of fracture surfaces in the same polymer, Kusy and Turner [29] could observe no interference colours for a viscosity average molecular mass of less than 90 000 daltons, concluding that there was a dramatic decrease in the size of the craze. Based on craze shape studies of polycarbonate, Pitman and Ward [23] reported a very high dependence of both craze stress and crack opening displacement on molecular mass and observed that both would be expected to become negligibly small for $\bar{M}_w < 10^4$. Berry speculated that the smallest molecule which could contribute to the surface energy would have its end on the boundaries of the craze region, on opposite sides of the fracture plane, and be fully extended between these points. Kusy and Turner [30] presented a

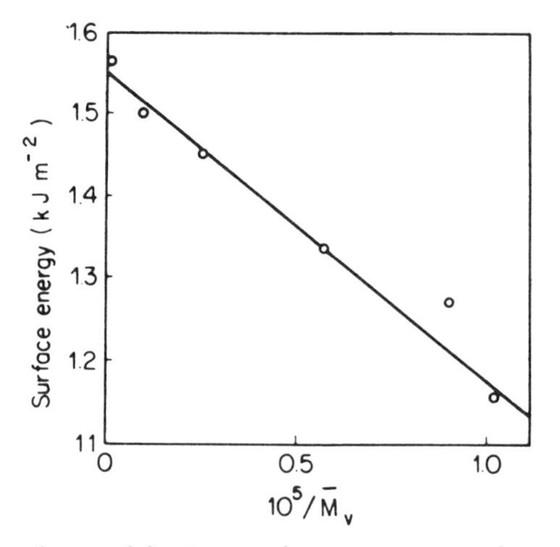

Figure 12.11 Dependence of fracture surface energy on reciprocal molecular mass ($\bar{M}_v$ is viscosity average molecular mass). (Reproduced with permission from Berry, *J. Polymer Sci.* **A2**, 4069 (1964))

fracture model for PMMA in which the surface energy measured, was determined by the number of chains above a critical length. Their data fitted well with their predictions, showing a limit to the surface energy at high molecular weight, but the model appeared inappropriate for the polycarbonate data of Pitman and Ward. Moreover, the extended molecular lengths, based on the extension of a random coil would be much less than the crack opening displacement (as discussed by Haward, Daniels and Treloar [31]) so that there is no direct correlation between the two quantities. The craze structure relates to the stretching of fibrils and the key molecular factors are the presence of random entanglements and the distance between these entanglements, not the extension of an isolated molecular chain.

12.5 THE STRUCTURE AND FORMATION OF CRAZES

We have seen how the craze at the crack tip in a glassy polymer plays a vital role in determining its fracture toughness. Crazing in polymers also manifests itself in another way. When certain polymers, notably PMMA and polystyrene, are subjected to a tensile test in the glassy state, above a certain tensile stress opaque striations appear in planes whose normals are the direction of tensile stress, as in Figure 12.12.

The interference bands on the fracture surfaces, which relate to the craze at the crack tip, were first observed by Berry [32] and by Higuchi [33]. Kambour confirmed that the PMMA fracture-surface layers were qualitatively similar to the internal crazes of this polymer, by showing that the refractive indices were the same [16]. Both surface layer and bulk crazes appear to be oriented polymer structures of low density, which are produced by orienting the polymer under conditions of abnormal constraint: it is not allowed to contract in the lateral direction, while being extended locally to strains of the order of unity, and so has undergone inhomogeneous cold-drawing.

Detailed studies have been made of the structure of crazes, the stress or strain criteria for their formation and environmental effects. These subjects will now be discussed in turn.

12.5.1 The Structure of Crazes

The structure of crazes in bulk specimens was studied by Kambour [16], who used the critical angle for total reflection at the craze/polymer interface to determine the refractive index of the craze, and showed that the craze was roughly 50% polymer and 50% void. Another investigation involved transmission electron microscopy of polystyrene crazes impregnated with an

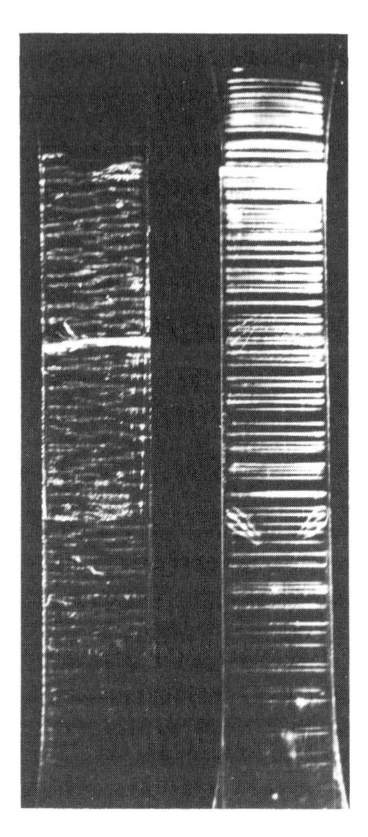

Figure 12.12 Craze formation in polystyrene

iodine–sulphur eutectic to maintain the craze in its extended state [34, 35]. The structure of the craze was clearly revealed as fibrils separated by the voids which are responsible for the overall low density.

Using a rather more direct technique Beahan, Bevis and Hull [36], examined the microstructure of microtomed thin sections of precrazed bulk polystyrene. Crazes were also examined which were formed by straining microtomed thin sections of polystyrene [37, 38]. In both cases a fibril structure was observed within the craze, it being rather finer in the thin specimens. Further detailed studies of craze structure with transmission electron microscopy [39, 40], selected area electron diffraction [41] and small angle X-ray scattering [42] have confirmed that the craze is constituted of cylindrical fibrils of highly oriented polymer (Figure 12.13). The fibril axes are parallel to the tensile stress direction, as expected if the fibrils consist essentially of drawn polymer.

It has been concluded [43] that the extension ratio of the fibrils relates to

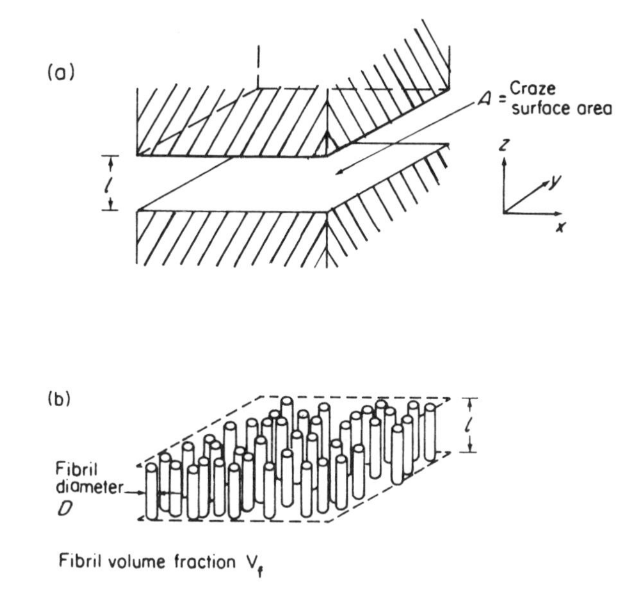

Figure 12.13 Schematic diagram of the forest of cylindrical fibrils oriented normal to the craze surface. (Reproduced with permission from Brown and Kramer, *J. Macromol. Sci. B*, **19**, 487 (1981) by courtesy of Marcel Dekker Inc.

the extensibility of a molecular network. Values for the extension ratios, estimated by Kramer and his colleagues [43, 44] and also by Ward and co-workers [19, 23], from analysis of optical interference patterns, compare reasonably well with estimates of the network extensibility from small angle neutron scattering data [43] or stress–optical measurements [45].

The studies of the craze structure by Kramer and his co-workers confirm the earlier findings of Hull and colleagues in showing that the craze structure is not uniform along its length. Although there are significant discrepancies between the displacement and the stress on the craze with those predicted by the Dugdale zone model, the latter nevertheless provides a good overall description of the mechanics of the craze and is quite adequate for most purposes.

12.5.2 Craze Initiation and Growth

The studies of craze formation and structure described above indicate that there are clear differences between crazing and yield. Yield is essentially a shear process where the deformation occurs at constant volume (ignoring structural changes such as crystallization), but crazing occurs at a crack tip or in a solid section with a very appreciable increase in volume. It therefore appears that tensile stresses and in particular, the hydrostatic tensile stress will be important in craze initiation and growth.

It would be desirable to obtain a stress criterion for craze initiation analogous to that for yield behaviour described in Chapter 11. Although all proposals made so far have not achieved general acceptance, it is of value to review the most important findings. Sternstein, Ongchin and Silverman [46] examined the formation of crazes in the vicinity of a small circular hole (1.59 mm diameter) punched in the centre of PMMA strips (12.7 mm × 50.8 mm × 0.79 mm) when the latter are pulled in tension. A typical pattern is shown in Figure 12.14(a). When the solutions for the elastic stress field in the vicinity of the hole were compared with the craze pattern it was found that the crazes grew parallel to the minor principal stress vector. As the contours of the minor principal stress vector are orthogonal to those of the major principal stress vector, this result shows that the major principal stress acts along the craze plane normal and therefore parallel to the molecular orientation axis of the crazed material.

The boundary of the crazed region coincided to a good approximation with contour plots showing lines of constant major principal stress σ_1, as shown in Figure 12.14(b) where the contour numbers are per unit of applied stress. At low applied stresses it is not possible to discriminate between the contours of constant σ_1 and contours showing constant values of the first stress invariant $I_1 = \sigma_1 + \sigma_2$. However, the consensus of the results is in accord with a craze-stress criterion based on the former rather than on the latter and, as we have seen, the direction of the crazes is consistent with the former.

Sternstein and Ongchin [47] extended this investigation by examining the formation of crazes under biaxial stress conditions, and found that the stress conditions for crazing involved both the principal stresses σ_1 and σ_2. The most physically acceptable explanation of these results was proposed by Bowden and Oxborough [48], who suggested that crazing occurs when the extensional strain in any direction reaches a critical value e_1, which depends on the hydrostatic component of stress.

For small strains, for the two-dimensional stress field, e_1 is given by

$$e_1 = \frac{1}{E}(\sigma_1 - \nu\sigma_2),$$

where E is Young's modulus and ν Poisson's ratio.

It was proposed that the crazing criterion was

$$Ee_1 = \sigma_1 - \nu\sigma_2 = A + B/I_1, \tag{12.19}$$

where $I_1 = \sigma_1 + \sigma_2$. Equation (12.19) predicts that the stress required to

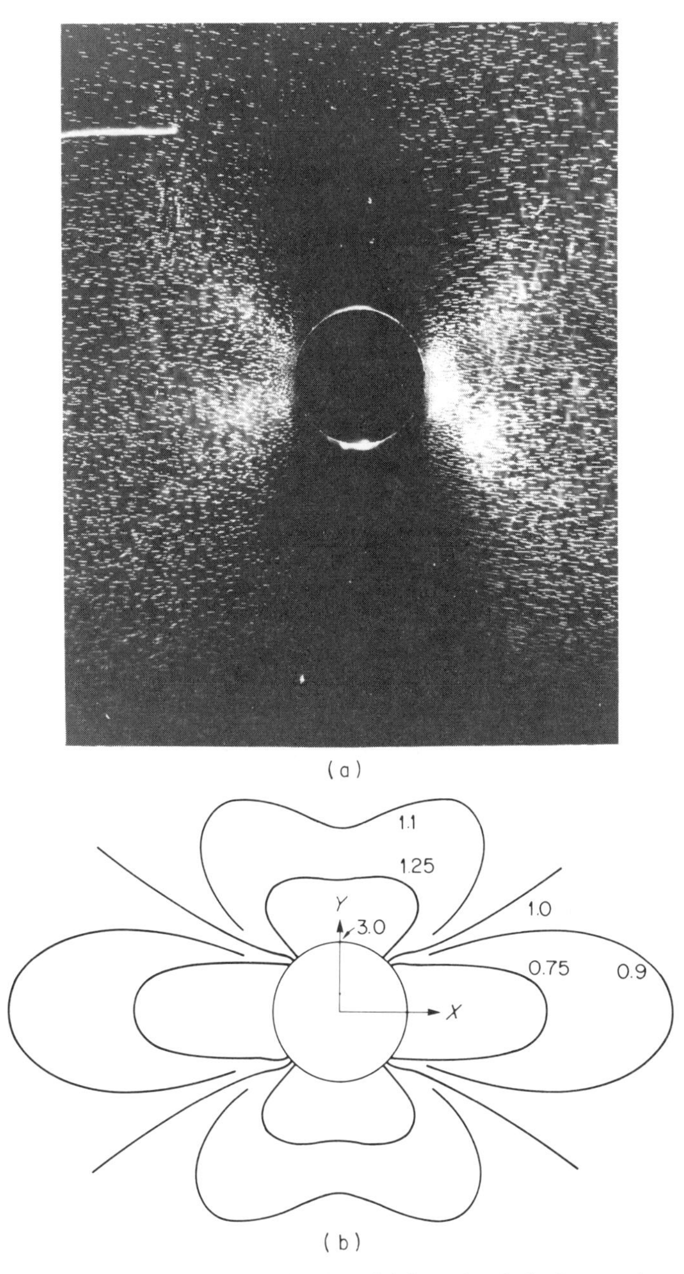

Figure 12.14 (a) Craze formation in the vicinity of a hole in a strip of PMMA loaded in tension. (Result obtained by L. S. A. Smith) (b) Major principal stress contours (σ_1) for an elastic solid containing a hole. The specimen is loaded in tension in the x direction. Contour numbers are per unit of applied tensile stress. (Reproduced with permission from Sternstein, Ongchin and Silverman, *Appl. Polymer Symp*. **7**, 175 (1968))

initiate a craze becomes infinite when $I_1 = 0$, i.e. crazing requires a dilational stress field. Unfortunately there are several pieces of experimental evidence [49–51] which contradict this assumption, so that there is still no completely satisfactory stress criterion for craze initiation.

There is, however, a theory for the growth of crazes which is consistent with all the experimental evidence. Argon, Hannoosh and Salama [52] have proposed that the craze front advances by a meniscus instability mechanism in which craze tufts are produced by the repeated break-up of the concave air/polymer interface at the crack tip, as illustrated in Figure 12.15. A theoretical treatment of this model predicted that the steady-state craze velocity would relate to the five-sixths power of the maximum principal tensile stress, and support for this result was obtained from experimental results on polystyrene and PMMA [52].

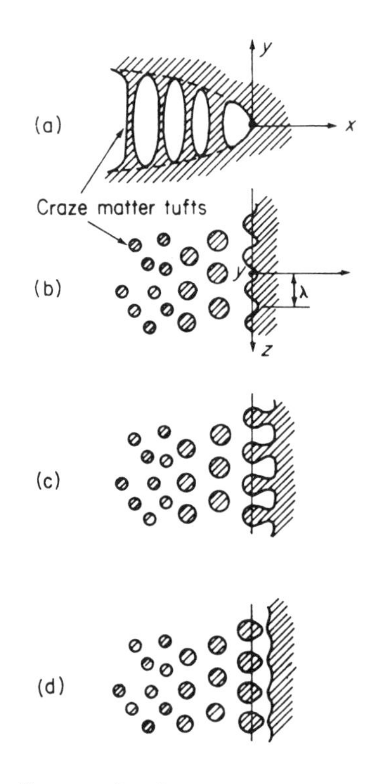

Figure 12.15 Schematic diagram showing craze matter production by the mechanism of meniscus instability: (a) outline of a craze tip; (b) cross-section in the craze plane across craze matter tufts; (c), (d) advance of the craze front by a completed period of interface convolution. (Reproduced with permission from Argon, Hannoosh and Salama, in *Fracture 1977*, Vol. 1, Waterloo, 1977, p. 445)

12.5.3 Crazing in the Presence of Fluids and Gases: Environmental Crazing

The crazing of polymers by environmental agents is of considerable practical importance and has been studied extensively, with notable contributions from Kambour [17, 53–55], Andrews and Bevan [56], Williams and co-workers [57, 58] and Brown [59–61]. The subject has been reviewed by Kambour [62] and by Brown [63]. In general environmental agents, which can be fluids or solids, reduce the stress or strain required to initiate crazing.

Kambour and co-workers [17, 53–55] showed that the critical strain for crazing decreased as the solubility of the environmental agent was increased. It was also found that the critical strain decreased as the glass transition temperature of the solvated polymer decreased. Andrews and Bevan [56] adopting a more formal approach, and applying the ideas of fracture mechanics, performed fracture tests on single-edge notched tensile specimens, where a central edge crack of length c is introduced into a large sheet of polymer which is then loaded in tension. The fracture stress is related to the surface work parameter $\mathfrak{J}$ of Andrews (or the strain energy release rate $G_c = 2\gamma$) by an equation identical in form to equation (12.2) above. The critical stress for crack and craze propagation σ_c was indeed proportional to $c^{-1/2}$, so that $\mathfrak{J}$ values could be determined. For constant experimental conditions, a range of values of $\mathfrak{J}$ was obtained from which a minimum value $\mathfrak{J}_0$, was estimated. From tests in a given solvent over a range of temperatures, it was found that values of $\mathfrak{J}_0$ decreased with increasing temperature up to a characteristic temperature $\mathfrak{J}_c$, above which $\mathfrak{J}_0$ remained constant at a value $\mathfrak{J}_0^*$. The values of $\mathfrak{J}_0^*$ for the different solvents were shown to be a smooth function of the difference between the solubility parameters of the solvent and the polymer, reaching a minimum when this difference was zero (Figure 12.16).

These findings were explained on the basis that the work done in producing the craze can be modelled by the expansion of a spherical cavity of radius r under a negative hydrostatic pressure p, which has two terms so that

$$p = \frac{2\gamma_\tau}{r} + \frac{2\sigma_Y}{3}\psi \tag{12.20}$$

where γ_τ is the surface tension between the solvent in the void and the surrounding polymer, σ_Y is the yield stress and ψ is a factor close to unity. The effect of temperature is to change the yield stress, so that with increasing temperature σ_Y falls, eventually to zero at T_c, which is the glass transition temperature of the plasticized polymer. Above T_c the fracture surface energy $\mathfrak{J}_0^*$ then relates solely to the intermolecular forces represented by the surface tension γ_τ.

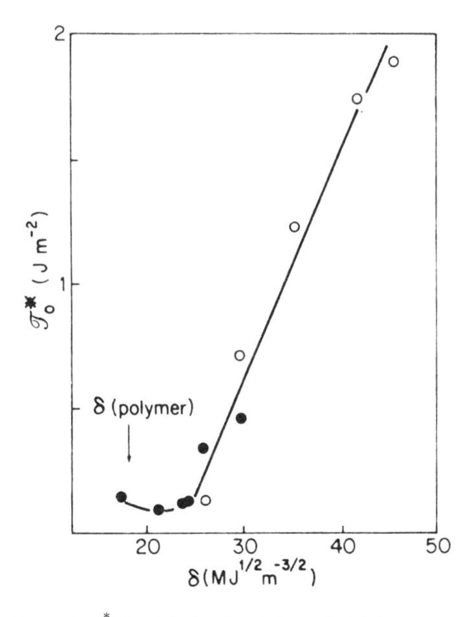

Figure 12.16 Variation of J_0^* for PMMA with solubility parameter of the solvent: (●) pure solvents; (○) water-isopropanol mixtures. (Reproduced with permission from Andrews and Bevan, *Polymer*, **13**, 337 (1972). (C) IPC Business Press Ltd.)

Brown has pointed out that gases at sufficiently low temperatures make almost all linear polymers craze [59–61, 63]. Parameters such as the density of the crazes and the craze velocity increase with the pressure of the gas and decrease with increasing temperature. It was concluded that the surface concentration of the absorbed gas was a key factor in determining its effectiveness as a crazing agent.

In a related, but somewhat different development, Williams and co-workers [57, 58] studied the rate of craze growth in PMMA in methanol. In all cases the craze growth depended on the initial stress-intensity factor K_0, calculated from the load and the initial notch length. Below a specific value of K_0 termed K_0^*, the craze would decelerate and finally arrest. For $K_0 > K_0^*$ the craze would decelerate initially and finally propagate at constant speed.

It was argued that the controlling factor determining craze growth was the diffusion of methanol into the craze. Where $K_0 < K_0^*$, the methanol is considered to diffuse along the length of the craze, and it may be shown that the length of the craze x is proportional to the square root of the time of growth (Figure 12.17). In the second type of growth, where $K_0 > K_0^*$, it is considered that the methanol diffuses through the surface of the specimens, maintaining the pressure gradient in the craze and producing craze growth at constant velocity.

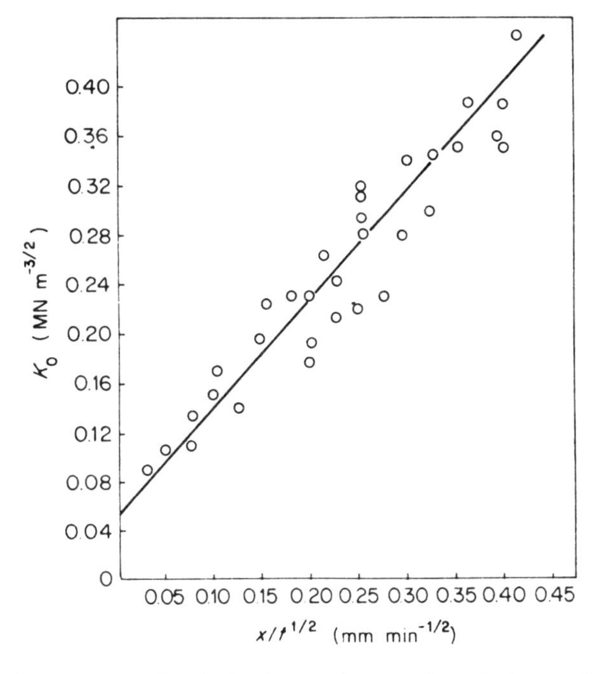

Figure 12.17 Craze growth behaviour for polymethyl methacrylate in methanol at 20 °C (Reproduced with permission from Williams and Marshall, *Proc. Roy. Soc.* **A342**, 55 (1975))

12.6 THE MOLECULAR APPROACH

It has long been recognized that oriented polymers (i.e. fibres) are much less strong than would be predicted on the basis of elementary assumptions that fracture involves simultaneously breaking the bonds in the molecular chains across the section perpendicular to the applied stress. Calculations of this nature were originally undertaken by Mark [64] and rather more recently by Vincent [65] on polyethylene. It was found that in both cases the measured tensile strength was at least an order of magnitude less than that calculated.

We have seen one possible explanation of this discrepancy—the Griffith flaw theory of fracture. It has also been considered that there may be a general analogy between this difference between measured and calculated strengths and the difference between measured and calculated stiffnesses for oriented polymers. A general argument for both discrepancies could be that only a small fraction of the molecular chains are supporting the applied load. In Chapter 8, we have discussed how the tie molecules or crystalline bridges which connect adjacent crystalline blocks play a key role in determining the axial stiffness of an oriented semi-crystalline polymer. There has therefore been considerable interest in examining chain fracture in oriented polymers, using electron paramagnetic resonance to observe free radicals produced, or infrared spectroscopy to identify such entities as aldehyde end groups which suggest chain scission. A very comprehensive survey of the results of such studies has been given by Kausch [66]. Kausch and Becht [67] have emphasized that the total number of broken chains is much too small for their load-carrying capacity to account for the measured reductions in macroscopic stress. We must therefore conclude that the tie molecules which eventually break are not the main source of strength of highly oriented polymers, a conclusion confirmed by the lack of any positive correlations between the strength of fibres and the radical concentration at break.

Although these strong reservations have to be borne in mind, the examination of chain fracture is not totally irrelevant to the deformation of polymers. Examination of the infrared spectra of oriented polymers under stress shows that there is a distinct shift in frequency from the unstressed state [68, 69] indicative of a distortion of bonds in the chain due to stress. It has also been proposed that there is a change in the shape of the absorption line, which has been interpreted as implying that the stress is inhomogeneous at a molecular level so that certain bonds are much more highly stressed than the average. The actual concentration of free radicals may therefore be primarily an indication of the stress distribution within the structure, for example between different microfibrils and hence tie molecules, and not relate to the strength *per se*.

A positive attempt to obtain a molecular understanding of fracture took as its starting-point the time and temperature dependence of the fracture processes. Zhurkov and his collaborators [70] measured the time for polymers to fracture as a function of tensile stress at various temperatures and proposed that the relationship between the lifetime, the tensile stress σ_B and the absolute temperature T could be represented by an Eyring-type equation:

$$\tau = \tau_0 \exp\left\{\frac{U_0 - \beta\sigma_B}{kT}\right\}$$

where τ_0, U_0 and β are constants determining the strength characteristics of a polymer.

The parameter U_0 has the dimensions of energy and it is suggested that it corresponds to the height of the activation barrier which has to be surmounted for fracture to occur. For a wide range of polymers U_0 was approximately equal to the activation energies obtained for thermal breakdown. The technique of electron spin resonance then showed that free radicals are produced in the fracture process in polymers and moreover that correlations can be established between the radical formation rate and the time to break.

The existence of submicrocracks in polymers has already been mentioned in connection with the Argon theory of craze initiation. Zhurkov, Kuksenko and Slutsker [71] have used small angle X-ray scattering to establish the presence of such submicroscopic cracks. Although it has been proposed by Zakrevskii [72] that the formation of these submicrocracks is associated with a cluster of free radicals and the associated ends of molecular chains, Peterlin [73] has argued that the cracks occur at the ends of microfibrils, and Kausch [66] has concluded that the submicrocrack formation is essentially independent of chain scission.

12.7 FACTORS INFLUENCING BRITTLE–DUCTILE BEHAVIOUR: BRITTLE–DUCTILE TRANSITIONS

12.7.1 The Ludwig–Davidenkov–Orowan Hypothesis

Many aspects of the brittle–ductile transition in metals, including the effect of notching, which we will discuss separately, have been discussed in terms of the Ludwig–Davidenkov–Orowan hypothesis, that brittle fracture occurs when the yield stress exceeds a critical value [74], as illustrated in Figure 12.18(a). It is assumed that brittle fracture and plastic flow are independent processes, giving separate characteristic curves for the brittle fracture stress σ_B and the yield stress σ_Y as a function of temperature at constant strain rate (as shown in Figure 12.18(b)). Changing strain rate will produce a shift in these curves. It is then argued that whichever process, either fracture or yield, that can occur at the lower stress will be the operative one. Thus the intersection of the σ_B/σ_Y curves defines the brittle–ductile transition and the material is ductile at all temperatures above this point.

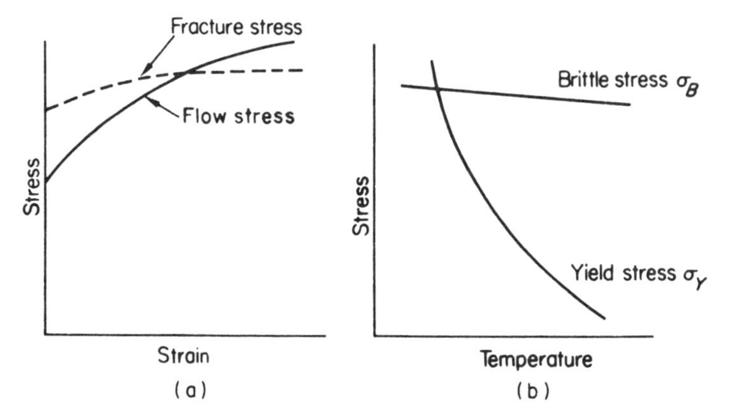

Figure 12.18 (a) and (b) Diagrams illustrating the Ludwig–Davidenkov–Orowan theories of brittle–ductile transitions

The influence of chemical and physical structure on the brittle–ductile transition can be analysed by considering how these factors affect the brittle stress curve and the yield stress curve respectively. As will be appreciated, this approach bypasses the relevance of fracture mechanics to brittle failure. If, however, we consider fracture initiation (as distinct from propagation of a crack) as governed by a fracture stress σ_B, the concept of regarding yield and fracture as competitive processes provides a useful starting-point.

Vincent and others [75–77] have shown that the brittle stress is not much affected by strain rate and temperature (e.g. by a factor of 2 in the temperature range −180 to +20 °C). The yield stress, on the other hand, is greatly affected by strain rate and temperature, increasing with increasing strain rate and decreasing with increasing temperature. (A typical figure would be a factor of 10 over the temperature range −180 to +20 °C.) These ideas are clearly illustrated by results for PMMA shown in Figure 12.19(a). The brittle–ductile transition will therefore be expected to move to higher temperatures with increasing strain rate (Figure 12.19(b)). The effect can be illustrated by varying the strain rate in a tensile test on a sample of nylon at room temperature: at low strain rates the sample is ductile and cold-draws, whereas at high strain rates it fractures in a brittle manner.

A further complication in varying strain rate occurs at low speeds, where within a certain temperature range cold-drawing occurs. It is possible that at high speeds the heat is not conducted away rapidly enough, so that strain hardening is prevented and the specimen fails in a ductile manner. Such an isothermal–adiabatic transition does not affect the yield stress and therefore does not affect the brittle–ductile transition; but it does cause a considerable reduction in the energy to break and may be operative in impact tests, even if brittle fracture does not intervene. It has therefore been proposed that there are two critical velocities at which the fracture energy drops sharply as the strain rate is increased. First there is the isothermal–adiabatic transition, and at higher strain rates, the brittle–ductile transition. Changes in ambient temperature have very little effect on the position of the isothermal–adiabatic transition, but a large effect on the brittle–ductile transition.

It was at first thought that the brittle–ductile transition was related to mechanical relaxation and in particular to the glass transition, which is true for natural rubber, polyisobutylene and polystyrene, but is not for most thermoplastics. It was then proposed [78] that where there is more than one mechanical relaxation, the brittle–ductile transition may be associated with a lower temperature relaxation. Although again it appeared that there might be cases where this is correct, it was soon shown that this hypothesis has no general validity. Because the brittle–ductile transition occurs at fairly high strains, whereas the dynamic mechanical behaviour is measured in the linear, low strain region, it is unreasonable to expect that the two can be directly linked. It is certain that fracture, for example, depends on several

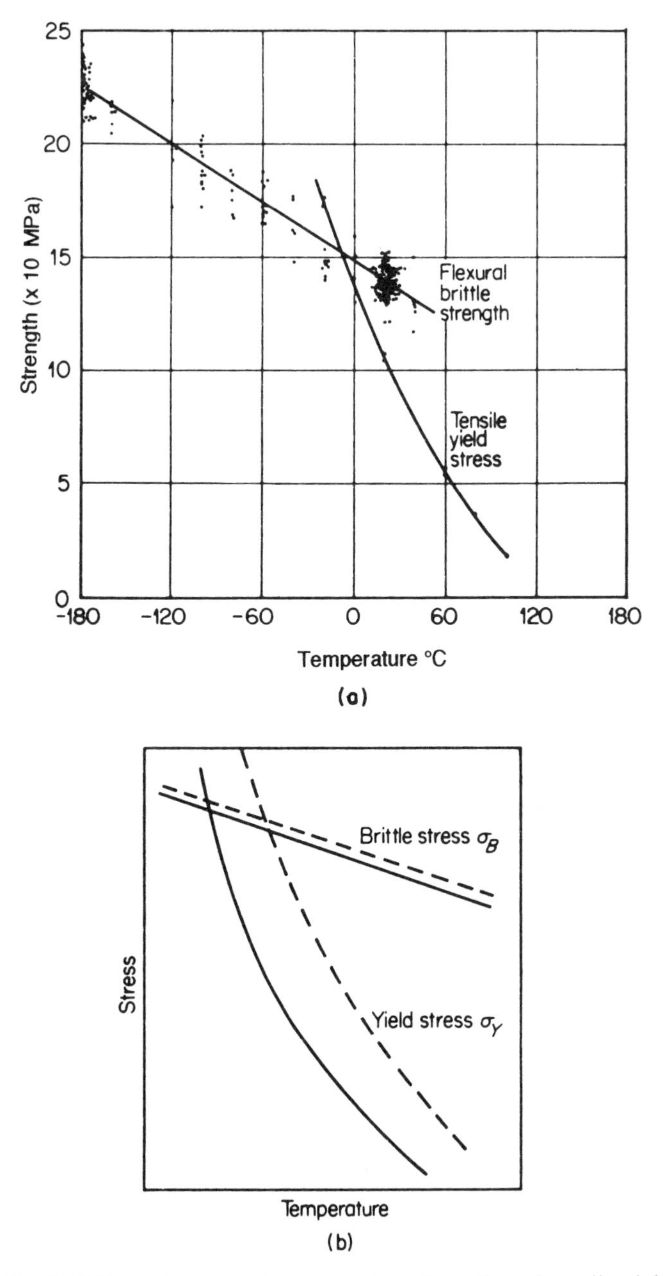

Figure 12.19 (a) Effect of temperature on brittle strength and tensile yield stress of PMMA. (Reproduced with permission from Vincent, *Plastics* **26**, 141 (1961)) (b) Diagram illustrating the effect of strain rate on the brittle–ductile transition: (——). low strain rate; (– – –), high strain rate

other factors such as the presence of flaws which will not affect the low-strain dynamic mechanical behaviour. The subject has been discussed extensively by Boyer [79] and by Heijboer [80].

12.7.2 Notch Sensitivity and Vincent's σ_B–σ_Y Diagram

As for metals the presence of a sharp notch can change the fracture of a polymer from ductile to brittle. For this reason a standard impact test for a polymer is the Charpy or Izod test, where a notched bar of polymer is struck by a pendulum and the energy dissipated in fracture calculated.

A very simple explanation of the effect of notching has been given by Orowan [74]. For a deep, symmetrical tensile notch, the distribution of stresses is identical with that for a flat frictionless punch indenting a plate under conditions of plane strain [81] (Figure 12.20). The compressive stress on the punch required to produce plastic deformation can be shown to be $(2 + \pi)K$, where K is the shear yield stress. For the Tresca yield criterion the value is $2.57\sigma_Y$ and for the von Mises yield criterion $2.82\sigma_Y$, where σ_Y is the tensile yield stress. Hence for an ideally deep and sharp notch in an infinite solid, the plastic constraint raises the yield stress to a value of approximately $3\sigma_Y$ which leads to the following classification for brittle–ductile behaviour first proposed by Orowan [1]:

1. If $\sigma_B < \sigma_Y$ the material is brittle;
2. If $\sigma_Y < \sigma_B < 3\sigma_Y$, the material is ductile in an unnotched tensile test, but brittle when a sharp notch is introduced;
3. If $\sigma_B < 3\sigma_Y$, the material is fully ductile, i.e. ductile in all tests, including those in notched specimens.

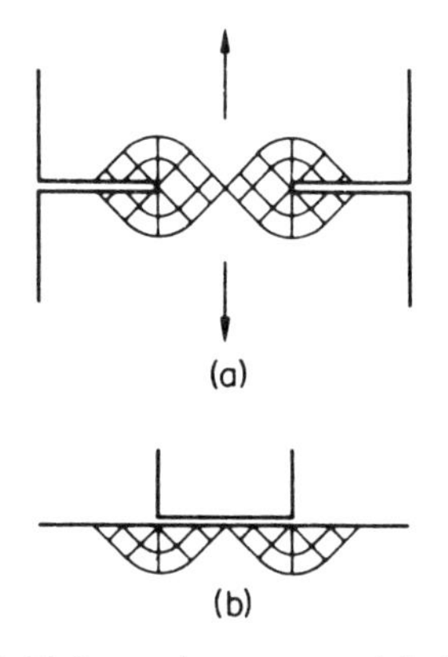

Figure 12.20 The slip-line field for a deep symmetrical notch (a) is identical with that for the frictionless punch indenting a plate under conditions of plane strain (b). (Reproduced with permission from Cottrell, *The Mechanical Properties of Matter*, Wiley, New York, 1964)

12.7.2.1 Vincents σ_B–σ_Y diagram

We may ask how relevant the above ideas are to the known behaviour of polymers. Vincent [82] has constructed a σ_B–σ_Y diagram which is very instructive in this regard (Figure 12.21).

Where possible, the value of σ_Y was taken as the yield stress in a tensile test at a strain rate of about 50% per minute; for polymers which were brittle in tension, σ_Y was the yield stress in uniaxial compression, and σ_B was the fracture strength measured in flexure at a strain rate of 18 min^{-1} at −180 °C.

The yield stresses were measured at +20 and −20 °C, the idea being that the −20 °C values would give a rough indication of the behaviour in impact at +20 °C, i.e. lowering the temperature by 40 °C is assumed to be equivalent to increasing the strain rate by a factor of about 10^5.

In the diagram the circles represent σ_B and σ_Y at +20 °C; the triangles σ_B and σ_Y at −20 °C. Both σ_Y and σ_B are affected by subsidiary factors such as

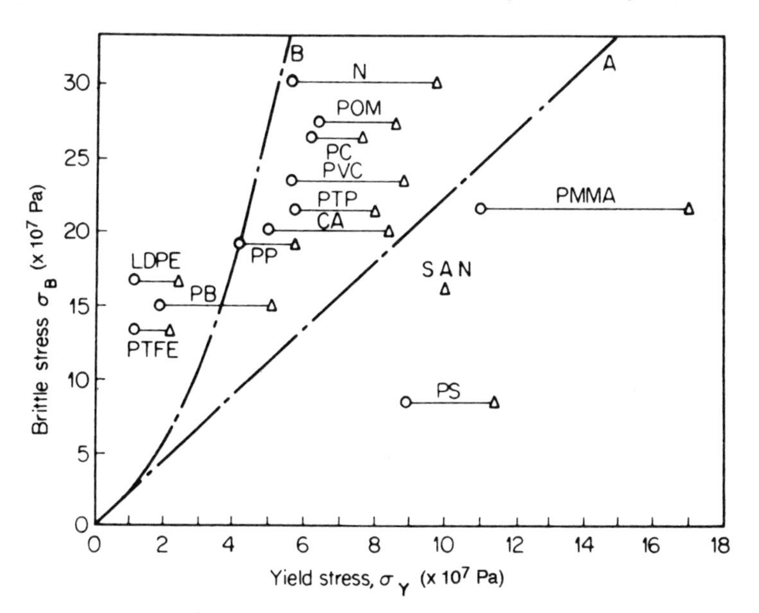

Figure 12.21 Plot of brittle stress at about −180 °C against a line joining yield-stress values at −20 °C (○) respectively for various polymers. Line A divides polymers which are brittle unnotched from those which are ductile unnotched but brittle notched, and line B divides polymers which are brittle notched, but ductile unnotched, from those which are ductile even when notched. PMMA, polymethyl methacrylate; PVC polyvinyl chloride; PS, polystyrene; PET polyethylene terephthalate; SAN, copolymer of styrene and acrylonitrile; CA cellulose acetate; PP, polypropylene; N, nylon 6:6; LDPE low-density polyethylene; POM, polyoxymethylene; PB, polybutene-1; PC, polycarbonate; PTFE polytetrafluoroethylene. (Reproduced with permission from Vincent, Plastics, **29**, 79 (1964))

molecular mass and the degree of crystallinity so that each point can only be regarded as of first-order significance.

From the known behaviour of the 13 polymers shown in this diagram, two characteristic lines can be drawn. Line A divides the brittle materials on the right which are brittle when notched from those on the left which are ductile even when notched. Both these lines are approximations, but they do summarize the existing knowledge.

For the line A the ratio $\sigma_B/\sigma_Y \sim 2$, rather than unity, but the difference may be accounted for by the measurement of σ_B at very low temperatures and possibly by the measurement of σ_B in flexure rather than in tension. (The latter may reduce the possibility of fracture at serious flaws in the surface.) It is encouraging that even an approximate relationship holds along the lines of the Ludwig–Davidenkov–Orowan hypothesis. Even more encouraging is the fact that the line B has a slope $\sigma_B/\sigma_Y \sim 6$, three times that of A, as expected on the basis of the plastic constraint theory.

The principal value of the σ_B–σ_Y diagram is that it may guide the development of modified polymers or new polymers. Together with the ideas of the previous section on the influence of material variables on the brittle strength and yield stress, it can lead to a systematic search for improvements in toughness.

12.8 THE IMPACT STRENGTH OF POLYMERS

The ability of a structural part to maintain its integrity and to absorb a sudden impact is often a relevant issue when selecting a suitable material. Impact testing of polymers is thus a subject of some importance, and extensively employed, although many of the results obtained are of an empirical and hence comparative nature.

The two major types of impact test are categorized as flexed beam and falling weight.

12.8.1 Flexed-beam Impact

Examples of flexed-beam impact are the Izod and Charpy impact test, in which a small bar of polymer is struck with a heavy pendulum. In the Izod test the bar is held vertically by gripping one end in a vice and the other free end is struck by the pendulum. In the Charpy test the bar is supported near its ends in a horizontal plane and struck either by a single-pronged or two-pronged hammer so as to simulate a rapid three-point or four-point bend test respectively (Figure 12.22(a)). It is customary to introduce a centre notch into the specimen so as to add to the severity of the test, as discussed in section 12.5.1 above. The standard Charpy impact specimen has a 90°

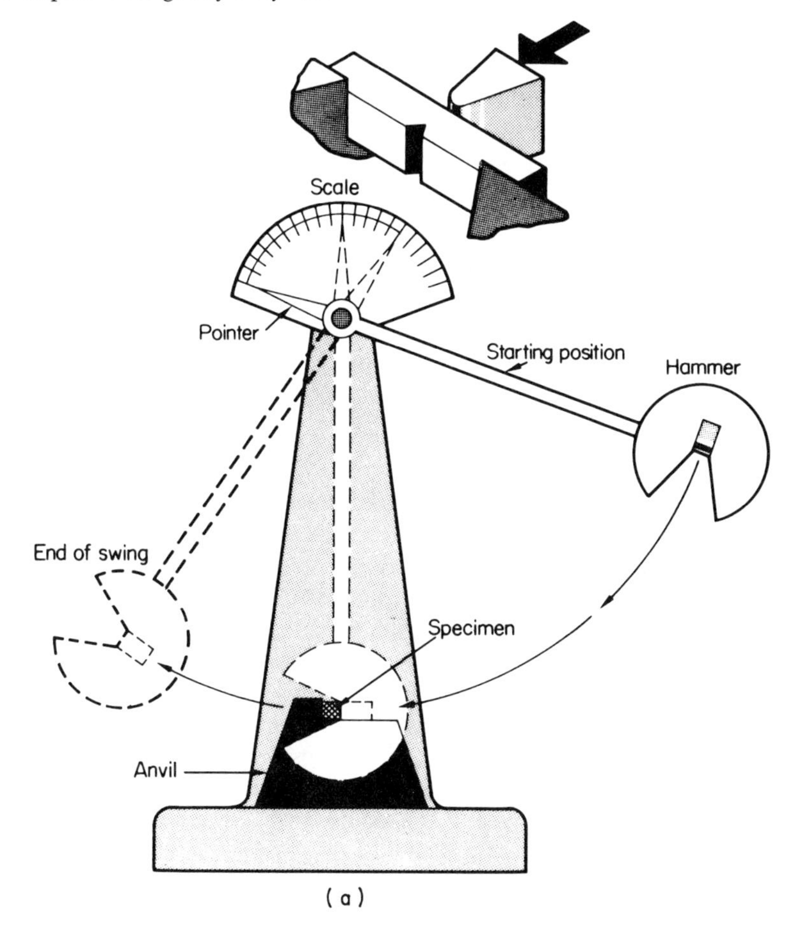

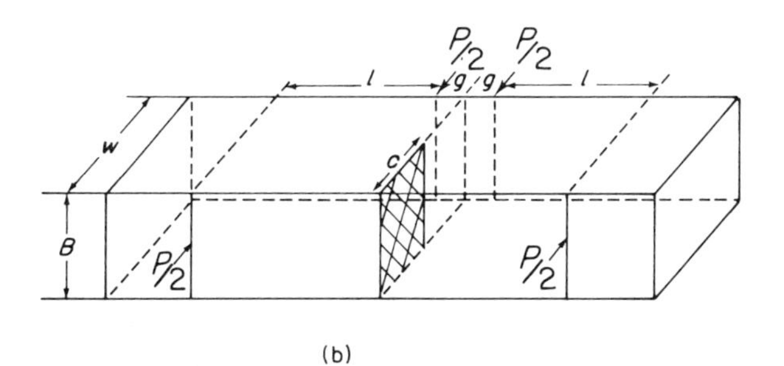

Figure 12.22 (a) Schematic drawing of a Charpy impact tester; (b) the notched Charpy impact specimen

V-notch with a tip radius of 0.25 mm. For polymers a very much sharper notch is often adopted by tapping a razor blade into a machined crack tip, which has important consequences for the interpretation of the subsequent impact test.

The interpretation of impact tests is not straightforward and it is necessary to consider several alternatives, as follows:

1. The most unsophisticated interpretation of the flexed bend impact test is that it is a measure of the energy required to propagate the crack across the specimen, irrespective of whether the specimen is notched or unnotched. Notch sensitivity is ignored and only the energy of propagation is involved. In this case

$$G_c = \frac{U_0}{A} = \frac{U_0}{BW(1 - c/W)} \tag{12.21}$$

   where the area of the uncracked cross section is $A = B(W - c)$. This is not the situation for polymers *per se*, but does apply to some rubber toughened polymers (e.g. acrylonitrile–butadiene–styrene (ABS) polymers).
2. Vincent [83] and others have recognized that the impact strength depends on the geometry of the notch, which led Fraser and Ward [84] to propose that for comparatively blunt notches (i.e. those not introduced by a razor blade or a sharp cutting tool) failure occurs when the stress at the root of the notch reaches a critical value. This stress, which in a glassy polymer marks the stress required to initiate a craze, can be calculated by assuming that the deformation is elastic. On this hypothesis, the Charpy test, as undertaken in the Hounsfield impact tester, can be regarded as a four-point bend test with the bending moment $M = Pl/2$, where P is the applied load and l is a sample dimension (Figure 12.22(b)). Immediately prior to fracture, $M = M_0$, $P = P_0$ and the elastically stored energy is $U_0 = \frac{1}{2}(2M_0/l)^2 C$, where C is the sample compliance. Hence

$$M_0 = \frac{l}{2}\sqrt{\frac{2U_0}{C}}$$

   where C is calculable from specimen geometry.

   For pure bending, the nominal stress at the root of the notch σ_n is given by $\sigma_n = (M/I)y$, where I is the second moment of area ($= Bt^3/12$ for a rectangular beam) and y is the distance to the neutral axis.

   Using the linear stress assumption the maximum stress at the root of the notch is the product of the nominal stress and the stress concentration factor α_k. Calculations of α_k for general shapes of notch are available in the literature, but when the crack length c is much greater than the notch tip radius ρ, α_k reduces to the simple expression $\alpha_k = 2\sqrt{c/\rho}$.

It has been shown that the impact behaviour of blunt notched specimens of PMMA is consistent with a critical stress at the root of the notch [84], and similar considerations apply to polycarbonate [85] and polyethersulphone [86] in the absence of shear lips. In these instances it appears therefore that the maximum local stress is the fracture criterion, independent of specimen geometry.

3. It was proposed independently by Brown [87], and by Marshall, Williams and Turner [88], that Charpy impact tests on sharply notched specimens can be quantitatively analysed in terms of linear elastic fracture mechanics. It is assumed that the polymer deforms in a linear elastic fashion up to the point of failure, which occurs when the change in stored elastic energy due to crack growth satisfies the Irwin–Kies relationship (equation 12.10) above). So that

$$G_c = \frac{K_c^2}{E^*} = \frac{P_0^2}{2B}\frac{dC}{dc}$$

where P_0 is the load immediately prior to fracture. Since the elastically stored energy in the specimens immediately prior to failure is $U_0 = P_0^2 C/2$,

$$G_c = \frac{U_0}{B}\frac{1}{C}\frac{dC}{dc} \qquad (12.22)$$

where U_0 is determined in a commercial impact tester from the potential energy lost due to impact. The total measured impact energy U_I must be reduced by the kinetic energy of the sample U_k to give $U_0 = U_I - U_k$.

The term $[(1/C)(dC/dc)]$ is calculable from the dimensions of the specimen, and details of such calculations have been given by Plati and Williams [89]. A plot of U_I versus

$$\left[\left(\frac{1}{C}\right)(dC/dc)\right]$$

for different specimen dimensions then provides a value of G_c. This approach has been shown to give values for G_c which are independent of specimen geometry for impact tests on razor-notched samples of several glassy polymers, including PMMA, polycarbonate [85] and polyether sulphone [86]. Similar results have also been obtained for razor-notched samples of polyethylene [90].

12.8.2 Falling-weight Impact

In the falling-weight impact test a circular disc of material (typically 6 cm diameter and 2 mm thickness, freely supported on an annulus of 4 cm diameter) is impacted by a metal dart with a hemispherical tip (typically of radius 1 cm). The tests are carried out either under conditions where the impact energy is far in excess of that required to break the specimen or at low levels of impact energy so that damage tolerance and the possible initiation of a crack can be observed.

Moore and his colleagues have described the application of such tests to polymers, and to polymer composites [91–93]. It is emphasized that for any reasonable attempt at interpretation the following must be carried out:

1. Measurement of the force–time curve so that the input energy to maximum force can be determined, as well as the total impact energy;
2. Photography of the tension surface during the impact event.

For fibre composites Moore and his colleagues showed that the peak on the force–time curve corresponds well with the energy required to initiate a crack. It was also shown that for both composites and polymers the total fracture energy corresponded quite well with that determined from notched Charpy tests.

Only for the Charpy test, and to a rather lesser extent the Izod test, has a satisfactory theoretical analysis been achieved. Even for these tests, however, there is still a gap between the engineering analysis and any accepted interpretation in physical terms. For example, although it seems likely that the brittle failure of razor-notched impact specimens is associated with the craze at the crack tip, there is no convincing numerical link between craze parameters and the fracture toughness K_{IC}, as exists for the cleavage fracture of compact tension specimens (see section 12.2 above). Again, although the mechanics point to a critical stress criterion for some blunt notched specimens and there is an empirical correlation with the craze stress determined in other ways, the magnitude of the critical stress is very great and suggests that a more sophisticated explanation may be required. For the brittle epoxy resins, which do not show a craze at the crack tip, Kinloch and Williams [94] have suggested that the fracture of both razor-notched and blunt-notched specimens can be described by a critical stress at a critical distance ($\sim 10\ \mu$m) below the root of the notch.

As temperature and strain rate in a polymer change, the nature of the stress–strain curve can alter remarkably. It is therefore natural to seek for correlations between the area beneath the stress–strain curve and the impact strength, and between dynamic mechanical behaviour and the impact strength. Attempts to make such correlations directly have met with a mixed success [95], which is not surprising in view of the complex quantitative interpretations of impact strength suggested above.

Vincent [96] has examined the statistical significance of a possible inverse correlation between impact strengths and dynamic modulus and concluded that, at best, this correlation only accounts for about two-thirds of the variance in impact strength. Factors such as the influence of molecular mass, and details of molecular structure such as the presence of bulky side groups, are not accounted for. He also reported impact tests over a wide temperature range on some polymers, notably polytetrafluorethylene and polysulphone, where peaks in brittle impact strength were observed at temperatures close to dynamic loss peaks, suggesting that in some instances it may be necessary to consider the relevance of a more generalized form of fracture mechanics [97], where the viscoelastic losses occurring during loading and unloading must be taken into account.

12.8.3 Toughened Polymers: High-impact Polyblends

The comparatively low impact strength of many well-known polymers such as PMMA, polystyrene and PVC led to the production of rubber-modified thermoplastics with high impact strength. The best known examples are high-impact polystyrene (HIPS) and ABS copolymer where the rubbery phase is dispersed throughout the polymer in the form of small aggregates or balls. Other polymers which have been toughened in this way include PMMA, PVC, polypropylene, polycarbonate, nylons and thermosets such as epoxies, polyesters and polyimides.

It is now generally accepted that the prinicipal reason for enhanced toughness in these polyblends is that the presence of the rubber particles produces an enhancement of the deformation mechanisms in the polymer phase, either by extensive shear yielding or by crazing or a combination of both.

Nielsen [9] lists three conditions which are required for an effective polyblend:

1. The glass temperature of the rubber must be well below the test temperature;
2. The rubber must form a second phase and not be soluble in the rigid polymer;
3. The two polymers should be similar enough in solubility behaviour for good adhesion between the phases.

Newman and Strella [99] suggested that, provided the rubber particle is well bonded to the polymer, a state of triaxial tension will be produced on the surface of the particle when the polymer is deformed. Transmission electron microscope studies by Donald and Kramer [100] showed that in ABS shear deformation of the matrix occurs when the particles are small, and crazing when the particles are large. In HIPS, Bucknall and Smith [101] showed that the improved toughness was related to crazing and stress whitening. Bucknall [102] and Bucknall and Smith compared the force–time curves for impact specimens over a range of temperatures, with both the notched Izod impact strength and the falling weight impact strength and the nature of the fracture surface. The force–time curves, such as in Figure 12.23(a) show regions similar to those observed for a homopolymer as discussed in the introduction above. Both impact strength tests also showed three regions (Figure 12.23(b) and (c)). The fracture surfaces at the lowest temperature were quite clear, whereas at high temperatures stress whitening or craze formation occurred. Three temperature regions were considered.

1. *Low temperatures*. The rubber is unable to relax at any stage of fracture. There is no craze formation and brittle fracture occurs.
2. *Intermediate temperatures*. The rubber is able to relax during the relatively slow build-up of stress at the base of the notch, but not during the fast crack propagation stage. Stress whitening occurs only in the first (precrack) stage of fracture, and is therefore confined to the region near the notch.
3. *High temperature*. The rubber is able to relax even in the rapidly forming stress field ahead of the travelling crack. Stress whitening occurs over the whole of the fracture surface. Bucknall and Smith [101] report similar results for other rubber-modified impact polymers.

12.8.4 Crazing and Stress Whitening

Bucknall and Smith [101] have remarked on the connection between crazing and stress whitening. It was observed that the fracture of high-impact polystyrene which incorporates rubber particles into the polystyrene is usually preceded by opaque whitening of the stress area. Figure 12.12 shows a stress-whitened bar of high-impact polystyrene which failed at an elongation of 35%. A combination of different types of optical measurements (polarized light to measure molecular orientation and phase contrast microscopy to determine refractive index) showed that these stress-whitened regions are similar to the crazes formed in unmodified polystyrene. They are birefringent, of low refractive index, capable of bearing load and are healed by annealing treatments. The difference between stress whitening and crazing exists merely in the size and concentration of the craze bands, which are of much smaller size and greater quantity in stress whitening. Thus the higher conversion of the polymer into crazes accounts for the high breaking

Figure 12.23 (a) Fracture surfaces of modified polystyrene notched Izod impact specimens: top, broken at −70 °C, type I fracture; centre, broken at 40 °C, type II fracture; bottom, broken at 150 °C, type III fracture. (b) Notched Izod impact strength of modified polystyrene as a function of temperature, showing the limits of the three types of fracture behaviour. (c) Drop weight impact strength of 2.03 mm high-impact polystyrene sheet as a function of temperature. (Reproduced with permission from Bucknall, *Brit. Plast.*, **40**, 84 (1967))

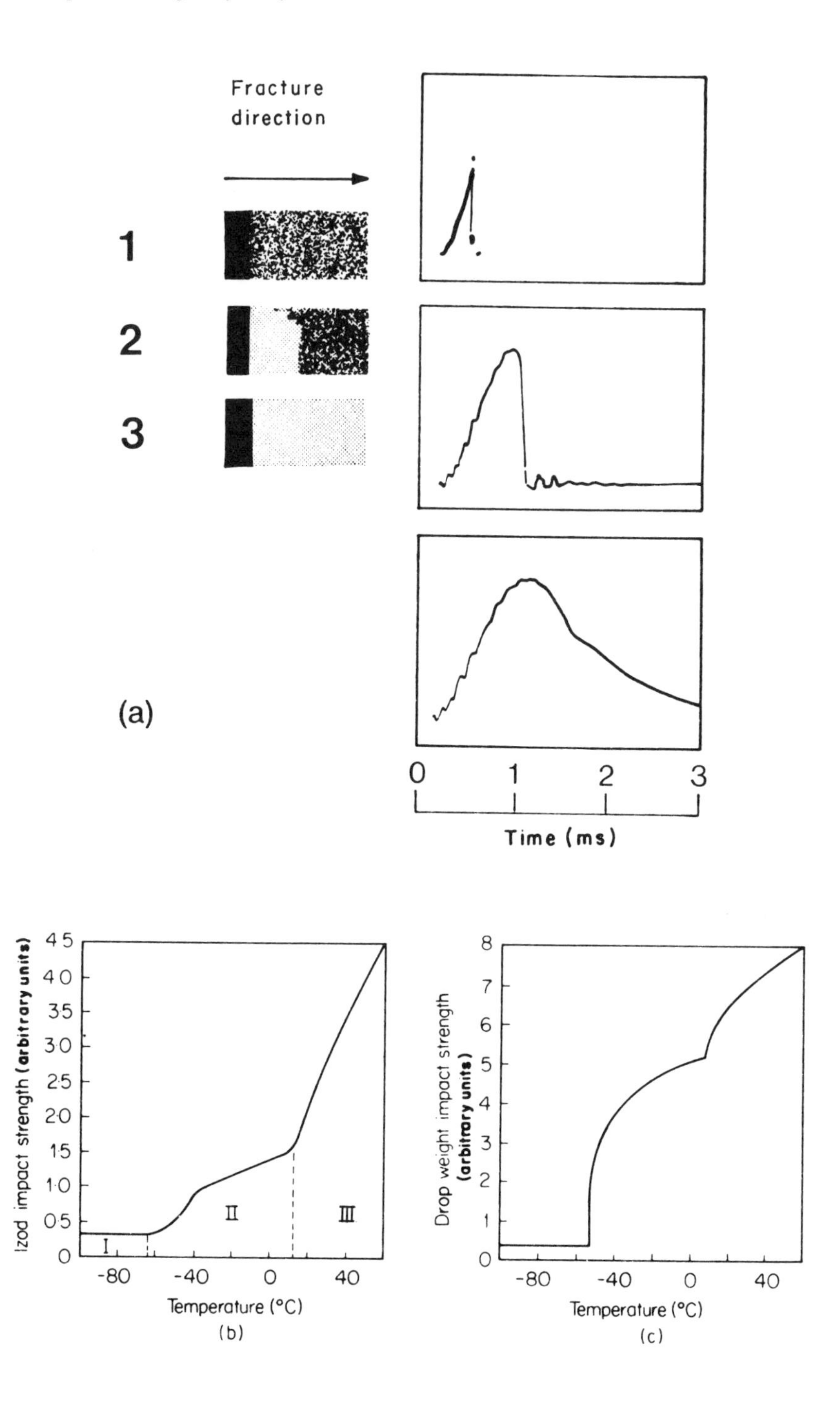
Fracture direction
1
2
3
(a)
0
1
2
3
Time (ms)
Izod impact strength (arbitrary units)
4·5
4·0
3·5
3·0
2·5
2·0
1·5
1·0
0·5
0
I
II
III
-80
-40
0
40
Temperature (°C)
(b)
Drop weight impact strength (arbitrary units)
8
7
6
5
4
3
2
1
0
-80
-40
0
40
Temperature (°C)
(c)

elongation of toughened polystyrene. It is suggested that the effect of the rubber particles is to lower the craze initiation stress relative to the fracture stress, thereby prolonging the crazing stage of deformation. The crazing stage appears to require the relaxation of the rubber phase, so that it behaves like a rubber and not a glass. The function of the rubber particles is not, however, merely to provide points of stress concentration, and there must be a good bond between the rubber and polystyrene, which is achieved by chemical grafting. The rubber must bear part of the load at the stage when the polymer has crazed but not fractured. Bucknall and Smith suggest that the rubber particles may be constrained by the surrounding polystyrene matrix so that their stiffness remains high. These ideas lead directly to an explanation of the three regimes for impact testing, as discussed above. At low temperatures there is no stress whitening because the rubber does not relax during the fracture process giving low impact strengths. At intermediate temperatures, stress whitening occurs near the notch, where the crack initiates and is travelling sufficiently slowly compared with the relaxation of the rubber. Here the impact strength increases. Finally at high temperatures, stress whitening is observed along the whole of the crack, and the impact strength is high. It seems likely that these ideas have a greater generality, and will apply to the fracture of other polymers including, for example, impact-modified PVC.

12.9 THE TENSILE STRENGTH AND TEARING OF POLYMERS IN THE RUBBERY STATE

12.9.1 The Tearing of Rubbers: Extension of Griffith Theory

The Griffith theory of fracture implies that the quasi-static propagation of a crack is a reversible process. Rivlin and Thomas [103, 104] recognized, however, that this may be unnecessarily restrictive, and also that the reduction in elastically stored energy due to the crack propagation may be balanced by changes in energy other than that due to an increase in surface energy. Their approach was to define a quantity termed the 'tearing energy', which is the energy expended per unit thickness per unit increase in crack length. The tearing energy includes surface energy, energy dissipated in plastic flow processes and energy dissipated irreversibly in viscoelastic processes. Providing that all these changes in energy are proportional to the increase in crack length and are primarily determined by the state of deformation in the neighbourhood of the tip of the crack, then the total energy will still be independent of the shape of the test piece and the manner in which the deforming forces are applied.

In formal mathematical terms, if the crack increases in length by an amount dc, an amount of work TB dc must be done, where T is the tearing energy per unit area and B is the thickness of the sheet. Assuming no external work is due, this can be equated to the change in elastically stored energy giving

$$-\left[\frac{\partial U}{\partial c}\right]_l = TB. \qquad (12.23)$$

The suffix l indicates that differentiation is carried out under conditions of constant displacement of the parts of the boundary which are not force-free. Equation (12.23) is similar in form to (12.1) above but T is defined for unit thickness of specimen and is therefore equivalent to 2γ in equation (12.1). As in the case of glassy polymers, T is not to be interpreted as a surface free energy, but involves the total deformation in the crack tip region as the crack propagates.

The so-called 'trouser tear' experiment shown in Figure 12.24 is a particularly simple case where the equation can be immediately evaluated. After making a uniform cut in a rubber sheet the sample is subjected to tear under the applied forces F. The stress distribution at the tip of the tear is complex, but providing that the legs are long, it is independent of the depth of the tear.

If the sample tears a distance Δc under the force F, and changes in extension of the material between the tip of the tear and the legs are ignored, the work done is given by $\Delta W = 2F\Delta c$.

Since the tearing energy $T = \Delta W/B\ \Delta c$, $T = 2F/B$ and can be measured easily.

Rivlin and Thomas [103] found that two characteristic tearing energies could be defined, one for very slow rates of tearing ($T = 37$ kJ m^{-2}) and one for catastrophic growth ($T = 130$ kJ m^{-2}) and that both these quantities were independent of the shape of the test peice.

The tearing energy is the energy required to extend the rubber to its maximum elongation and does not relate directly to tensile strength, but

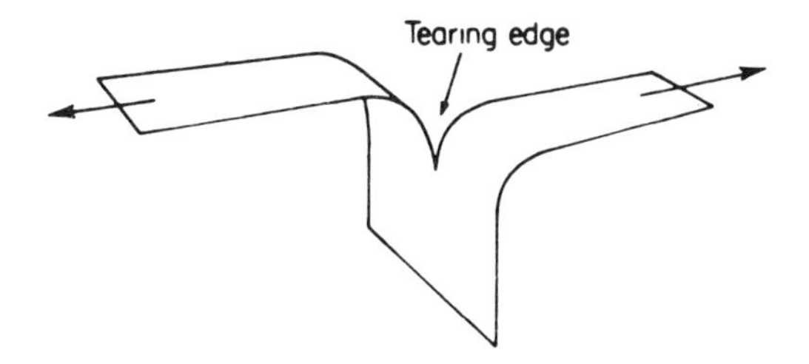

Figure 12.24 The standard 'trouser tear' experiment

depends on the shape of the stress–strain curve together with the viscoelastic nature of the rubber. For example, we may contrast two different rubbers, the first possessing a high tensile strength but a very low elongation to fracture and very low viscoelastic losses, and the second possessing a low tensile strength but a high elongation to fracture and high viscoelastic losses. In spite of its comparatively low tensile strength the second rubber may still possess a high tearing energy.

12.9.2 Molecular Theories of the Tensile Strength of Rubbers

Most molecular theories of the strength of rubber treat rupture as a critical stress phenomenon. It is accepted that the strength of the rubber is reduced from its theoretical strength in a perfect sample by the presence of flaws. Moreover, it is assumed that the strength is reduced from that of a flawless sample by approximately the same factor for different rubbers of the same basic chemical composition. It is then possible to consider the influence on strength of such factors as the degree of cross-linking and the primary molecular mass.

Bueche [105] has considered the tensile strength of a model network consisting of a three-dimensional net of cross-linked chains. Figure 12.25 illustrates a unit cube whose edges are parallel to the three chain directions in the idealized network. Assume that there are N chains in this unit cube and that the number of chains in each strand of the network is n. There are then n^2 strands passing through each face of the cube. To relate the number n to the number of chains per unit volume of the network (and so form a link with rubber elasticity theory) we note that the product of the number of strands passing through each cube face and the number of chains in each strand will be $\frac{1}{3}N$ since there are three strand directions. Thus

$$n^3 = \tfrac{1}{3} N, \qquad n = (N/3)^{1/3}. \tag{12.24}$$

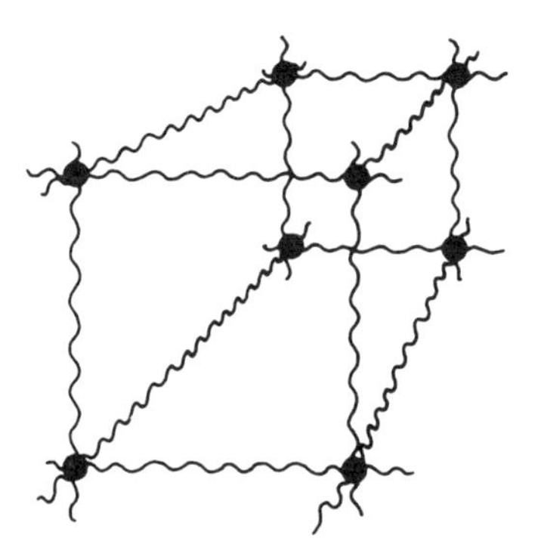

Figure 12.25 Model network of cross-linked chains

Apply a stress σ parallel to one of the three strand directions and assume that the strands break simultaneously at an individual fracture stress σ_c. Then

$$\sigma_B = n^2 \sigma_c,$$

which from equation (12.24) can be written as

$$\sigma_B = (N/3)^{2/3} \sigma_c.$$

For a real network N is the number of effective chains per unit volume, and is given in terms of the actual number of chains per unit volume N_a by the Flory relationship

$$N = N_a[1 - 2\bar{M}_c/M_n],$$

where $\bar{M}_c$ and $\bar{M}_n$ are the average molecular mass between cross-links and the number averge molecular mass of the polymer, respectively. (Note that for a network there must be at least two cross-links per chain, i.e. $\bar{M}_n > 3\bar{M}_c$.)

This substitution gives

$$\sigma_B \alpha [1 - 2\bar{M}_c/\bar{M}_n]^{2/3} \sigma_c.$$

Flory [106] found that the variation of tensile strength with the polymer molecular mass $\bar{M}_n$, for butyl rubber, follows the predicted $[1 - 2\bar{M}_c/\bar{M}_n]^{2/3}$ relationship, but for natural rubber [107] an initial increase in tensile strength with increasing degree of cross-linking was followed by a decrease at very high degrees of cross-linking. Flory attributed this decrease to the influence of cross-links in the crystallization of the rubber. However, a similar effect was observed for the non-crystallizing SBR rubber by Taylor and Darin [108], which led Bueche [109] to propose that the simple model described above fails because of the assumption that each chain holds the load at fracture, which may be a good approximation at low degrees of cross-linking but is less probable at high degrees of cross-linking.

It is of considerable technological importance that the tensile strength of rubbers can be much increased by the inclusion of reinforcing fillers such as carbon black and silicone, which increase the tensile strength by allowing the applied load to be shared among a group of chains, thus decreasing the chance that a break will propagate [110].

12.10 EFFECT OF STRAIN RATE AND TEMPERATURE

The influence of strain rate and temperature on the tensile properties of elastomers and amorphous polymers has been studied extensively, particularly by Smith and his co-workers [111–113], who measured the variation of tensile strength and ultimate strain as a function of strain rate for a number of elastomers. The results for different temperatures could be superimposed, by shifts along the strain rate axis, to give master curves for tensile strength and ultimate strain as a function of strain rate. Results of this nature are shown in Figure 12.26 which summarizes Smith's data for an unfilled SBR rubber. Remarkably, the shift factors obtained from superposition of both tensile strength and ultimate strain took the form predicted by the WLF equation (see Section 6.3.2) for the superposition of low strain linear viscoelastic behaviour of amorphous polymers (Figure 12.27). The actual value for T_g agreed well with that obtained from dilatometric measurements.

This result suggests that, except at very low strain rates and high temperatures, where the molecular chains have complete mobility, the fracture process is dominated by viscoelastic effects. Bueche [114] has treated this problem theoretically and obtained the observed form of the dependence of tensile strength on strain rate and temperature. Later theories have attempted to obtain the time dependence for both tensile strength and ultimate strain, or the time to break at a constant strain rate [115, 116].

Smith plotted log σ_B/T against log e for the above and similar data to obtain a unique curve for all strain rates and test temperatures, which he termed the 'failure envelope' for elastomers. It was also found [113] that the failure envelope can represent failure under more complex conditions such as creep and stress relaxation. In Figure 12.28 such failure can take place by starting from the initial stage G and progressing parallel to the abscissa (constant stress, i.e. creep) or parallel to the ordinate (constant strain, i.e. stress relaxation) until a point is reached on the failure envelope ABC, as indicated by the progress along the dotted lines.

12.11 FATIGUE IN POLYMERS

Materials frequently fail by fatigue due to the cyclic application of stresses which are below that required to cause yield or fracture when a continuously rising stress is applied. The effect of such cyclic stresses is to initiate microscopic cracks at centres of stress concentration within the material or on the surface, and subsequently to enable these cracks to propagate, leading to eventual failure.

Early studies of fatigue in polymers concentrate on stress cycling of

Figure 12.26 Variation of (a) tensile strength and (b) ultimate strain of a rubber with reduced strain rate $\dot{e}a_T$. Values were measured at various temperatures and rates and reduced to a temperature of 263 K. (Reproduced with permission from Smith, *J. Polymer. Sci*, **32**, 99 (1958))

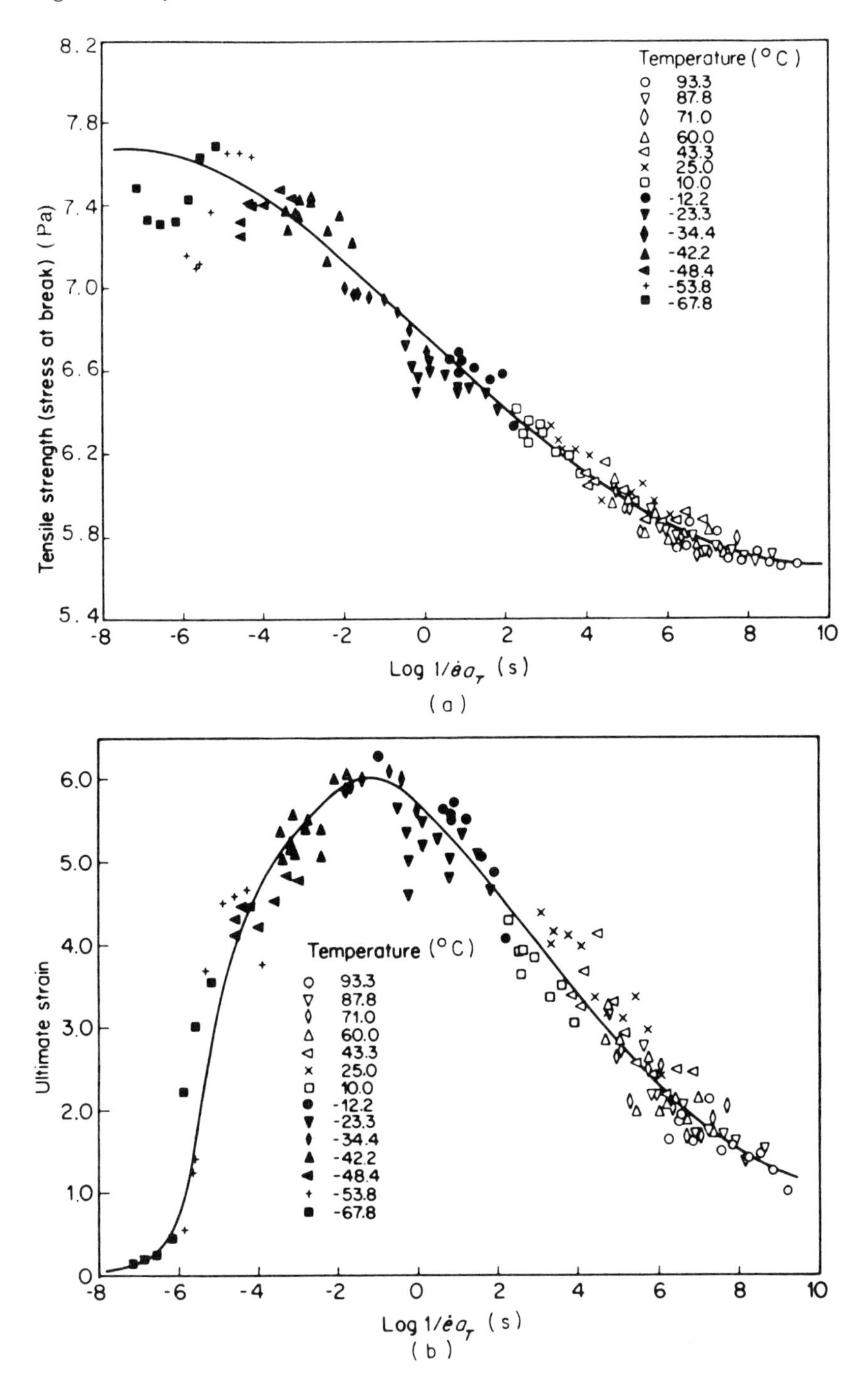
Temperature (°C)
93.3
87.8
71.0
60.0
43.3
25.0
10.0
-12.2
-23.3
-34.4
-42.2
-48.4
-53.8
-67.8
Tensile strength (stress at break) (Pa)
8.2
7.8
7.4
7.0
6.6
6.2
5.8
5.4
-8
-6
-4
-2
0
2
4
6
8
10
Log 1/ėa_T (s)
(a)
Temperature (°C)
93.3
87.8
71.0
60.0
43.3
25.0
10.0
-12.2
-23.3
-34.4
-42.2
-48.4
-53.8
-67.8
Ultimate strain
6.0
5.0
4.0
3.0
2.0
1.0
0
-8
-6
-4
-2
0
2
4
6
8
10
Log 1/ėa_T (s)
(b)

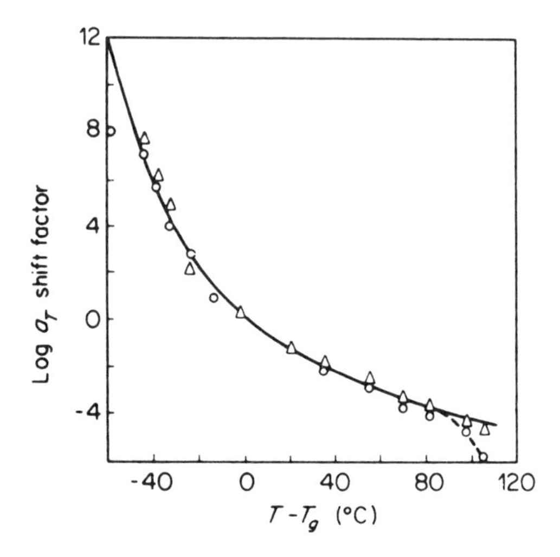

Figure 12.27 Experimental values of log a_T shift factor obtained from measurement of ultimate properties compared with those predicted using the WLF equation. (△) From tensile strength; (○) from ultimate strain; (——) WLF equations with $T_g = 263$ K (Reproduced with permission from Smith, *J. Polymer. Sci.*, **32**, 99 (1958))

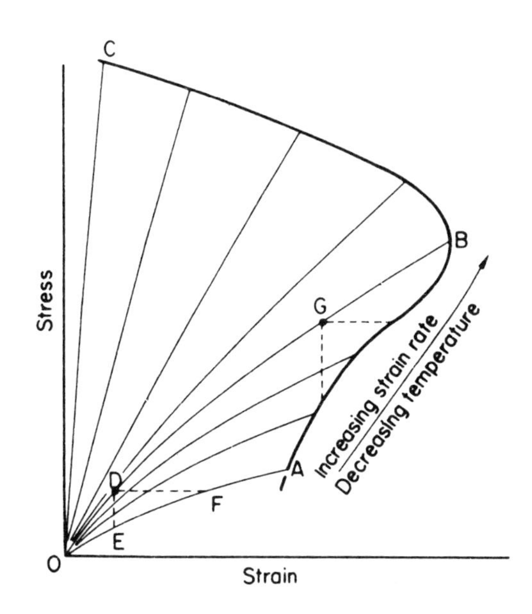

Figure 12.28 Schematic representation of the variation of stress–strain curves with the strain rate and temperature. Envelope connects rupture point and the dotted lines illustrate stress relaxation and creep under different conditions. (Reproduced with permission from Smith and Stedry, *J. Appl. Phys.*, **31**, 1892 (1960))

unnotched samples, to produce S versus N plots similar to those which have proved so useful for characterizing fatigue in metals (S being the maximum loading stress, N the number of cycles to failure). An example of this type of plot for PVC [117] is shown in Figure 12.29. A major aspect of such a test is the question of adiabatic heating, which can lead to failure by thermal melting. Clearly there will be a critical frequency above which thermal effects become important.

Stress cycling tests in unnotched samples do not readily distinguish between crack intiation and crack propagation. Further progress requires a similar approach to that adopted in fracture studies, namely the introduction of very sharp initial cracks so as to examine crack propagation utilizing fracture mechanics concepts.

The first quantitative studies of fatigue in polymers, which concentrated on rubbers [118–120], applied the tearing energy concept of fracture proposed by Rivlin and Thomas to fatigue crack propagation. Thomas [118] showed that the fatigue crack growth rate could be expressed in the form of an empirical relationship

$$\frac{dc}{dN} = A\mathfrak{I}^n, \tag{12.25}$$

where c is the crack length, N the number of cycles, and $\mathfrak{I}$ the surface work parameter, which is analogous to the strain energy release rate G in linear elastic fracture mechanics. For a single edge notch specimen

$$\mathfrak{I} = 2k_1cU, \tag{12.26}$$

where $U = \sigma^2/2E$ is the stored energy density for a linear elastic material,

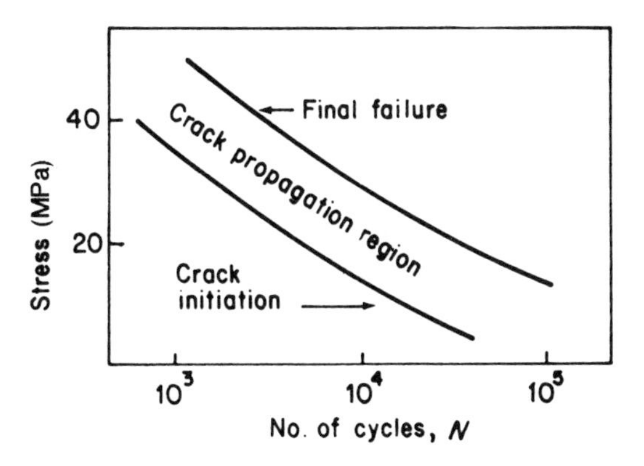

Figure 12.29 Fatigue response of PVC: relationship between applied stress σ and number of cycles to failure N, for both initiation of fatigue cracks and final failure. (Reproduced with permission from Manson and Hertzberg, *CRC Crit. Rev. Macromol. Sci.*, **1**, 433 (1973))

and k_1 a constant which varies from π at small extensions (the linear elastic value) to approximately unity at large extensions [121]. Here A and n are constants which are dependent on the material and generally vary with test conditions such as temperature. The exponent n usually lies between 1 and 6 and for rubber is approximately 2 for anything other than very small $\mathrm{d}c/\mathrm{d}N$.

As expressed in equation (12.25) $\mathfrak{J}$ is essentially a positive quantity and can be considered to vary during the test cycle from zero ($\mathfrak{J} = \mathfrak{J}_{\min} = 0$) to a finite value ($\mathfrak{J} = \mathfrak{J}_{\max}$). It has been found that where $\mathfrak{J}_{\min}$ is increased, there is a corresponding decrease in A, which has been attributed to reduced crack propagation where strain-induced crystallization occurs. Furthermore, it has been shown that there is a fatigue limit $\mathfrak{J} = \mathfrak{J}_0$, below which a fatigue crack will not be propagated. Lake and Thomas showed that $\mathfrak{J}_0$ corresponds to the minimum energy required per unit area to extend the rubber at the crack tip to its breaking point. Andrews [122] pointed out that initiation requires either intrinsic flaws of magnitude c_0 or that flaws of this size are produced during the test itself, with c_0 defined by equation (12.26), where $\mathfrak{J}_0 = k_1 c_0 U$. Andrews and Walker [123] carried this approach one stage further, incorporating a generalized form of fracture mechanics, to analyse the fatigue behaviour of low-density polyethylene, which was viscoelastic in the range of interest, so that the more generalized fracture mechanics was required to deal with unloading as well as loading during crack propagation. The fatigue characteristics were predicted from the crack growth data using a single fitting constant, the intrinsic flaw size c_0, which it was suggested corresponded to the spherulite dimensions, so that interspherulite boundary cracks constituted the intrinsic flaws.

For glassy polymers, fracture mechanics has been the usual starting-point [124–127], with the fatigue crack growth rate usually expressed as an empirical relationship

$$\frac{\mathrm{d}c}{\mathrm{d}N} = A'(\Delta K)^m, \tag{12.27}$$

where c is the crack length, N the number of cycles, ΔK the range of the stress intensity factor (i.e. $K_{\max} - K_{\min}$, where $K_{\min}$ is generally zero), and A' and m are constants depending on the material and test conditions.

For $K_{\min} = 0$, equation (12.27) is clearly identical in form to equation (12.25), which is generally adopted for rubbers. Recall from section 12.2.3 that the strain energy release rate $G = K^2/E$ for plane stress. Then

$$G = 2\mathfrak{J} = K_{\max}^2/2E = (\Delta K)^2/2E,$$

and equations (12.25) and (12.27) are formally equivalent if $m = 2n$.

Equation (12.27) is also the most general form of the law proposed by Paris [128, 129] for predicting fatigue crack growth rates in metals. The general situation for glassy polymers is illustrated in Figure 12.30(a) with

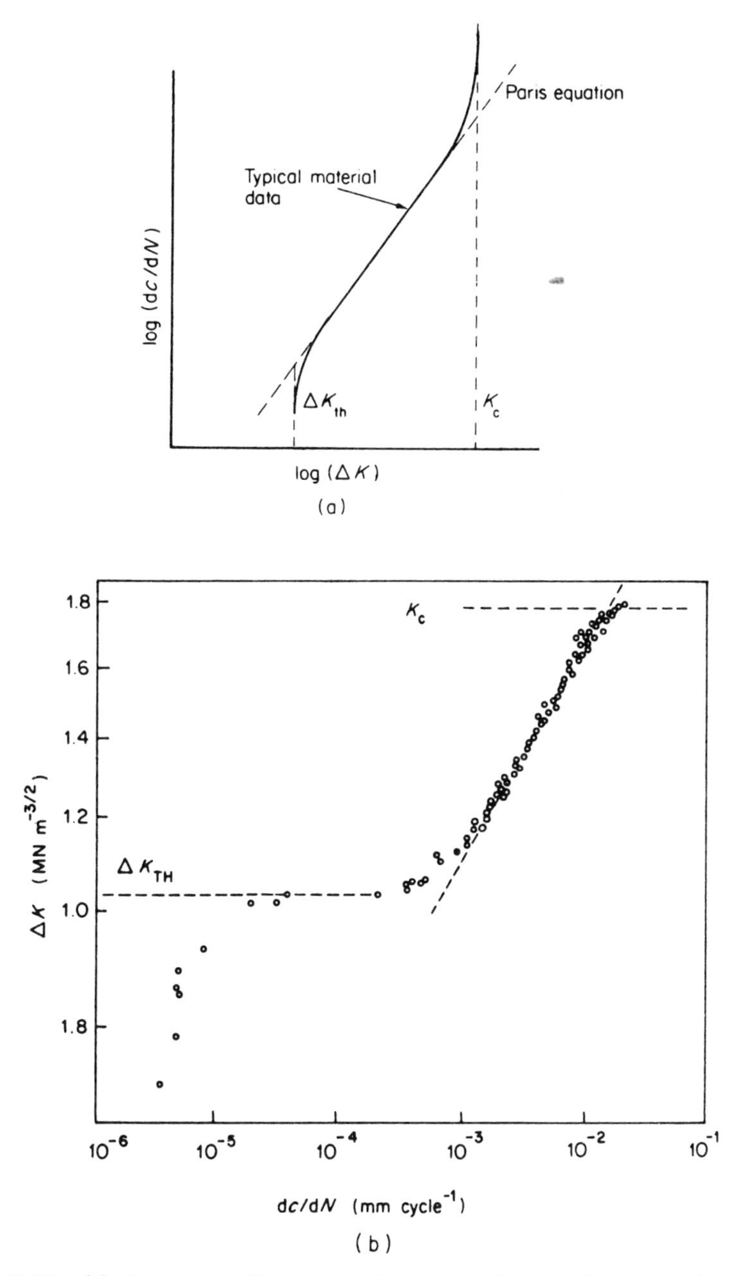

Figure 12.30 (a) Schematic diagram of fatigue crack growth rate dc/dN as a function of the range of stress intensity factor ΔK. (b) Fatigue crack growth characteristics for a vinyl urethane polymer. (Reproduced with permission from Harris and Ward, *J. Mater. Sci.*, **8**, 1655 (1973))

some typical results shown in Figure 12.30(b). The data differ in two respects from the Paris equation: first, analogous to the case of rubbers, there is a distinct threshold value of ΔK, denoted by ΔK_{th} below which no crack growth is observed; second, as ΔK approaches the critical stress intensity factor K_c, the crack accelerates. A further criticism of equation (12.27) is that it allows for the influence of the range of the stress intensity factor but not for the mean stress which usually has an important influence on the crack growth rate. The latter consideration led Arad, Radon and Culver [130] to suggest an equation of the form

$$\frac{\mathrm{d}c}{\mathrm{d}N} = \beta\lambda^n, \tag{12.28}$$

where $\lambda = (K_{\max}^2 - K_{\min}^2)$. This relation is equivalent to equation (12.27) because the cycle strain energy release rate ΔG is given by

$$\Delta G = \frac{1}{E}(K_{\max}^2 - K_{\min}^2).$$

A comprehensive review of the application of the Paris equation and its modified form (12.28) to the fatigue behaviour of polymers has been given by Manson and Hertzberg [131], who considered the effect of physical variables such as crystallinity and molecular mass. They noted a strong sensitivity of fatigue crack growth to molecular mass: in polystyrene a fivefold increase in molecular mass resulted in a more than tenfold increase in fatigue life. A general correlation was observed between the fracture toughness K_c and the fatigue behaviour expressed as the stress intensity range ΔK corresponding to an arbitrary value of $\mathrm{d}c/\mathrm{d}N$ (chosen as 7.6×10^{-7} m cycle^{-1}), as is shown in Figure 12.31. A study of fatigue behaviour in polycarbonate by Pitman and Ward [132] also brought out the similarity between fatigue and fracture, so that the fatigue behaviour can be analysed in terms of mixed mode failure. Similar to the fracture behaviour described in section 12.2, changing molecular mass again changed the balance between energy dissipated in propagating the craze and shear lips respectively. A development by Williams [133, 134] attempts to model fatigue crack propagation behaviour in terms of the Dugdale plastic zone analysis of the crack tip. Each fatigue cycle is considered to reduce the craze stress in one part of craze, so that a two-stage plastic zone is established, leading to an equation for crack growth of the form.

$$\frac{\mathrm{d}c}{\mathrm{d}N} = \beta'[K^2 - \alpha K_c^2] \tag{12.29}$$

which gives a good fit to experimental data for polystyrene over a substantial range of temperatures.

Both Williams and Pitman and Ward conclude that it is difficult to assign physical significance to the parameters in the Paris equation. Further developments in this area will require a more distinctly physical approach.

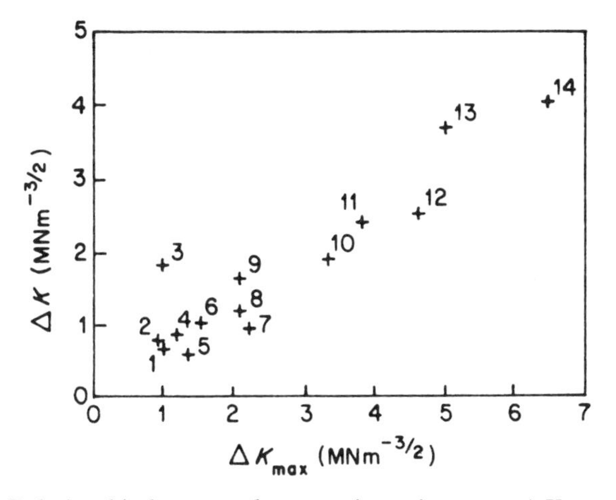

Figure 12.31 Relationship between the stress intensity range ΔK, corresponding to an arbitary value of dc/dN. 7.6×10^{-7} m cycle^{-1} and the maximum stress intensity factor range $\Delta K_{\max}$, observed at failure for a group of polymers. The polymers are (1) cross-linked polystyrene, (2) PMMA, (3) PVC, (4) LDPE, (5) polystyrene, (6) polysulphone, (7) high-impact polystyrene, (8) ABS resin, (9) chlorinated polyether, (10) polyphenylene oxide, (11) nylon 6, (12) polycarbonate, (13) nylon 6:6, (14) polyvinylidene fluoride (Reproduced with permission from Manson and Hertzberg, *CRC Crit. Rev. Macromol. Sci.*, **1**, 433 (1973))

REFERENCES

1. A. A. Griffith, *Phil. Trans. Roy. Soc.*, **221**, 163 (1921).
2. G. E. Inglis, *Trans. Inst. Naval Architect.*, **55**, 219 (1913).
3. G. R. Irwin, *J. Appl. Mech.*, **24**, 361 (1957).
4. G. R. Irwin and J. A. Kies, *Welding J. Res. Suppl.*, **33**, 1935 (1954).
5. W. F. Brown and J. F. Srawley, *ASTM STP 410*, 1966.
6. J. F. Srawley and B. Gross, *NASA Report E-3701*, 1967.
7. J. G. Williams, *Stress Analysis of Polymers*, 2nd edn, Ellis Horwood, Chichester, 1980.
8. J. J. Benbow and F. C. Roesler, *Proc Phys Soc.*, **B70** 201 (1957).
9. B. Cottrell, *Int J. Fract. Mech.*, **2**, 526 (1966).
10. J. P. Berry, *J. Appl. Phys.*, **34**, 62 (1963).
11. J. J. Benbow, *Proc. Phys. Soc.*, **78**, 970 (1961).
12. N. L. Svensson, *Proc. Phys. Soc.*, **77**, 876 (1961).
13. J. P. Berry, *J. Polymer Sci.*, **50**, 313 (1961)
14. E. H. Andrews, in *Proceedings of the Conference on the Physical Basis of Yield and Fracture,* Oxford, 1966, p. 127.
15. J. P. Berry in *Fracture* (eds B. L. Auerbach et al.), Wiley, New York, 1959, p.263.
16. R. P. Kambour, *Polymer*, **5**, 143 (1964).
17. R. P. Kambour, *J. Polymer Sci., A2*, **4**, 349 (1966).
18. R. P. Kambour, *Macromolecular Reviews*, **7**, 1 (1973).

19. H. R. Brown and I. M. Ward, *Polymer*, **14**, 469 (1973).
20. W. Döll and G. W. Weidmann, *Colloid Polymer Sci.*, **254**, 205 (1976).
21. D. S. Dugdale, *J. Mech. Phys. Solids*, **8**, 100 (1960).
22. J. R. Rice, in *Fracture—An Advanced Treatise* (ed. H. Liebowitz.), Academic Press, New York and London, 1968, Ch.3.
23. G. L. Pitman and I. M. Ward., *Polymer*, **20**, 895 (1979).
24. P. J. Hine, R. A. Duckett and I. M. Ward., *Polymer*, **22**, 1745 (1981).
25. J. G. Williams and M. Parvin., *J. Mater. Sci.*, **10**, 1883 (1975).
26. J. P. Berry, *J. Polymer Sci., A2*, **2**, 4069 (1964).
27. P. J. Flory, *J. Amer. Chem. Soc.*, **67**, 2048 (1945).
28. W. Döll and G. W. Weidmann. *Progr. Colloid Polymer Sci.*, **66**, 291 (1979).
29. R. P. Kusy and D. T. Turner, *Polymer*, **18**, 391 (1977).
30. R. P. Kusy and D. T. Turner, *Polymer*, **17**, 161 (1976).
31. R. N. Haward, H. E. Daniels and L. R. G. Treloar, *J. Polymer Sci. Polymer Phys. Ed.*, **16**, 1169 (1978).
32. J. P. Berry, in *Fracture* (eds B. L. Averback et al.), Wiley, New York, 1959, p. 263.
33. M. Higuchi, *Rept. Res. Inst. Appl. Mech. (Japan)*, **6**, 173 (1959).
34. R.P. Kambour and A. S. Holik, *J. Polymer Sci., A2*, **7**, 1393 (1969).
35. R. P. Kambour and R. R. Russell, *Polymer*, **12**, 237 (1971).
36. P. Beahan, M. Bevis and D. Hull, *Phil. Mag.*, **24**, 1267 (1971).
37. P. Beahan, M. Bevis and D. Hull, *J. Mater. Sci.*, **8**, 169 (1972).
38. P. Beahan, M. Bevis and D. Hull, *Polymer,* **14**, 96 (1973).
39. B. D. Lauterwasser and E. J. Kramer, *Phil. Mag.*, **39**, 469 (1979).
40. A. M. Donald and E. J. Kramer, *Phil. Mag.*, **43**, 857 (1981).
41. H. R. Brown, *J. Polymer Sci. Polymer Phys. Ed.*, **17**, 143 (1979).
42. H. R. Brown and E. J. Kramer, *J. Macromol Sci B.*, **19**, 487 (1981).
43. A. M. Donald, E. J. Kramer and R. A. Bubeck, *J. Polymer. Sci., Polymer Phys. Ed.* **20**, 1129 (1982).
44. A. M. Donald and E. J. Kramer, *Polymer*, **23**, 457 (1982).
45. F. Rietsch, R. A. Duckett and I. M. Ward, *Polymer*, **20**, 2235 (1979).
46. S. S. Sternstein, L. Ongchin and A. Silverman, *Appl. Polymer Symp.*, **7**, 175 (1968).
47. S. S. Sternstein and L. Ongchin, Amer. Chem. Soc. *Polymer Preprints*, **10**, 1117 (1969).
48. P. B. Bowden and R. J. Oxborough, *Phil. Mag.*, **28**, 547 (1973).
49. K. Matsushige, S. V. Radcliffe and E. Baer, *J. Mater Sci.*, **10**, 833 (1974).
50. R. A. Duckett, B. C. Goswami, L. S. A. Smith, I. M. Ward and A. M. Zihlif, *Brit. Polymer J.*, **10**, 11 (1978).
51. M. Kitagawa, *J. Polymer Sci. Polymer Phys Ed.*,**14**, 2095 (1976).
52. A. S. Argon, J. C. Hannoosh and M. M. Salama, in *Fracture 1977*, Vol. 1, Waterloo, Canada, 1977, p. 445.
53. G. A. Bernier and R. P. Kambour, *Macromolecules*, **1**, 393 (1968).
54. R. P. Kambour, C. L. Gruner and E. E. Romagosa, *J Polymer. Sci.*, **11**, 1879 (1973).
55. R. P. Kambour, C. L. Gruner and E. E. Romagosa, *Macromolecules* **7**, 248 (1974).
56. E. H. Andrews and L. Bevan., *Polymer*, **13**, 337 (1972).
57. G. P. Marshall, L. E. Culver and J. G. Williams, *Proc. Roy. Soc.*, **A319** 165 (1970).
58. J. G. Williams and G. P. Marshall, *Proc. Roy. Soc.*, **A342**, 55 (1975).

59. N. Brown and Y. Imai, *J. Appl. Phys.*, **46**, 4130 (1975).
60. Y. Imai and N. Brown, *J. Mater. Sci.*, **11**, 417 (1976).
61. N. Brown, B. D. Metzger and Y. Imai, *J. Polymer. Sci. Polymer. Phys. Ed.*, **16**, 1085 (1978).
62. R. P. Kambour, in *Proceedings of the International Conference on the Mechanics of Environment Sensitive Cracking Materials*, 1977, p. 213.
63. N. Brown, in *Methods of Experimental Physics*, Vol. 16, Part C (ed. R. A. Fava), Academic Press, New York, 1980, p. 233.
64. H. Mark, *Cellulose and its Derivatives,* Interscience, New York, 1943.
65. P. I. Vincent, *Proc. Roy. Soc.*, **A282**, 113 (1964).
66. H. H. Kausch, *Polymer Fracture*, Springer-Verlag, Berlin, 1978.
67. H. H. Kausch and J. Becht, *Rheol. Acta.* **9**, 137 (1970).
68. S. N. Zhurkov, I. I. Novak, A. I. Slutsker, V. I. Vettegren, V. S. Kuksenko, S. I. Veliev, M. A. Gezalov and M. P. Vershina, in *Proceedings of the Conference on the Yield, Deformation and Fracture of Polymers*, Cambridge, 1970.
69. R. P. Wool, *J. Polymer Sci.*, **13**, 1795 (1975).
70. S. N. Zhurkov and E. E. Tomashevsky, in *Proceedings of the Conference on the Physical Basis of Yield and Fracture*, Oxford, 1966, p. 200.
71. S. N. Zhurkov, V. S. Kuksenko and A. I. Slutsker, in *Proceedings of the Second International Conference on Fracture*, Brighton, 1969, p. 531.
72. V. A. Zakrevskii and V. Ye. Korsukov, *Polymer Sci USSR*, **14**, 1064 (1972).
73. A. Peterlin, *Int. J. Fracture*, **11**, 761 (1975).
74. E. Orowan, *Rept. Prog. Phys.*, **12**, 185 (1949).
75. P. I. Vincent, *Polymer*, **1**, 425 (1960).
76. J. M. Stearne and I. M. Ward, *J. Mater. Sci.*, **4**, 1088 (1969).
77. P. L. Clarke, PhD thesis, Leeds University, 1981.
78. E. A. W. Hoff and S. Turner, *Bull. Amer. Soc. Test. Mater.*, **225**, TP208 (1957).
79. R. F. Boyer, *Polymer Eng. Sci.*, **8**, 161 (1968).
80. J. Heijboer, *J. Polymer Sci., C*, **16**, 3755 (1968).
81. A. H. Cottrell, *The Mechanical Properties of Matter*, Wiley, New York, 1964, p. 327.
82. P. I. Vincent, *Plastics*, **29**, 79 (1964).
83. P. I. Vincent, *Impact Tests and Service Performance of Thermoplastics,* Plastics and Rubber Institute, London, 1971.
84. R. A. W. Fraser and I. M. Ward, *J. Mater. Sci.*, **9**, 1624 (1974).
85. R. A. W. Fraser and I. M. Ward, *J. Mater. Sci.* **12**, 459 (1977).
86. P. J. Hine, PhD thesis, Leeds University, 1981.
87. H. R. Brown, *J. Mater. Sci.*, **8**, 941 (1973).
88. G. P. Marshall, J. G. Williams and C. E. Turner, *J. Mater. Sci.*, **8**, 949 (1973).
89. E. Plati and J. G. Williams, *Polym. Eng. Sci.*, **15**, 470 (1975).
90. R. W. Truss, R. A. Duckett and I. M. Ward, *Polym. Eng. Sci.*, **23** 708 (1983).
91. A. E. Johnson, D. R. Moore, R. S. Prediger, P. E. Reed and S. Turner, *J. Mater. Sci.*, **21**, 3153 (1986); **22**, 1724 (1987).
92. D. R. Moore and R. S. Prediger, *Polymer Composites*, **9**, 330 (1988).
93. D. P. Jones, D. C. Leach and D. R. Moore, *Plastics and Rubber, Proc. Appl.*, **6**, 67 (1986).
94. A. J. Kinlock and J. G. Williams, *J. Mater. Sci.*, **15**, 987 (1980).
95. R. M. Evans, H. R. Nara and R. G. Bobalek, *Soc. Plast. Engrs J.*, **16**, 76 (1960).
96. P. I. Vincent, *Polymer*, **15**, 111 (1974).

97. E. H. Andrews, *J. Mater. Sci.*, **9**, 887 (1974).
98. L. E. Nielsen, *Mechanical Properties of Polymers*, Reinhold, New York, 1962.
99. S. Newman and S. Strella, *J. Appl. Poly, Sci.*, **9**, 2297 (1965).
100. A. M. Donald and E. J. Kramer., *J. Mater Sci.*, **17**, 1765 (1982).
101. C. B. Bucknall and R. R. Smith, *Polymer*, **6**, 437 (1965).
102. C. B. Bucknall, *Brit. Plast.*, **40**, 84 (1967).
103. R. S. Rivlin and A. G. Thomas, *J Polymer Sci.*, **10**, 291 (1953).
104. A. G. Thomas, *J. Polymer Sci.*, **18**, 177 (1955).
105. F. Bueche, *Physical Properties of Polymers,* Interscience, New York, 1962, p. 237.
106. P. J. Flory, *Ind. Eng. Chem.*, **38**, 417 (1946).
107. P. J. Flory, N. Rabjohn and M. C. Shaffer, *J. Polymer Sci.*, **4**, 435 (1949).
108. G. R. Taylor and S. Darin, *J. Polymer Sci.*, **17**, 511 (1955).
109. F. Bueche, *J. Polymer Sci.*, **24**, 189 (1957).
110. F. Bueche, *J. Polymer Sci.*, **33**, 259 (1958).
111. T. L. Smith, *J. Polymer Sci.* **32**, 99 (1958).
112. T. L. Smith, *Soc. Plast. Engrs J.*, **16**, 1211 (1960).
113. T. L. Smith and P. J. Stedry, *J. Appl Phys.*, **31**, 1892 (1960).
114. F. Bueche, *J. Appl. Phys.*, **26**, 1133 (1955).
115. F. Bueche and J. C. Halpin, *J. Appl Phys.*, **35**, 36 (1964).
116. J. C. Halpin., *J. Appl. Phys.*, **35**, 3133 (1964).
117. S. J. Hutchinson and P. P. Benham., *Plast. Polymer,* April 1970, 102.
118. A. G. Thomas., *J. Polymer Sci.*, **31**, 467 (1958).
119. G. J. Lake and A. G. Thomas., *Proc. Roy. Soc.* **A300**, 108 (1967).
120. G. J. Lake and P. B. Lindley, in *Proceedings of the Conference on the Physical Basis of Yield and Fracture*, Oxford, 1966, p. 176.
121. H. W. Greensmith, *J. Appl. Polymer Sci.*, **7**, 993 (1963).
122. E. H. Andrews, in *Testing of Polymers*, Vol. 4 (ed. W. E. Brown,), Wiley, New York, 1968, p. 237.
123. E. H. Andrews and B. J. Walker, *Proc Roy. Soc.*, **A325**, 57 (1971).
124. H. F. Borduas, L. E. Culver and D. J. Burns, *J. Strain Analysis*, **3**, 193 (1968).
125. R. W. Hertzberg, H. Nordberg and J. A. Manson, *J. Mater Sci.*, **5**, 521 (1970).
126. S. Arad, J. C. Radon and L. E. Culver, *J. Mech. Eng. Sci.*, **13**, 75 (1971).
127. J. S. Harris and I. M. Ward, *J. Mater. Sci.*, **8**, 1655 (1973).
128. P. C. Paris, in *Fatigue, an Interdisciplinary Approach*, Syracuse University Press, Syracuse, NY, 1964, p. 107.
129. P. C. Paris and F. Erdogan, *J. Basic. Eng., Trans. ASME*, **85**, 528 (1963).
130. S. Arad, J. C. Radon and L. E. Culver, *Polymer Eng. Sci.*, **12**, 193 (1972).
131. J. A. Manson and R. W. Hertzberg, *CRC Crit Rev. Macromol. Sci.*, **1**, 433 (1973).
132. G. L. Pitman and I. M. Ward, *J. Mater Sci.*, **15**, 635 (1980).
133. J. G. Williams, *J. Mater. Sci.*, **12**, 2525 (1980).
134. Y. W. Mai and J. G. Williams, *J. Mater. Sci.*, **14**, 1933 (1979).

PROBLEMS FOR CHAPTERS 11 AND 12

1. The critical shear stress τ for yielding of a certain polymer is given by $\tau = \tau_0 + \mu\sigma_N$, where σ_N is the compressive stress on the yield plane and τ_0 and μ are constants.

If $\tau_0 = 10^7$ Pa and $\mu = 0.1$, calculate the magnitude of the yield stress in compression, showing that yield occurs on the plane whose normal makes an angle 47°51′ with the compressive stress direction.

2. A thin-walled cylinder with closed ends of radius r and wall thickness t is fabricated from a polymer with a yield stress in pure shear of k. Calculate the internal pressure required to produce yielding of the cylinder walls if the yield criterion under appropriate conditions of temperature and strain rate may be written as

$$(\sigma_1 - \sigma_2)^2 + (\sigma_2 - \sigma_3)^2 + (\sigma_3 - \sigma_1)^2 = 6k^2 + \mu(\sigma_1 + \sigma_2 + \sigma_3).$$

(Hint: the hoop stress, radial stress and axial stress due to an internal pressure p are pr/t, 0 and $pr/2t$ respectively.)

3. A batch of isotropic polymer is observed to yield at 40 MPa in compression and 35 MPa in tension, both tests at a uniaxial strain rate of $10^{-3}\ \mathrm{s}^{-1}$. Suggest a reason for the difference in these two yield stresses and estimate the yield stress in equal biaxial tension ($\sigma_1 = \sigma_2 > 0$, $\sigma_3 = 0$), explaining carefully the assumptions you have made. What strain rate ($\dot{\varepsilon}_1 = \dot{\varepsilon}_2 = -\dot{\varepsilon}_3/2$) would you use in biaxial tension to obtain the same effective strain rate as in the uniaxial tests?

4. The compressive yield stress of a fictional polymer decreases linearly from 200 MPa at 50 K to 100 MPa at room temperature when tested at a strain rate of $10^{-3}\ \mathrm{s}^{-1}$. The room temperature compressive yield stress increases to 120 MPa at a strain rate of $10^{-1}\ \mathrm{s}^{-1}$.

   According to the Eyring theory of flow these observations can be analysed in terms of the activation volume $V = kT\,\Delta(\ln\dot{\varepsilon})/\Delta\sigma$, the activation energy ΔU and a pre-exponential factor $\dot{\varepsilon}_0$. Estimate the values of these parameters using the given data. (Boltzmann constant = $1.38 \times 10^{-23}\ \mathrm{J\,K^{-1}}$).

5. A large sheet of PMMA has a central elliptical crack of length 1 cm. Given that K_{IC} is 1 $\mathrm{MNm^{-3/2}}$ calculate the fracture stress.

6. Show how the Irwin–Kies relationship can be derived by considering the load–extension curve for a specimen with a central crack. Use this relationship to calculate the load which must be applied to a double cantilever beam specimen of half-width 1 cm and thickness 3 mm to cause a crack of initial length 8 cm to begin to propagate, given that the fracture toughness K_{IC} is 2 $\mathrm{MN\,m^{-3/2}}$.

7. A craze at the crack tip in glassy polymer has a crack opening displacement of 4 μm. Given that the fracture toughness is 1 $\mathrm{MN\,m^{-3/2}}$ and the plane strain modulus is 1.5 GPa, calculate the length of the craze and the craze stress.

Appendix 1

A1.1 SCALARS, VECTORS AND TENSORS

Quantities such as mass and temperature are scalars, whose magnitude does not depend on direction. Force, however, is a vector, which relative to a given set of axes may be resolved into three components parallel to the coordinate axes. Referred to a new set of axes obtained by rotating the initial set, the three axial components of the force will be changed. It may be advantageous to perform a rotation of axes so that the total force is directed along one axis. Area may also be represented as a vector, whose direction is that of the outward-facing normal to the surface and whose length is proportional to the area.

When each component of one vector is linearly related to each component of another vector the coefficients of proportionality are the components of a second-rank tensor. Stress, defined as force per unit area, is the quotient of two vectors, and is an example of a second-rank tensor. Note that the condition defining positive in the direction of the outward normal means that a hydrostatic pressure must be negative. Strain is also a tensor, and in the most general case both stress and strain can be expressed in terms of nine tensor components. Stress and strain are both examples of second-rank tensors, which have nine components; vectors, which have three components, are sometimes referred to as first-rank tensors; single-valued scalars are zero rank tensors.

A1.2 TENSOR COMPONENTS OF STRESS

The components of stress have already been defined in Section 2.1 in terms of the equilibrium of a cube and these form the elements of a symmetric second-rank tensor

$$\sigma_{ij} = \begin{bmatrix} \sigma_{xx} & \sigma_{xy} & \sigma_{xz} \\ \sigma_{xy} & \sigma_{yy} & \sigma_{yz} \\ \sigma_{xz} & \sigma_{yz} & \sigma_{zz} \end{bmatrix}$$

A1.3 TENSOR COMPONENTS OF STRAIN

It was emphasized in section 2.2 above that the engineering components of strain were defined. The nine tensor components of strain are defined by the symmetric strain tensor

$$\varepsilon_{ij} = \begin{bmatrix} \dfrac{\partial u}{\partial x} & \dfrac{1}{2}\left(\dfrac{\partial v}{\partial x} + \dfrac{\partial u}{\partial y}\right) & \dfrac{1}{2}\left(\dfrac{\partial w}{\partial x} + \dfrac{\partial u}{\partial z}\right) \\ \dfrac{1}{2}\left(\dfrac{\partial v}{\partial x} + \dfrac{\partial u}{\partial y}\right) & \dfrac{\partial v}{\partial y} & \dfrac{1}{2}\left(\dfrac{\partial v}{\partial z} + \dfrac{\partial w}{\partial y}\right) \\ \dfrac{1}{2}\left(\dfrac{\partial w}{\partial x} + \dfrac{\partial u}{\partial z}\right) & \dfrac{1}{2}\left(\dfrac{\partial v}{\partial z} + \dfrac{\partial w}{\partial y}\right) & \dfrac{\partial w}{\partial z} \end{bmatrix}$$

It can be seen that this follows from a general definition of the strain ε_{ij} as

$$\varepsilon_{ij} = \frac{1}{2}\left(\frac{\partial u_i}{\partial x_j} + \frac{\partial u_j}{\partial x_i}\right)$$

where i, j, take the values 1, 2 and 3 in turn and we write $x_1 = x$, $x_2 = y$, $x_3 = z$ and $u_1 = u$, $u_2 = v$, $u_3 = w$.

In terms of the engineering components of strain e_{xx}, etc. defined in section 2.2 we have

$$\varepsilon_{ij} = \begin{bmatrix} e_{xx} & \frac{1}{2}e_{xy} & \frac{1}{2}e_{xz} \\ \frac{1}{2}e_{xy} & e_{yy} & \frac{1}{2}e_{yz} \\ \frac{1}{2}e_{xz} & \frac{1}{2}e_{yz} & e_{zz} \end{bmatrix}$$

i.e. the engineering and tensor extensional strains are identical but the engineering shear strains are twice the tensor shear strains.

A1.4 GENERALIZED HOOKE'S LAW

The generalized Hooke's law assumes that each of the nine components of the stress tensor is linearly related to each of the nine components of the strain tensor and vice versa. For example

$$\sigma_{xx} = a\varepsilon_{xx} + b\varepsilon_{xy} + c\varepsilon_{xz} + d\varepsilon_{yx} + e\varepsilon_{yy} + f\varepsilon_{yz} + g\varepsilon_{zx} + h\varepsilon_{zy} + i\varepsilon_{zz}$$

and

$$\varepsilon_{xx} = a'\sigma_{xx} + b'\sigma_{xy} + c'\sigma_{xz} + d'\sigma_{yx} + e'\sigma_{yy} + f'\sigma_{yz} + g'\sigma_{zx} + h'\sigma_{zy} + i'\sigma_{zz},$$

where a, b, ... and a', b' ... are constants.

In one dimension Hooke's law defines stiffness (or modulus) as stress divided by strain; alternatively, compliance is strain divided by stress. In three dimensions the stress and strain tensors are related through stiffness $[c]$ and compliance $[s]$ tensors **(note the confusing nomenclature)**. As stress and strain are each second-rank symmetric tensors, $[c]$ and $[s]$ are each fourth-rank symmetric tensors: each component of strain is linearly related to all nine components of stress, and vice versa, so there are 81 components in the stiffness and compliance tensors, which when written out in full form a 9×9 array.

$$\sigma_{ij} = c_{ijkl}\varepsilon_{kl}$$

or equivalently

$$\varepsilon_{ij} = s_{ijkl}\sigma_{kl}.$$

In terms of earlier nomenclature $\sigma_{ij} = \sigma_{xx}, \sigma_{yy}$, etc. and $\varepsilon_{ij} = \varepsilon_{xx}, \varepsilon_{yy}$, etc.

The fourth-rank tensors s_{ijkl} and c_{ijkl} define the compliance and stiffness constants, with i, j, k, l taking values 1, 2, 3 in turn. In these equations the use of 1, 2, 3 is synonymous with the x, y, z used to define stress and strain components.

In a fourth-rank symmetric tensor not all 81 terms are independent: in general

$$s_{ijkl} = s_{ijlk} = s_{jilk} = s_{jikl},$$

and similarly for stiffness components. When crystalline symmetry is taken into account the number of independent terms is reduced still further.

A1.5 ENGINEERING STRAINS AND MATRIX NOTATION

It is often more convenient to work in terms of engineering strains rather than use tensor strain components. Such an approach leads to a more compact notation, in which a generalized Hooke's law relates the six independent components of stress to the six independent components of engineering strain:

$$\sigma_p = c_{pq}e_q \quad \text{and} \quad e_p = s_{pq}\sigma_q,$$

where σ_p represents $\sigma_{xx}, \sigma_{yy}, \sigma_{zz}, \sigma_{xz}, \sigma_{yz}$ or σ_{xy} and e_q represents $e_{xx}, e_{yy}, e_{zz}, e_{xz}, e_{yz}$ or e_{xy}; c_{pq} and s_{pq} now form 6×6 matrices, in which p and q take the value 1, 2, . . . 6.

For stiffness constants the following substitution is used for obtaining p and q in terms of i, j, k and l:

Tensor subscript	11	22	33	23 or 32	31 or 13	12 or 21
Matrix subscript	1	2	3	4	5	6

For compliance constants the substitution is based on the above conversion for stiffness constants, but additional rules apply because of the factor 2 difference between the definition of engineering and tensor shear strains:

$$s_{ijkl} = s_{pq} \text{ when } p \text{ and } q \text{ are 1, 2 or 3,}$$

$$2s_{ijkl} = s_{pq} \text{ when either } p \text{ or } q \text{ are 4, 5 or 6,}$$

$$4s_{ijkl} = s_{pq} \text{ when both } p \text{ and } q \text{ are 4, 5 or 6.}$$

A typical relation between strain and stress is thus changed from

$$\varepsilon_{xx} = s_{1111}\sigma_{xx} + s_{1112}\sigma_{xy} + s_{1113}\sigma_{xz} + s_{1121}\sigma_{yx} + s_{1122}\sigma_{yy}$$
$$+ s_{1123}\sigma_{yz} + s_{1131}\sigma_{zx} + s_{1132}\sigma_{zy} + s_{1133}\sigma_{zz},$$

which equals

$$\varepsilon_{xx} = s_{1111}\sigma_{xx} + s_{1122}\sigma_{yy} + s_{1133}\sigma_{zz} + 2s_{1112}\sigma_{xy}$$
$$+ 2s_{1113}\sigma_{xz} + 2s_{1123}\sigma_{yz},$$

because of the symmetries already discussed, to

$$e_{xx} = s_{11}\sigma_{xx} + s_{12}\sigma_{yy} + s_{13}\sigma_{zz} + s_{14}\sigma_{xz} + s_{15}\sigma_{yz} + s_{16}\sigma_{xy},$$

which is sometimes abbreviated further as

$$e_1 = s_{11}\sigma_1 + s_{12}\sigma_2 + s_{13}\sigma_3 + s_{14}\sigma_4 + s_{15}\sigma_5 + s_{16}\sigma_6.$$

The existence of a strain-energy function ([2], p. 49) means that

$$c_{pq} = c_{qp} \quad \text{and} \quad s_{pq} = s_{qp},$$

so that the number of independent constants is reduced from 36 to 21, before taking crystal symmetry into account, i.e.

$$c_{pq} = \begin{bmatrix} c_{11} & c_{12} & c_{13} & c_{14} & c_{15} & c_{16} \\ c_{12} & c_{22} & c_{23} & c_{24} & c_{25} & c_{26} \\ c_{13} & c_{23} & c_{33} & c_{34} & c_{35} & c_{36} \\ c_{14} & c_{24} & c_{34} & c_{44} & c_{45} & c_{46} \\ c_{15} & c_{25} & c_{35} & c_{45} & c_{55} & c_{56} \\ c_{16} & c_{26} & c_{36} & c_{46} & c_{56} & c_{66} \end{bmatrix}$$

Similarly

$$s_{pq} = \begin{bmatrix} s_{11} & s_{12} & s_{13} & s_{14} & s_{15} & s_{16} \\ s_{12} & s_{22} & s_{23} & s_{24} & s_{25} & s_{26} \\ s_{13} & s_{23} & s_{33} & s_{34} & s_{35} & s_{36} \\ s_{14} & s_{24} & s_{34} & s_{44} & s_{45} & s_{46} \\ s_{15} & s_{25} & s_{35} & s_{45} & s_{55} & s_{56} \\ s_{16} & s_{26} & s_{36} & s_{46} & s_{56} & s_{66} \end{bmatrix}$$

Because the simplified notation involves a matrix rather than a tensor it is necessary to convert back into tensor notation in order to calculate, for instance, a compliance in terms of a rotated set of coordinate axes, using the tensor relationship

$$s'_{mnop} = a_{mi}a_{nj}a_{ok}a_{pl}s_{ijkl},$$

where a_{mi}, etc. represent direction cosines of the angles between the two sets of axes. An example is given in A1.7.

A1.6 THE ELASTIC MODULI OF ISOTROPIC MATERIALS

Most of the present book is concerned with isotropic polymers, for which measured properties, such as the Young's modulus E, Poisson's ratio ν and the shear modulus G, relate directly to the constants of the compliance matrix.

For an isotropic solid, the matrix s_{pq} reduces to

$$s_{pq} = \begin{pmatrix} s_{11} & s_{12} & s_{12} & 0 & 0 & 0 \\ s_{12} & s_{11} & s_{12} & 0 & 0 & 0 \\ s_{12} & s_{12} & s_{11} & 0 & 0 & 0 \\ 0 & 0 & 0 & 2(s_{11} - s_{12}) & 0 & 0 \\ 0 & 0 & 0 & 0 & 2(s_{11} - s_{12}) & 0 \\ 0 & 0 & 0 & 0 & 0 & 2(s_{11} - s_{12}) \end{pmatrix}$$

It can be shown that the Young's modulus is given by

$$E = 1/s_{11},$$

the Poisson's ratio by

$$\nu = -s_{12}/s_{11}$$

and the torsional modulus by

$$G = \frac{1}{2(s_{11} - s_{12})}.$$

Thus we obtain the stress–strain relationships derived more simply in section 2.3:

$$e_{xx} = \frac{1}{E}\sigma_{xx} - \frac{\nu}{E}(\sigma_{yy} + \sigma_{zz}),$$

$$e_{yy} = \frac{1}{E}\sigma_{yy} - \frac{\nu}{E}(\sigma_{xx} + \sigma_{zz}),$$

$$e_{zz} = \frac{1}{E}\sigma_{zz} - \frac{\nu}{E}(\sigma_{xx} + \sigma_{yy}),$$

$$e_{xz} = \frac{1}{G}\sigma_{xz},$$

$$e_{yz} = \frac{1}{G}\sigma_{yz},$$

$$e_{xy} = \frac{1}{G}\sigma_{xy},$$

where

$$G = \frac{E}{2(1 + \nu)}$$

Another basic quantity is the bulk modulus K, which determines the dilation $\Delta = e_{xx} + e_{yy} + e_{zz}$ produced by a uniform hydrostatic pressure. Using the stress–strain relationships above, it can be shown that the strains produced by a uniform hydrostatic pressure p are given by

$$e_{xx} = (s_{11} + 2s_{12})p,$$

$$e_{yy} = (s_{11} + 2s_{12})p,$$

$$e_{zz} = (s_{11} + 2s_{12})p.$$

Then

$$K = \frac{p}{\Delta} = \frac{1}{3(s_{11} + 2s_{12})} = \frac{E}{3(1 - 2\nu)}$$

A1.7 TRANSFORMATION OF TENSORS FROM ONE SET OF COORDINATE AXES TO ANOTHER

If the stress or strain components are known in terms of one set of initially perpendicular axes, they can readily be determined for another set of perpendicular axes by the tensor transformation rule

$$\sigma'_{ij} = a_{ik}a_{jl}\sigma_{kl},$$

$$\varepsilon'_{ij} = a_{ik}a_{jl}\varepsilon_{kl},$$

where $i, j, k, l = x, y, z$ in turn and a_{ik}, a_{jl}, etc. represent the direction cosines relating the two sets of axes.

Similarly if the compliances and stiffnesses are defined for one set of axes (usually the symmetry axes of the polymer) their values can be determined in another set of axes by the tensor transformation rule

$$s'_{\mathrm{mnop}} = a_{mi}a_{nj}a_{ok}a_{pl}s_{ijkl},$$

$$c'_{\mathrm{mnop}} = a_{mi}a_{nj}a_{ok}a_{pl}c_{ijkl}.$$

For example, consider a film of polymer with orthorhombic symmetry, with the x and z axes lying in the plane of the film and the y axis normal to that plane. The compliance matrix is

$$\begin{bmatrix} s_{11} & s_{12} & s_{13} & 0 & 0 & 0 \\ s_{12} & s_{22} & s_{23} & 0 & 0 & 0 \\ s_{13} & s_{23} & s_{33} & 0 & 0 & 0 \\ 0 & 0 & 0 & s_{44} & 0 & 0 \\ 0 & 0 & 0 & 0 & s_{55} & 0 \\ 0 & 0 & 0 & 0 & 0 & s_{66} \end{bmatrix}$$

Determine Young's modulus (E_θ) for a strip of material cut in a direction making an angle θ with the z axis (Figure A1.1). Using $E_\theta = 1/s_\theta$ take the direction of the strip as the z' axis of a rotated set of Cartesian axes. Then $s_\theta = s_{3'3'3'3'}$.

In the most general case of rotation of axes the expansion will involve 81 terms. Here, however, the z' axis is perpendicular to the y axis, so that $a_{3'2}$ will be zero, immediately reducing the number of terms to 16. Crystal symmetry is now used to determine which of these 16 terms can be neglected because the relevant compliance constant is zero. The compliance constants are of the following forms:

$$s_{1111}, s_{1113}, s_{1133}, s_{1333}, s_{3333}.$$

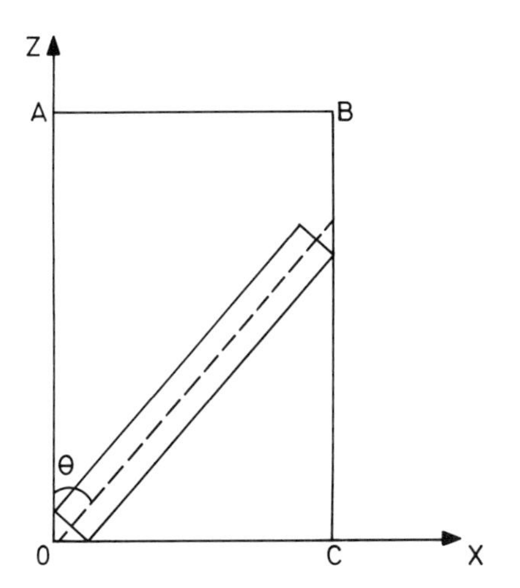

Figure A1.1 Strip cut at θ to the z axis from a polymer sheet with orthorhombic symmetry

Terms which include an odd number of threes relate to terms in the compliance matrix of the form s_{p5}, where $p = 5$, and so are zero. The complete expansion therefore contains only eight terms, which correspond with the compliance matrix constants s_{11}, s_{33}, s_{13} and s_{55} (Table A2.1).

$$\begin{aligned} s_{3'3'3'3'} &= a_{3'1}a_{3'1}a_{3'1}a_{3'1}s_{1111} + a_{3'3}a_{3'3}a_{3'3}a_{3'3}s_{3333} \\ &\quad + a_{3'1}a_{3'1}a_{3'3}a_{3'3}s_{1133} + a_{3'3}a_{3'3}a_{3'1}a_{3'1}s_{3311} \\ &\quad + a_{3'1}a_{3'3}a_{3'3}a_{3'1}s_{1331} + a_{3'3}a_{3'1}a_{3'1}a_{3'3}s_{3113} \\ &\quad + a_{3'3}a_{3'1}a_{3'3}a_{3'1}s_{3131} + a_{3'1}a_{3'3}a_{3'1}a_{3'3}s_{1313}, \end{aligned}$$

but

$$a_{3'3} = \cos\theta \quad \text{and} \quad a_{3'1} = \cos(\pi/2 - \theta) = \sin\theta.$$

$$\begin{aligned} \therefore \quad s_{3'3'3'3'} &= \sin^4\theta\, s_{1111} + \cos^4\theta\, s_{3333} + 2\sin^2\theta\cos^2\theta\, s_{1133} \\ &\quad + 4\sin^2\theta\cos^2\theta s_{1313}, \end{aligned}$$

as

$$s_{1133} = s_{3311} \quad \text{and} \quad s_{1331} = s_{3113} = s_{3131} = s_{1313}.$$

Converting back to the abbreviated matrix notation

$$s_\theta = s_{3'3'} = \sin^4\theta\, s_{11} + \cos^4\theta\, s_{33} + \sin^2\theta\cos^2\theta\,(2s_{13} + s_{55}).$$

Note the factor of 4 that is required when converting s_{1313} to the matrix notation s_{55}, as in this case both p and q are greater than 3.

Important practical cases are those for strips cut along the z axis and at 45° and 90°

Table A2.1 Conversion from tensor to matrix format

Tensor term	Matrix equivalent	Does it occur?
1111	11	Yes
1113	15	No
1131	15	No
1311	51	No
3111	51	No
1133	13	Yes
1331	55	Yes
3311	31	Yes
1313	55	Yes
3131	55	Yes
3113	55	Yes
1333	53	No
3331	35	No
3313	35	No
3133	53	No
3333	33	Yes

to the z axis. Substituting in the general expression above we obtain the appropriate Young's moduli:

$$E_0 = \frac{1}{s_0} = \frac{1}{s_{33}}, \qquad E_{90} = \frac{1}{s_{90}} = \frac{1}{s_{11}}$$

and

$$\frac{1}{E_{45}} = s_{45} = \frac{1}{4}[s_{11} + s_{33} + (2s_{13} + s_{55})].$$

A1.8 THE MOHR CIRCLE CONSTRUCTION

This construction enables the components of a two-dimensional stress (or strain), expressed in terms of a given set of perpendicular axes, to be converted into the components relative to any other set of perpendicular axes. In particular it provides a simple method for determining the principal axes of stress and strain. For fibre symmetry all planes that contain the fibre axis as one of the perpendicular axes are identical: thus the construction is again applicable.

Let the components of stress be σ_{xx}, σ_{yy}, $\sigma_{xy} = \sigma_{yx}$ in terms of perpendicular axes OX, OY. Now consider an identical total stress in terms of axes OX′, OY′, obtained by rotating OX and OY through an angle θ in the anticlockwise direction. It can be shown that the shear stress components σ'_{xy} and σ'_{yx} in terms of the new axes are

$$\sigma'_{xy} = \sigma'_{xy} = \tfrac{1}{2}(\sigma_{yy} - \sigma_{xx})\sin 2\theta + \sigma_{xy}\cos 2\theta. \tag{A1.1}$$

For a proof of (A1.1) using elementary mathematics see Hall [1]. The expression may be obtained directly from the general tensor relation $\sigma'_{ij} = a_{ik}a_{jl}\sigma_{kl}$, where a_{ik}, etc. represent direction cosines: $a_{11} = \cos\theta$, $a_{12} = \sin\theta$, $a_{21} = -\sin\theta$, $a_{22} = \cos\theta$.

A particular angle θ, where $0 < \theta < \pi/2$, can always be found for which σ'_{xy} in (A1.1) is zero. The condition is

$$\tan 2\theta = \frac{2\sigma_{xy}}{\sigma_{xx} - \sigma_{yy}} \tag{A1.2}$$

When stress components are referred to axes rotated through that value of θ compared with the original axes, the shear stress components vanish, and we have defined the principal axes of stress. Principal axes of strain are defined similarly. In Figure A1.2 extensional stresses σ_{xx}, σ_{yy} are plotted along the OX axis, and shear stresses $\sigma_{xy} = \sigma_{yx}$ are plotted along the OY axis.

Here P is the point $(\sigma_{xx}, -\sigma_{xy})$; Q is the point $(\sigma_{yy}, \sigma_{xy})$. A circle with PQ as diameter will cut the OX axis at A. Call the angle between PQ and the OX axis 2θ. Then $\tan 2\theta = MP/AM$.

$$AM = \sigma_{xx} - OA = \sigma_{xx} - \tfrac{1}{2}(\sigma_{xx} + \sigma_{yy}) = \tfrac{1}{2}(\sigma_{xx} - \sigma_{yy}). \qquad MP = \sigma_{xy}.$$

Hence

$$\tan 2\theta = \frac{2\sigma_{xy}}{(\sigma_{xx} - \sigma_{yy})},$$

which is identical with equation (A1.2), used above to define the principal axes of stress. Hence θ (*half* the above angle) gives the angle between the original axes and the principal axes. Similarly the stress components referred to any set of axes rotated

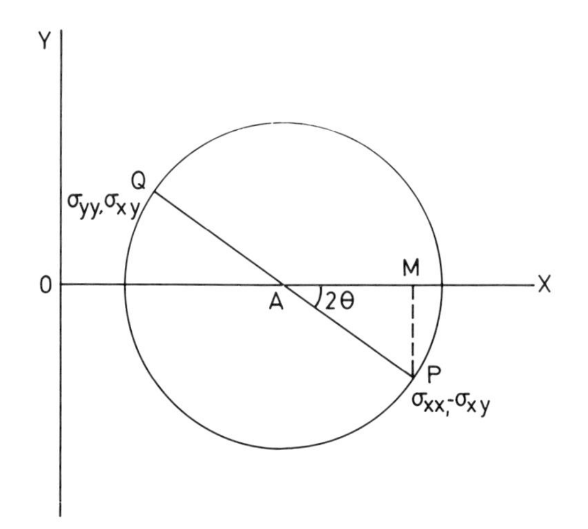

Figure A1.2 The Mohr circle construction

through an angle ϕ compared with the original axes, can be determined from the coordinates of the ends of the diameter of the Mohr circle obtained by rotating PQ through an angle 2ϕ. The construction holds also for strain components.

REFERENCES

1. I. H. Hall, *The Deformation of Solids*, Nelson, London 1968.
2. A. E. H. Love, *A Treatise on the Mathematical Theory of Elasticity*, 4th Edn, Macmillan, New York, 1944.

Appendix 2

A2.1 RIVLIN, MOONEY, OGDEN

A very useful and simple form for the strain-energy function has already been given as follows:

$$U = \tfrac{1}{2}NkT(\lambda_1^2 + \lambda_2^2 + \lambda_3^2 - 3), \tag{3.23a}$$

but more complex forms have been proposed. Because a rubber is an isotropic material whose properties are the same in all directions, it is plausible that any function of λ_1, λ_2, λ_3 which is invariant to a permutation of λ_1, λ_2, λ_3 and also becomes zero for $\lambda_1 = 1$, $\lambda_2 = 1$, $\lambda_3 = 1$, might provide a satisfactory form for the strain energy function. Functions of λ_1, λ_2, λ_3 which are independent of the choice of axes are termed strain invariants. Rivlin [1], and Rivlin and Saunders [2] examined the behaviour of vulcanized rubber and explored the use of a strain energy function of the form

$$U = C_1(I_1 - 3) + f(I_2 - 3), \tag{A2.1}$$

where $I_1 = \lambda_2^2 + \lambda_2^2 + \lambda_3^2$ is called the first strain invariant so that the first term in equation (A2.1) is identical to equation (3.24), $I_2 = \lambda_1^2\lambda_2^2 + \lambda_2^2\lambda_3^2 + \lambda_3^2\lambda_1^2$ is called the second strain invariant and f indicates that we can envisage a power law series

$$f(I_2 - 3) = C_2(I_2 - 3) + C_3(I_2 - 3)^2 + C_4(I_2 - 3)^3,$$

where C_2, C_3 and C_4 are constants.

A further simple strain invariant is $I_3 = \lambda_1^2\lambda_2^2\lambda_3^2$, which for an incompressible rubber is always unity, because $\lambda_1\lambda_2\lambda_3 = 1$. Therefore I_2 can also be written as

$$I_2 = \frac{1}{\lambda_1^2} + \frac{1}{\lambda_2^2} + \frac{1}{\lambda_3^2}$$

This leads us to a practically useful form of the strain-energy function, originally proposed by Mooney [3], and shown by him to give a good fit to experimental data for the uniaxial extension of rubbers. The Mooney equation is

$$U = C_1(\lambda_1^2 + \lambda_2^2 + \lambda_3^2 - 3) + C_2\left(\frac{1}{\lambda_1^2} + \frac{1}{\lambda_2^2} + \frac{1}{\lambda_3^2}\right) \tag{A2.2}$$

Using the incompressibility relationship $\lambda_1\lambda_2\lambda_3 = 1$, for uniaxial extension with $\lambda_1 = \lambda$, we have $\lambda_2^2 = \lambda_3^2 = 1/\lambda$ and

$$U = C_1(\lambda^2 - 2/\lambda - 3) + C_2\left(\frac{1}{\lambda^2} + 2\lambda - 3\right). \tag{A2.3}$$

Differentiation with respect to λ gives the force per unit unstrained area as

$$f = 2\left(\lambda - \frac{1}{\lambda^2}\right)\left(C_1 + \frac{C_2}{\lambda}\right) \tag{A2.4}$$

This equation can be rearranged to give

$$\frac{f}{2(\lambda - 1/\lambda^2)} = C_1 + \frac{C_2}{\lambda}$$

A plot of

$$\frac{f}{2(\lambda - 1/\lambda^2)}$$

(often called the 'reduced stress') versus $1/\lambda$ is called a Mooney plot, and should be linear with a gradient of C_2 and an ordinate of $(C_1 + C_2)$ at $\lambda = 1$.

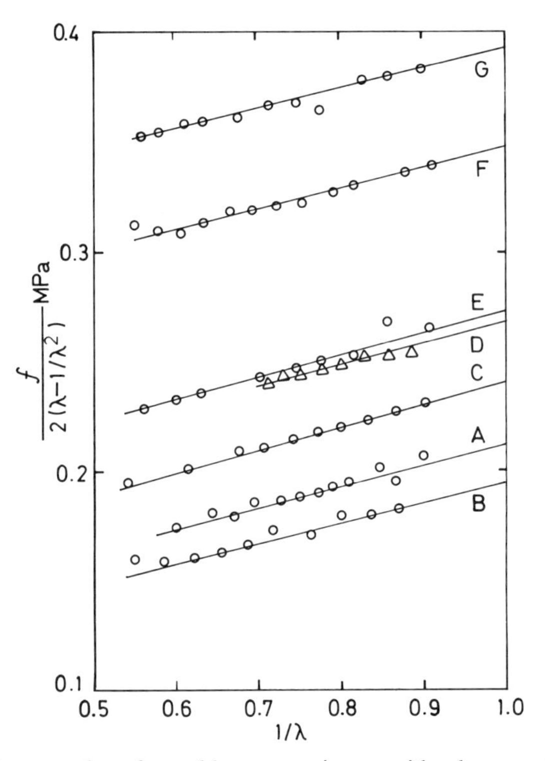

Figure A2.1 Mooney plots for rubbers covering a wide degree of vulcanization (Gumbrell, Mullins, and Rivlin 1953). Reproduced with permission from Treloar, *The Physics of Rubber Elasticity*, 3rd edn., Clarendon Press, Oxford, 1975

In practice, it is found that the results for rubbers are typically similar to those shown in Figure A2.1 where the Mooney equation is adequate for $1/\lambda > 0.45$. In spite of this limitation, Mooney plots are a very useful way of dealing with data for the deformation of rubber networks. Such plots are often used as a first step to fitting results to more complicated relationships such as those proposed by Edwards and Vilgis [4], which provide a molecular understanding of rubber elasticity.

Rivlin's choice of a strain-energy function U that involved the squares of the extension ratios arose because he assumed that negative values of the extension ratios λ were a mathematical possibility, whereas it was necessary for U always to be greater than zero. We have seen, however, that by choosing suitable rotations of coordinate axes the most general deformation can be described in terms of pure strain, i.e. three principal extension ratios λ_1, λ_2, λ_3 which are all positive (although some are necessarily less than unity, because $\lambda_1\lambda_2\lambda_3 = 1$).

In addition to these theoretical considerations, which suggest that we need not be restricted to squares of extension ratios in formulating the strain energy function, it has been found by experimentalists that there is high sensitivity to experimental error when small values of the strain invariants I_1 and I_2 are involved. It is therefore natural to postulate that the only constraint on the form U is that imposed by material isotropy, which implies that $U(\lambda_1, \lambda_2, \lambda_3)$ should be a symmetric function of the extension ratio, i.e. invariant to any permutation of the indices 1, 2, 3.

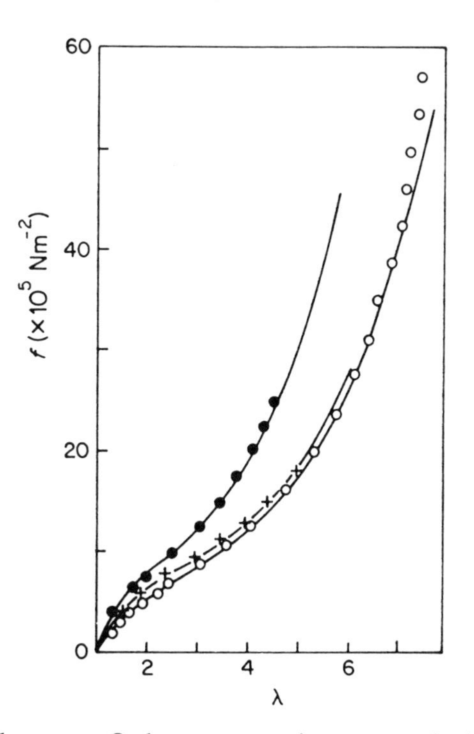

Figure A2.2 The three-term Ogden representation compared with the Treloar data for simple tension (○); pure shear (+) and equibiaxial tension (○). (Reproduced with permission from Treloar, *Proc. Roy Soc.* **A351**, 301 (1976) and Ogden, *Proc. Roy. Soc.* **A326**, 565 (1972))

Ogden, Chadwick and Haddon [5] have proposed that a useful form is

$$U = \sum_n \frac{\mu_n}{\alpha_n}(\lambda_1^{\alpha_n} + \lambda_2^{\alpha_n} + \lambda_3^{\alpha_n} - 3).$$

They showed that experimental data of Treloar [6] for tension, pure shear and equibiaxial tension could be fitted very well indeed (Figure A2.2) by assuming a three-form expression for U with $\alpha_1 = 1.3$, $\alpha_2 = 5.0$, $\alpha_3 = -2.0$, $\mu_1 = 6.2 \times 10^5$ Pa, $\mu_2 = 1.2 \times 10^3$ Pa, $\mu_3 = -1 \times 10^3$ Pa.

REFERENCES

1. R. S. Rivlin, *Phil. Trans. Roy. Soc.*, **A240**, 459, 491 (1948), **241**, 379 (1949).
2. R. S. Rivlin and D. W. Saunders., *Phil. Trans. Roy. Soc.*, **A243**, 251 (1951); *Trans. Faraday. Soc.*, **48**, 200 (1952).
3. M. Mooney, *J. Appl. Phys.*, **11**, 582 (1940)
4. S. F. Edwards and Th. Vilgis, *Polymer*, **27**, 483 (1986).
5. R. W. Ogden, P. Chadwick and E. W. Haddon, *Quart. J. Mech. Appl. Math.*, **26**, 23 (1973).
6. L. R. G. Treloar, *Trans Faraday. Soc.*, **40**, 59 (1944).

Answers to Problems

Chapters 2 and 3

1. An ideal rubber is assumed to be composed of very long molecules which can adopt a variety of conformations. The individual chains are interlinked by cross-links or junction points. The number of cross-links per unit volume is low and there is no interference with the motion of the chains.

   In the absence of external stress the chain between junction points adopts a maximum entropy state. On straining, the junction points deform affinely resulting in the orientation of the molecular segments joining them and hence a reduction in entropy.

   The entropy of a single chain can be calculated from the number of conformations which are possible for the given position of the junction points. The entropy of the whole system is obtained by summing over the network.

   The dependence of the entropy on strain in the network leads directly to the dependence of the free energy on strain and the differential of the free energy with respect to strain is the force developed in the network.

   For an ideal rubber, the shear modulus G at a temperature T is given by $G = NkT$, where N is the number of chains per unit volume so

$$N = \frac{G}{kT}$$

   Also we have $\rho = N \times$ mean mass of chain, $\rho = N \times n_{\mathrm{m}} \times$ mass of monomer (M), where n_{m} is the mean number of monomers per chain, i.e.

$$N = \frac{\rho N_{\mathrm{A}}}{n_{\mathrm{m}} M}$$

   Combining these equations gives

$$\begin{aligned} n_{\mathrm{m}} &= \frac{\rho N_{\mathrm{A}}}{M} \frac{kT}{G} \\ &= \frac{1300 \times 6.022 \times 10^{26} \times 1.38 \times 10^{-23} \times 293}{68 \times 4 \times 10^{5}} \\ &= 116. \end{aligned}$$

2. Volume before deformation $V_0 = \pi r_0^2 l$. Volume after deformation

$$V_0 = \pi r_0^2 l(1 + \varepsilon_x)^2(1 + \varepsilon_z)$$
$$= \pi r_0^2 l(1 + 2\varepsilon_x + \varepsilon_z + \varepsilon_x{}^2 + 2\varepsilon_x\varepsilon_z + \varepsilon_x{}^2\varepsilon_z).$$

At low strains second-order terms can be neglected, i.e.

$$\pi r_0^2 l = \pi r_0^2 l(1 + 2\varepsilon_x + \varepsilon_z),$$
$$\therefore \quad \varepsilon_x = -\tfrac{1}{2}\varepsilon_z \quad \text{or} \quad \nu = 0.5000.$$

At higher strains second-order terms cannot be neglected.

$$\therefore \quad \pi r_0^2 l = \pi r_0^2 l(1 + \varepsilon_x)^2(1 + \varepsilon_z)$$

or

$$(1 + \varepsilon_x) = (1 + \varepsilon_z)^{-1/2} = 1 - \varepsilon_{z/2} + \tfrac{3}{8}\varepsilon_z^2$$
$$\nu = -\varepsilon_x/\varepsilon_z = \tfrac{1}{2} - \tfrac{3}{8}\varepsilon_z.$$

Required value is given by

$$0.49 = \tfrac{1}{2} - \tfrac{3}{8}\varepsilon_z \quad \text{or} \quad \tfrac{3}{8}\varepsilon_z = 0.01$$

Therefore $\varepsilon_z = 0.027$ or 2.7%.

3. The root mean square distance

$$(\bar{r}^2)^{1/2} = \sqrt{n}l,$$

where n is the number of links, length l, in the chain

$$(\bar{r}^2)^{1/2} = \sqrt{1000} \times 1.53 \text{ Å},$$
$$= 48.4 \text{ Å}.$$

4.
$$U = C_1(\lambda_1^{1.3} + \lambda_2^{1.3} + \lambda_3^{1.3} - 3).$$

For this deformation let λ be the extension ratio in the direction of the applied force, i.e. $\lambda_1 = \lambda$ and $\lambda_2 = \lambda_3 = \lambda^{-1/2}$ from conservation of volume. We can rewrite the free energy function as

$$U = C_1(\lambda^{1.3} + 2\lambda^{-1.3/2} - 3).$$

The force per unit undeformed area is

$$f = \frac{\partial U}{\partial \lambda} = C_1 1.3\lambda^{0.3} - 2C_1\frac{1.3}{2}\lambda^{-3.3/2}.$$

For $\lambda = 3$

$$f = 1.3C_1(3^{0.3} - 3^{-1.65}).$$

The load required to produce this extension is just

$$Af = 1.3C_1A(1.390 - 0.163)$$
$$= 382 \text{ N}$$
$$= 39 \text{ kg}.$$

5. $$U = C_1(\lambda_1^2 + \lambda_2^2 + \lambda_3^2 - 3).$$

For simple extension

$$\lambda_1 = \lambda \quad \text{and} \quad \lambda_2 = \lambda_3 = \lambda^{-1/2},$$

$$\therefore \quad U = C_1(\lambda^2 + 2/\lambda - 3).$$

The force per unit undeformed area is

$$f = \frac{\partial U}{\partial \lambda} = 2C_1\lambda - \frac{2C_1}{\lambda^2}$$

The force per unit deformed area, or the stress, σ, is

$$\sigma = f\lambda = 2C_1(\lambda^2 - 1/\lambda)$$

or in terms of strain e, where $1 + 2e = \lambda^2$

$$\begin{aligned}\sigma &= 2C_1\{1 + 2e - (1 + 2e)^{-1/2}\} \\ &= 2C_1\{1 + 2e - \{1 - \tfrac{1}{2}2e + \tfrac{1}{2}\cdot\tfrac{1}{2}\cdot\tfrac{3}{2}(2e)^2 \ldots\}\} \\ &= 2C_1\{1 + 2e - 1 + e\} \text{ for small strains,}\end{aligned}$$

i.e. $\sigma = 6e$

For biaxial stretching we have

$$\lambda_1 = \lambda_2 = \lambda \quad \text{and} \quad \lambda_3 = 1/\lambda^2$$

$$\therefore \quad U = C_1(2\lambda^2 + 1/\lambda^2 - 3).$$

The force per unit undeformed area

$$f = \frac{\partial U}{\partial \lambda} = C_1\left(4\lambda - \frac{2}{\lambda^3}\right)$$

and the stress σ is given by

$$\begin{aligned}\sigma &= f\lambda = 2C_1\left(2\lambda^2 - \frac{1}{\lambda^2}\right) \\ &= 2C_1[2(1 + 2e) - (1 + 2e)^{-1}] \text{ in terms of strain } e \\ &= 2C_1[2 + 4e - [1 - 2e + \tfrac{1}{2}2(2e)^2 \ldots]] \\ &= 2C_1[2 + 4e - (1 - 2e)] \text{ at small strains}\end{aligned}$$

$$\sigma = 2C_1(1 + 6e).$$

6. $$U = C_1(\lambda_1^2 + \lambda_2^2 + \lambda_3^2 - 3) + C_2(\lambda_1^2\lambda_2^2 + \lambda_2^2\lambda_3^2 + \lambda_3^2\lambda_1^2 - 3)$$

For simple extension

$$\lambda_1 = \lambda \quad \text{and} \quad \lambda_2 = \lambda_3 = \lambda^{-1/2}.$$

Hence

$$U = C_1\left(\lambda^2 + \frac{1}{\lambda^2} + \frac{1}{\lambda^2} - 3\right) + C_2\left(\lambda + \frac{1}{\lambda^2} + \lambda - 3\right)$$

$$= C_1\left(\lambda^2 + \frac{2}{\lambda} - 3\right) + C_2\left(2\lambda + \frac{1}{\lambda^2} - 3\right)$$

Force f per unit undeformed area is given by

$$f = \frac{\partial U}{\partial \lambda} = C_1\left(2\lambda - \frac{2}{\lambda^2}\right) + C_2\left(2 - \frac{2}{\lambda^3}\right)$$

$$= 2C_1\left(\lambda - \frac{1}{\lambda^2}\right) + 2C_2\left(1 - \frac{1}{\lambda^3}\right)$$

The stress, σ (referred to the deformed area) is

$$\sigma = f\lambda = 2C_1\left(\lambda^2 - \frac{1}{\lambda}\right) + 2C_2\left(\lambda - \frac{1}{\lambda^2}\right)$$

or in terms of the strain e where $1 + 2e = \lambda^2$

$$\sigma = 2C_1[1 + 2e - \{1 - \tfrac{1}{2}2e + \tfrac{1}{2}\cdot\tfrac{1}{2}\cdot\tfrac{3}{2}(2e)^2 \ldots\}]$$

$$+ 2C_2\left[1 + \frac{1}{2}2e + \frac{1}{2}\frac{1}{2}\left(-\frac{3}{2}\right)\left(2e\right)^2 \ldots\right) - \left(1 - 2e + \frac{1.2}{2}(2e)^2 \ldots\right)\bigg]$$

$$= 2C_1[1 + 2e - 1 + e] + 2C_2[1 + e - 1 + 2e] \quad \text{for small strains}$$

$$= 2C_1(3e) + 2C_2(3e)$$

$$= 6e(C_1 + C_2).$$

Hence the modulus $= \sigma/e = 6(C_1 + C_2)$.

7. For a rubber we can use the equation

$$f = NkT\left(\lambda - \frac{1}{\lambda^2}\right)$$

where f is the force per unit undeformed area, N is the number of chains per unit volume (terminated by cross-links) and λ is the extension ratio.

Here a load of 1 kg on an initial cross-sectional area of 10 mm^2 is equivalent to a value f given by

$$f = \frac{9.81}{10^{-5}} \text{ Pa},$$

$$\lambda = 2 \quad \text{and} \quad T = 300 \text{ K},$$

$$\therefore \quad N = \frac{f}{kT}\left(\frac{1}{2 - \frac{1}{4}}\right) = \frac{4f}{7kT}$$

$$= \frac{4 \times 9.81 \times 10^5}{7 \times (1.38 \times 10^{-23}) \times 300}$$

$$= 1.35 \times 10^{26} \text{ chains per cubic metre.}$$

8. The force f per unit area of undeformed material is given by

$$f = NkT\left(\lambda - \frac{1}{\lambda^2}\right).$$

The density ρ is given by

$$\rho = \text{no. of chains per unit vol} \times \text{mean mass of the chains} = N \times M/N_A,$$

where N_A is Avogadro's number and M the mean molecular mass between cross-links, i.e.

$$\begin{aligned} M &= \frac{\rho N_A}{N} = \frac{\rho N_A kT(\lambda - 1/\lambda^2)}{f} \\ &= \frac{900 \times 6.022 \times 10^{26} \times 1.38 \times 10^{-23} \times 300 \times (3 \times \frac{1}{9})}{9.81/(15 \times 10^{-3} \times 1.5 \times 10^{-3})} \\ &= 14\,900. \end{aligned}$$

Chapter 4

1. The form of the stress relaxation is that for a Maxwell element

$$G(t) = E \exp\left(\frac{-t}{\tau}\right)$$

where $\tau = \eta/E$.

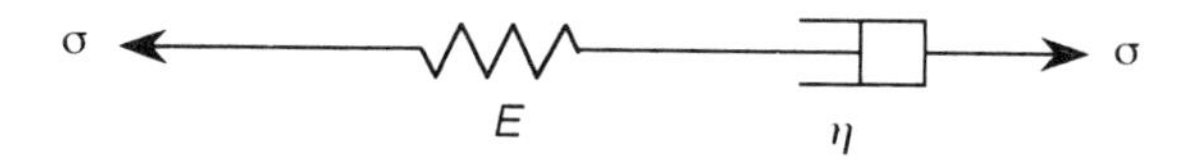

At $t = 0$

$$G(t) = E = 2 \text{ GPa}.$$

At $t = 10^4$

$$G(t) = 2 \exp\left(\frac{-10^4}{\tau}\right) = 1.0 \text{ GPa},$$

i.e.

$$\exp\left(\frac{10^4}{\tau}\right) = 2 \quad \text{or} \quad \frac{10^4}{\tau} = \ln_e 2 = 0.693,$$

$$\therefore \quad \tau = \frac{10^4}{0.693} = \frac{\eta}{E} \text{ s}, \ \eta = \frac{2 \times 10^4}{0.693} \text{ GPa s}.$$

Total strain in the Maxwell element

$$\varepsilon = \varepsilon_E + \varepsilon_v.$$

For the viscous element we can write

$$\frac{d\varepsilon_v}{dt} = \frac{\sigma}{\eta} \quad \text{or} \quad \int_0^{\varepsilon_v} d\varepsilon_v = \frac{\sigma}{\eta}\int_0^t dt$$

giving $\varepsilon_v = \dfrac{\sigma t}{\eta}$. Total strain therefore is

$$\varepsilon(t) = \frac{\sigma}{E} + \frac{\sigma t}{\eta} = \sigma\left(\frac{1}{E} + \frac{t}{\eta}\right)$$

putting $J_0 = 1/E$.

$$\therefore \quad J(t) = J_0 + \frac{t}{\eta}$$

$$\varepsilon(1000) = 0.1\left(\frac{1}{2} + \frac{10^3 \times 0.693}{2 \times 10^4}\right) = 0.1(0.5 + 0.0347)$$

$$= 5.35\%.$$

2.

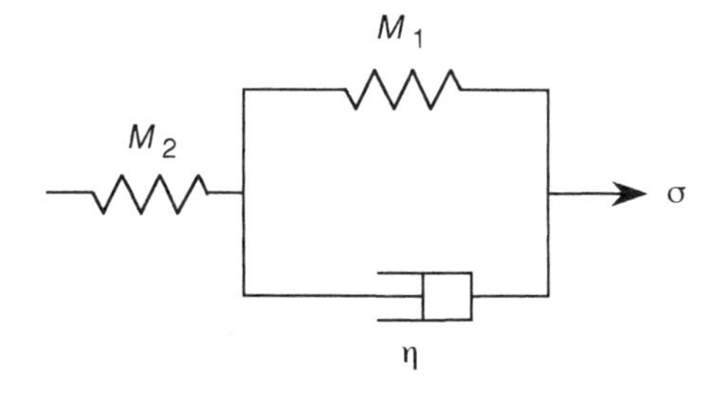

When the load is applied spring 2 extends by σ/M_2 and remains stretched. Time dependency is due entirely to the Kelvin unit, in which the total strain ε = viscous strain, ε_v = elastic strain ε_E.

In the Kelvin unit the stress is shared between the two components: $\sigma = \sigma_E + \sigma_v$

$$\sigma_E = M_1\varepsilon; \qquad \sigma_v = \eta\frac{d\varepsilon}{dt}$$

$$\therefore \quad \sigma = M_1\varepsilon + \eta\frac{d\varepsilon}{dt} \quad \text{or} \quad \frac{\sigma}{M_1} = \varepsilon + \frac{\eta}{M_1}\frac{d\varepsilon}{dt}$$

Rearranging

$$\int_0^{\varepsilon}\frac{d\varepsilon}{(\sigma/M_1 - \varepsilon)} = \frac{M_1}{\eta}\int_0^t dt$$

$$-\ln\left[\frac{\sigma}{M_1} - \varepsilon\right]_0^{\varepsilon} = \frac{M_1}{\eta}t$$

or

$$\ln\left(1 - \frac{\varepsilon}{\sigma/M_1}\right) = -\frac{M_1}{\eta}t,$$

where η/M_1 has the dimensions of time. It is defined as the retardation time τ_2.

$$\therefore \quad 1 - \frac{\varepsilon}{\sigma/M_1} = \exp(-t/\tau_2).$$

Rearranging

$$\varepsilon = \frac{\sigma}{M_1}[1 - \exp(-t/\tau_2)],$$

for total strain add on the instantaneous deformation

$$\varepsilon_0 = \frac{\sigma}{M_2}$$

$$\therefore \quad \varepsilon = \frac{\sigma}{M_2} + \frac{\sigma}{M_1}[1 - \exp(-t/\tau_2)].$$

At $t = 0$

$$\varepsilon = \frac{\sigma}{M_2} = 0.002.$$

As t tends to infinity

$$0.006 = 0.002 + \frac{\sigma}{M_1},$$

$$\therefore \quad \sigma/M_1 = 0.004.$$

Therefore at $t = 1000$ s

$$0.004 = 0.002 + 0.004[1 - \exp(-t/\tau_2)]$$

$$\therefore \quad 0.002 = 0.004[1 - \exp(-t/\tau_2)],$$

$$1 - \exp(-t/\tau_2) = \tfrac{1}{2}, \qquad \therefore \quad \exp(-t/\tau_2) = \tfrac{1}{2},$$

$$\exp(t/\tau_2) = 2, \qquad t/\tau_2 = \ln_e 2 = 0.693.$$

$$\therefore \quad \tau_2 = \frac{1000}{0.693} = 1440 \text{ s}.$$

3. The maximum stress occurs at 1% extension and will be $(2 \times 10^9 \div 100) = 20$ MPa. Because of the initial extension the amplitude of the vibrating stress (σ_0) is $\pm$ 10 MPa, and the amplitude of the vibrating strain (e_0) is $1/200 = 0.005$.

   The stored elastic energy will depend on the dimensions of the specimen. Its maximum value is $W_{st} = \frac{1}{2}$ (area) (length) $\sigma_0 \varepsilon_0 \cos\delta$. $A = 10^{-5}$ m^2, $L = 0.2$ m

$$\tan\delta = 0.1 \sim \sin\delta \qquad \therefore \quad \cos\delta = (1 - 0.01)^{1/2} \sim 0.995.$$

$$\therefore \quad W_{st} = \tfrac{1}{2} \times 10^{-5} \times 0.2 \times 10^7 \times 5 \times 10^{-3} \times 0.995 \sim 5 \times 10^{-2} \text{ J}.$$

Energy dissipated per cycle is

$$\Delta W = \pi A L \sigma_0 \varepsilon_0 \sin\delta = \pi \times 10^{-5} \times 0.2 \times 10^7 \times 5 \times 10^{-1} \sim 3.1 \times 10^{-2} \text{ J}.$$

At 0.5 Hz, logarithmic decrement $\Lambda = 0.2 = \pi \tan\delta_{0.5}$.

$$\therefore \quad \tan\delta_{0.5} = \frac{0.2}{\pi} = 0.064.$$

Therefore phase lag = 0.064 radians.

The damping is less at a lower frequency. Lower frequency is equivalent to testing at a higher temperature. Hence at 5 Hz the damping is less at a temperature somewhat above 20 °C, which implies the presence of a relaxation maximum somewhat below 20 °C.

4. Linear viscoelasticity—hence at any time stress is linearly related to strain, e.g. if extension is x for load m, t minutes after loading, then extension for load am will be ax, at the same time after loading.

   For step loading: m at $t = 0$, plus am at t_0; the subsequent creep is that for m for time t *plus am* for time $(t - t_0)$.

   Unloading is equivalent to adding a negative load; hence subsequent effect of unloading at time t_0 is that of m for time t, *minus m* for time $(t - t_0)$.

   (i) Residual extension at $T = 240$.
   Extension due to 0.1 kg for 240 min = 0.600%.
   Extension due to −0.1 kg for 200 min = −0.585%.
   Extension due to 0.1 kg for 160 min = 0.555%.
   Extension due to −0.1 kg for 120 min = −0.514%.
   Therefore net residual extension = 0.056%.

   (ii) (a) Extension due to 0.1 kg for 80 min = 0.462%.
   Extension due to 0.2 kg for 40 min = 2 × 0.390%.
   Therefore total extension after 80 min = 1.242%.

   (b) Immediate recovery is the same at any time: 0.300% due to 0.1 kg, plus 0.600% due to 0.2 kg.
   Therefore total immediate recovery = 0.900%.

   (c) Extension due to 0.1 kg for 240 min = 0.600%.
   Extension due to 0.2 kg for 200 min = 2 × 0.585%.
   Extension due to −0.3 kg for 40 min = −3 × 0.390%.
   Therefore residual extension at $T = 240$ is 0.600 + 1.170 − 1.170 = 0.600%.

5.

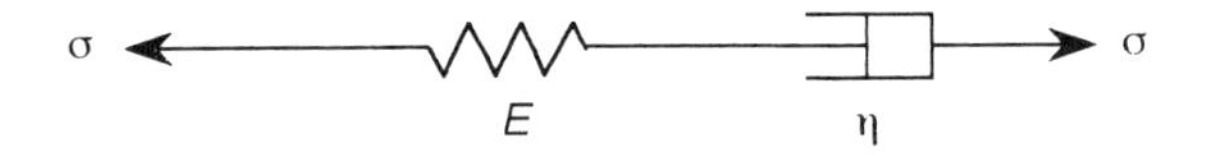

The relaxation time is the time taken for the stress to fall to a value $1/e$ times the initial value $\sigma = \sigma_0 \exp(-t/\tau)$, where

$$\tau = \frac{\eta}{E} = \frac{10^{10}}{10^8} = 100 \text{ s}.$$

At time $t = 0$ a strain of 1% is applied

$$\varepsilon_v = 0 \quad \text{and} \quad \varepsilon_E = \varepsilon_{total} = 1\%$$

$$\therefore \quad \sigma_0 = 0.01 \times 10^8 \text{ Pa} = 10^6 \text{ Pa}.$$

At time $t = 25$ s the stress σ_0 has dropped to

$$\sigma_0 \exp(-25/100) = \sigma_0 e^{-1/4}.$$

On application of a further 2% strain an additional stress of

$$0.02 \times E = 2 \times 10^6 \text{ Pa}$$

is developed in the element. Therefore the total stress is

$$2 \times 10^6 + 10^6 e^{-1/4}$$

which has decayed after a further 25 s to

$$\begin{aligned}(2 + e^{-1/4})10^6 \times \exp(-25/\tau) &= (2 + e^{-1/4}) \times e^{-1/4} \times 10^6 \text{ Pa} \\ &= 2.78 \times 0.78 \times 10^6 \\ &= 2.17 \text{ MPa}.\end{aligned}$$

6. Strain due to stress σ_1 applied at $t = 0$ is given by

$$\varepsilon_1(t) = \sigma_1 J_0(1 - \exp[-t/\tau]).$$

Strain due to stress $(\sigma_2 - \sigma_1)$ applied at $t = t_1$ is given by

$$\varepsilon_2(t) = (\sigma_2 - \sigma_1)J_0(1 - \exp[-(t - t_1)/\tau]).$$

Therefore total strain at time $t > t_1$ is

$$\begin{aligned}\varepsilon(t) &= \sigma_1 J_0(1 - \exp[-t/\tau]) + (\sigma_2 - \sigma_1)J_0(1 - \exp[-(t - t_1)/\tau]) \\ &= \sigma_2 J_0(1 - \exp[-(t - t_1)/\tau]) + \sigma_1 J_0(\exp[-(t - t_1)/\tau] - \exp[-t/\tau]) \\ &= \sigma_2 J_0(1 - \exp[-(t - t_1)/\tau]) + \sigma_1 J_0 \exp(-t/\tau)(\exp[t_1/\tau] - 1) \\ &= \sigma_2 J_0 - \sigma_2 J_0 \exp(-t/\tau)\exp(t_1/\tau) + \sigma_1 J_0 \exp(-t/\tau)(\exp[t_1/\tau] - 1) \\ &= \sigma_2 J_0 + [\sigma_1(\exp(t_1/\tau) - 1) - \sigma_2 \exp(t_1/\tau)]J_0 \exp(-t/\tau) \\ &= \sigma_2 J_0[1 - K\exp(-t/\tau)]\end{aligned}$$

where

$$K = \left[\exp(t_1/\tau) - \frac{\sigma_1}{\sigma_2}(\exp[t_1/\tau] - 1)\right]$$

7. For the Maxwell element

$$\tau = \frac{\eta}{E} = \frac{10^{11}}{10^9} = 100 \text{ s}.$$

In a Maxwell element the instantaneously applied strain appears across the spring. With time the dashpot strains, but since the applied strain is constant the strain developed across the spring drops.

At time $t = 0$, $\varepsilon_E = 1\%$ and hence $\sigma = \sigma_E = \sigma_v = 0.01 \times E$. At time $t = 30$ s this stress has dropped to

$$\sigma(30) = 0.01E\exp(-30/100) = 0.01Ee^{-0.3}.$$

At this stage an additional strain of 1% is applied to the element. This strain appears across the spring giving an additional stress of

$$\sigma_a = 0.01 \times E.$$

The total stress at time $t = 30$ s is therefore

$$\sigma_t(30)0.01E(1 + e^{-0.3}).$$

The stress at a total time of 100 s is then

$$\sigma(70) = 0.01E(1 + e^{-0.3})\exp(-70/100)$$

$$= 0.01E(1 + e^{-0.3})e^{-0.7}$$
$$= (0.01E)(1.74)(0.50)$$
$$= 8.7 \text{ MPa}.$$

8. For a Maxwell element

$$\tau = \frac{\eta}{E}$$

For a rubber $E \propto T$

$$\therefore \quad E_{100\,°\text{C}} = \frac{373}{293} E_{\text{room}} = \frac{373}{293} \times 10^6 \text{ Pa}.$$

The viscosity at 100 °C is

$$\eta_{100\,°\text{C}} = \eta_{\text{room}} \exp\left(\frac{-4000}{8.3 \times 373}\right) \text{Pa s},$$

$$\therefore \quad \tau_{100\,°\text{C}} = \frac{10^8}{10^6}\frac{293}{373} e^{-1.29}$$
$$= 21.6 \text{ s}$$
$$\text{cf } \tau_{\text{room}} = \frac{10^8}{10^6} = 100 \text{ s}$$

Chapters 7 and 8

1. (a) $E_3 = 10$ GPa, $E_1 = 1$ GPa(10^{10} N m^{-2}, 10^9 N m^{-2}); (b) $E_{45} = 1.59$ GPa (1.59×10^9 N m^{-2}).
2. $\varepsilon_{11} = S_{11}\sigma_{11} + S_{12}\sigma_{22}$; $\varepsilon_{22} = S_{12}\sigma_{11} + S_{22}\sigma_{22}$; $S_{22}\varepsilon_{11} - S_{12}\varepsilon_{22} = S_{11}S_{22}\sigma_{11} - S_{12}^2\sigma_{11}$.

$$\sigma_{11} = \frac{S_{22}}{S_{11}S_{22} - S_{12}^2}\varepsilon_{11} \quad \frac{-S_{12}}{S_{11}S_{22}S_{12}^2}\varepsilon_{22}.$$

For composites

$$\sigma_{11} = \frac{57}{2}\varepsilon_{11} + 7\varepsilon_{22}$$

and $S_{22} = S_{11}$. Equate terms

$$\left.\begin{aligned} S_{11}^2 - S_{12}^2 &= \frac{2S_{11}}{57} \quad (1) \\ S_{11}^2 - S_{12}^2 &= \frac{-S_{12}}{57} \quad (2) \end{aligned}\right\} \Rightarrow S_{12} = \frac{-2.7}{57} S_{11}.$$

In (1),

$$S_{11}^2\left(1-\left(\frac{2.7}{57}\right)^2\right)=\frac{2S_{11}}{57}$$

$$\Rightarrow S_{11}=\frac{2}{57(1-(2.7/57)^2)}$$

$$\Rightarrow E_1=\frac{1}{S_{11}}=26.8.$$

3. Take A and B in series. Call the combination X.

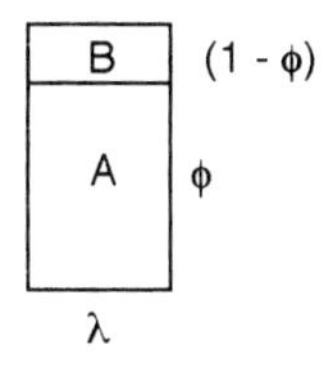

Here λ is common to A and B, therefore $\sigma_x = \sigma_a = \sigma_b$. For the same reason ϕ is the volume fraction of A and $(1-\phi)$ the volume fraction of B.

$$\Delta l_x = \Delta l_a + \Delta l_b \qquad \varepsilon_x = \frac{\Delta l_x}{1}$$

$$\therefore \quad \varepsilon_x = \phi\varepsilon_a + (1-\phi)\varepsilon_b \qquad \varepsilon_a = \frac{\Delta l_a}{\phi}$$

$$\text{Hooke's law: } E = \frac{\sigma}{\varepsilon} \qquad \varepsilon_b = \frac{\Delta l_b}{(1-\phi)}$$

or $\varepsilon = \sigma/E$.

$$\therefore \quad \frac{\sigma}{E_x}=\frac{\sigma\phi}{E_a}+\frac{\sigma(1-\phi)}{E_b} \quad \text{or} \quad E_x=\left(\frac{\phi}{E_a}+\frac{1-\phi}{E_b}\right)^{-1}$$

Now take X and B in parallel. Length is the same for each; strain is the same for each; but force is shared

$$F_c = F_B + F_X$$

$$\sigma_c = F_{c/1}; \qquad \sigma_B = F_{B/(1-\lambda)};\ \sigma_X = F_{X/\lambda}$$

Areas, and volume fractions of X and B are λ and $(1-\lambda)$.

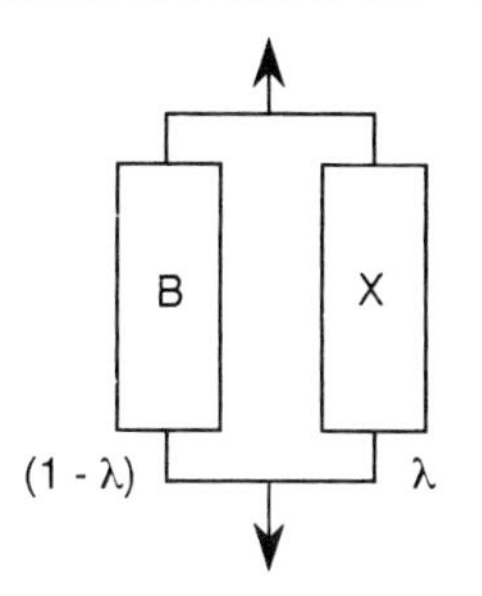

$$\frac{F_c}{1} = \sigma_c = \sigma_B(1-\lambda) + \sigma_X\lambda.$$

Modulus $= \sigma/\varepsilon$, but ε is common.

$$\therefore \quad E_c = (1-\lambda)E_b + \lambda E_X$$
$$= \lambda\left(\frac{\phi}{E_a} + \frac{1-\phi}{E_b}\right)^{-1} + (1-\lambda)E_b.$$

Calculation

(i) $\phi = 1, \lambda = 0.5, E_a = 10$ GPa; $E_b = 1$ GPa.

$$E_c = \frac{0.5}{(1/E_a + 0)} + 0.5E_b = 0.5 \times 10 + 0.5$$
$$E_c = 5.5 \text{ GPa}.$$

(ii)

$$E_c = \frac{0.5}{(0.8/10 + 0.2/1)} + 0.5 \times 1 = \frac{0.5}{(0.28)} + 0.5 = 2.29 \text{ GPa}.$$

A relatively small amount of the weaker component will reduce the modulus severely when it is in series with the stronger component.

Chapters 11 and 12

1. Keep compressive normal stress positive, take positive shear stress; $\sigma_N = \sigma\cos^2\theta$, shear and shear stress on plane $\tau_N = \sigma\sin\theta\cos\theta$, yield stress $\tau_Y = \tau_0 + \mu\sigma\cos^2\theta$.

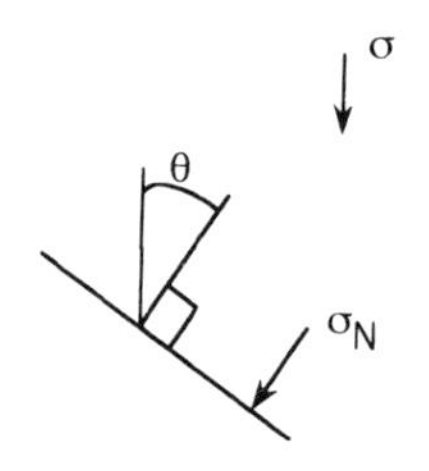

Yielding when $\tau_N - \tau_Y$ is a maximum,

$$\frac{d}{d\theta}\left\{\sigma\sin\theta\cos\theta - \tau_0 - \mu\sigma\cos^2\theta\right\} = 0,$$

$$\tfrac{1}{2}2\cos 2\theta + 2\mu\cos\theta\sin\theta = 0,$$

i.e.

$$\cos 2\theta + \mu\sin 2\theta = 0,$$

$$\tan 2\theta = -\frac{1}{\mu}$$

By assumption $\theta > 0$. Values of 2θ for $\tan 2\theta = -1/\mu$ are, with $\mu = 0.1$, $-84.29° = 180 - 84.29°$, etc. Here $2\theta = 180 - 84.29°$ gives $\theta = 47.86°$, i.e. $47° 51'$.

Yield occurs when $\tau_Y = \tau_N$, i.e. $\sigma \sin\theta \cos\theta = \tau_0 + \mu\sigma\cos^2\theta$, i.e.

$$\sigma = -\frac{\tau_0}{\sin\theta\cos\theta - \mu\cos^2\theta}$$

$$\tau_0 = 10 \text{ MPa gives } \sigma = 22.1 \text{ MPa}.$$

2. Put

$$\sigma_1 = \frac{pr}{t}, \qquad \sigma_2 = 0, \qquad \sigma_3 = \frac{pr}{2t}$$

The equation becomes

$$\left(\frac{pr}{t}\right)^2 + \left(\frac{pr}{2t}\right)^2 + \left(\frac{pr}{2t}\right)^2 = 6k^2 + \mu\frac{3pr}{2t}$$

$$6\left(\frac{pr}{2t}\right)^2 - 3\mu\left(\frac{pr}{2t}\right) - 6k^2 = 0$$

$$2\left(\frac{pr}{2t}\right)^2 - \mu\left(\frac{pr}{2t}\right) - 2k^2 = 0$$

$$\frac{pr}{2t} = \frac{\mu \pm 1(\mu^2 + 4.2.2k^2)}{4}$$

take positive root to maintain positive p:

$$p = \frac{t}{2r}(\mu + \sqrt{}(\mu^2 + 16k^2)).$$

3. Dependence of yield stress on hydrostatic pressure. Assume $\sigma_y = \sigma_{y0} - \alpha p$ where $\alpha > 0$ and $p > 0$ for tension. In tension,

$$35 = \sigma_{y0} - \tfrac{1}{3}\alpha \times 35.$$

In compression

$$40 = \sigma_{y0} + \tfrac{1}{3}\alpha \times 40.$$

Subtract

$$5 = \tfrac{1}{3}\alpha(40 + 35) \Rightarrow \alpha = \tfrac{1}{5}$$

$$\Rightarrow \sigma_{y0} = 37\tfrac{1}{3} \text{ MPa}$$

In biaxial tension, $P = \frac{2}{3}\sigma$

$$\sigma_y = \sigma_{y0} - \frac{\alpha 2\sigma_y}{3} \Rightarrow \sigma_y = \frac{\sigma_{y0}}{(1 + 2/15)} = 32.9 \text{ MPa}.$$

Uniaxial strain $(\varepsilon_u, -\frac{1}{2}\varepsilon_u, -\frac{1}{2}\varepsilon_u)$ assuming incompressibility. Biaxial strain $(\varepsilon_b, \varepsilon_b, -2\varepsilon_b)$. Equate maximum shear rates

$$\tfrac{3}{2}\dot{\varepsilon}_u = 3\dot{\varepsilon}_b \quad \text{i.e.} \quad \dot{\varepsilon}_b = \tfrac{1}{2}\dot{\varepsilon}_u$$

so use half the strain rate in biaxial tension

$$\dot{\varepsilon}_b = 0.5 \times 10^{-3}\ \mathrm{s}^{-1}.$$

4. Given that $V = \mathrm{kT}\Delta(\ln \dot{\varepsilon})/\Delta\sigma$,

$$\sigma(300) = 120\ \mathrm{MPa\ at\ } 10^{-1}\ \mathrm{s}^{-1}$$

$$\sigma(300) = 100\ \mathrm{MPa\ at\ } 10^{-3}\ \mathrm{s}^{-1}$$

then

$$V = 1.38 \times 10^{-23} \times 300 \times \ln(100)/20 \times 10^6 = 9.5 \times 10^{-28}\ \mathrm{m}^3$$

$$\sigma = (\Delta U + kT \ln(\dot{\varepsilon}/\dot{\varepsilon}_0))/V$$

so

$$\mathrm{d}\sigma/\mathrm{d}T = k \ln(\dot{\varepsilon}/\dot{\varepsilon}_0)/V$$

Since

$$\mathrm{d}\sigma/\mathrm{d}T = (200 - 100) \times 10^6/(50 - 300)\ \mathrm{Pa\,K}^{-1}$$

$$k \ln(\dot{\varepsilon}/\dot{\varepsilon}_0)/V = -4 \times 10^5\ \mathrm{Pa\,K}^{-1}$$

At $T = 50$ K,

$$\sigma = 200\ \mathrm{MPa} = \Delta U/V + 50k \ln(\dot{\varepsilon}/\dot{\varepsilon}_0)/V$$

so

$$\Delta U/V = 2 \times 10^8 + 2 \times 10^7 = 2.2 \times 10^8\ \mathrm{Pa}.$$

giving

$$\Delta U = 2.2 \times 10^8 \times 9.5 \times 10^{-28} = 2.09 \times 10^{19}\ \mathrm{J}.$$

$$\dot{\varepsilon} = \dot{\varepsilon}_0 \exp(-(\Delta U - \sigma V)/kT)$$

giving $\dot{\varepsilon}_0 = 9.1 \times 10^8\ \mathrm{s}^{-1}$.

5.

$$K_{\mathrm{IC}} = \sigma_{\mathrm{B}}(\pi c)^{1/2},$$

$$K_{\mathrm{IC}} = 1\ \mathrm{MN\,m}^{-3/2} \qquad c = 10^{-2}\ \mathrm{m},$$

$$\sigma_{\mathrm{B}} = 1\ \mathrm{MN\,m}^{-3/2}/(\pi \times 10^{-2})^{1/2} = \frac{10}{1.77} = 5.64\ \mathrm{MPa}.$$

6.

$$G_c = \frac{P^2\,\mathrm{d}C}{2B\,\mathrm{d}c}$$

$$K_{\mathrm{IC}} = 4\sqrt{6}\,\frac{P_c}{Bb^{3/2}} \qquad P = \frac{K_{\mathrm{IC}}Bb^{3/2}}{4\sqrt{6}c} = 0.433\ \mathrm{N}$$

$$= \frac{2 \times 0.003 \times 0.02^{3/2}}{4\sqrt{6} \times 8 \times 10^{-2}}$$

$$= 21.65\ \mathrm{N} = 2.21\ \mathrm{kg}.$$

7.

$$\sigma_c \delta_t = \frac{K_{IC}^2}{E^*} \qquad \sigma_c = \frac{10^3}{6}\,\text{MPa}$$

$$= 166\,\text{MPa}$$

$$R = \frac{\pi}{8}\frac{K_{IC}^2}{\sigma_c^2} \qquad R = \frac{\pi}{8}\frac{36}{10^6} = 1.414 \times 10^{-5}\,\text{m}$$

$$= 14\,\mu\text{m}.$$

Index